ENVIRONMENTAL RESEARCH ADVANCE

# PLANTATIONS

# BIODIVERSITY, CARBON SEQUESTRATION, AND RESTORATION

# ENVIRONMENTAL RESEARCH ADVANCE

ENVIRONMENTAL RESEARCH ADVANCE

# PLANTATIONS

## BIODIVERSITY, CARBON SEQUESTRATION, AND RESTORATION

HAI REN
EDITOR

nova publishers
*New York*

For permission to use material from this book please contact us:
Telephone 631-231-7269; Fax 631-231-8175
Web Site: http://www.novapublishers.com



**Library of Congress Cataloging-in-Publication Data**

ISBN: 978-1-62808-090-2

*Published by Nova Science Publishers, Inc. † New York*

# PREFACE

Food and Agriculture Organization of the United Nations (FAO, 2011) reported that the estimated loss of global forest area declined from 16 million hectares per year in the 1990s to an estimated 13 million hectares per year between 2000 and 2010. At the same time, large areas of plantations had been established worldwide because of timber, tree crop production, or restoration projects. It was estimated that there were currently about 230 million hectares of total plantations in the world, or about 6.7% of the total forest area. Planted species in plantations are primarily selected to produce timber and tree crops with high yield and productivity. As a novel ecosystem, plantations provide numerous ecosystem services to humans. However, plantations also have some undesirable characteristics, including low biodiversity, loss of original endemic, rare, and endangered species, poor structure due to extensive planting of fast-growing coniferous and exotic species, low ecosystem heterogeneity, an overuse of ornamental species rather than functional species, a lack of mature trees, frequent outbreaks of insect pests and diseases, and maintaining low soil fertility. Fortunately, many of these issues are being recognized to a greater extent, and efforts to manage their forests and plantations on the premise of multi-usability and sustainability has begun.

Today, the challenges of plantations are even greater than in the past because of accelerated climate change, changes in land use and land cover, changes in biogeochemical cycles, population growth, urbanization, as well as the loss of traditional knowledge and cultural diversity. Responding to these challenges and reversing the decline in plantation ecosystem quality will require more detailed knowledge and experience in plantation ecology and management.

This book focuses on plantations: biodiversity, carbon sequestration, and restoration. The primary content explores plantation and non-plantation biodiversity values, mushrooms and woodlands, the roles of arbuscular mycorrhizal in tropical forestry and agro-forestry, the impacts on the abundance and diversity of soil legume-nodulating rhizobia and arbuscular mycorrhizal fungal communities in semiarid regional plantations. The book also investigates carbon biosequestration and bioenergy options of plantations, lessons from plantation forestry carbon sequestration, carbon sequestration in agro forestry-block plantation or mangrove plantations. The book includes targeted chapters on forest restoration and management of plantations for restoring degraded landscapes, alluvial-gold-mining soil reclamation using *Acacia mangium* plantations, effects of forest practices on wild populations of *Testudo hermanni*, effects of incorporation of native species biomass in an acidic andisol to control water contamination from plant residues,and ecohydrology of Amazonian rain forest ecosystems.

As the book developed, we were consistently astounded at the remarkable insights and contributions of the authors of the chapters in this book. The international list of authors were carefully selected, and through a rigorous peer review process, this edited book aims to fill some of the information gaps in vegetation restoration and nutrient mobilization, particularly in plantation ecosystems around the world. Our intended audience includes planners of projects to restore and manage forests or plantations; practitioners who implement those plans; forest resource managers who oversee the sites; forestry consultants; tree farms; environmental authorities; conservationists and students of forestry. We also hope that researchers and the public can find valued information for their future use and efforts. We hope that our work can bring scientists and policy-makers together to envision a sustainable future for woodlands' health and productivity management.

The reality of this book was facilitated by many people who worked tirelessly to edit and select contributors. We express our thanks to the President Nadya Gotsiridze-Columbus, Carra Feagaiga, Jennifer Ramirez and the staff of Nova Science Publishers, Inc. for their encouragement and commitment to excellence in publishing. We also thank our reviewers: Bruce Jaffee, Qinfeng Guo, Osemwegie Osarekhoe Omorefosa, Miguel Taboada, Ryszard Mazurek, Walter Osorio, Juan Carlos Loaiza, J. C. Loaiza-Usuga, Mark P McHenry, Hai Ren and contributors (authors) who were generous with their time, ideas, and comments which improved each of the chapters and helped broaden the scope of the book.

Editor: Prof. Hai Ren, Ph.D.
South China Botanical Garden, Chinese Academy of Sciences, China

Co-editor: Mark P. McHenry, Ph.D.
Faculty of Minerals and Energy, Murdoch University, Western Australia

Co-editor: Prof. J. C. Loaiza-Usuga, Ph.D.
National University of Colombia - Medellin Campus, Colombia
Forest Sciences Center of Catalonia (CTFC), Spain

In: Plantations
Editor: Hai Ren

ISBN: 978-1-62808-090-2


*Chapter 1*

# Plantation and Non-Plantation Biodiversity Values: Distinctions of Economic Theories and Market-Based Mechanisms to Value Ecosystems and Utilization within an Australian Context

***Mark P. McHenry[1]* and Katinka X. Ruthrof[2]***

[1]Faculty of Minerals and Energy, Murdoch University, Western Australia
[2]Centre of Excellence for Climate Change Woodland and Forest Health, School of Biological Sciences and Biotechnology, Murdoch University, Western Australia

## Abstract

Decisions regarding plantation development, either implicitly or explicitly, assign a value to ecosystems. Whilst implicit valuation simply ignores biodiversity values in plantation decision-making, explicit valuation introduces a representative value of biodiversity losses or gains. This work explores the functional components of biodiversity, the existing economic theory of biodiversity, and both advantages and disadvantages of various mechanisms that drive ecosystem valuation to further the development of market-based biodiversity policy and markets. This theoretical refinement enables both public and private decisionmakers to clarify the data requirements that underpin uncertainties in what values of biodiversity exist, to whom, and discuss options to develop a comprehensive market-based mechanism that internalises biodiversity values into everyday plantation investment decisions in the Australian context. This work suggests a hybridisation of existing valuation methods are a bridge towards functional biodiversity valuation in both plantation and non-plantation land use, These new 'non-commodity' markets may close the economic and market 'externality gap' between ecosystem conservation and exploitation, achieving conservation objectives at little cost with thoughtful land use planning.

* Corresponding Author address Murdoch University, 90 South St Murdoch, Western Australia, 6150. Email: mpmchenry@gmail.com.

**Keywords**: Biodiversity; markets; ecosystems; plantations; native vegetation

## INTRODUCTION

Article 2 in the United Nations Convention on Biological Diversity defines biological diversity as "the variability among living organisms from all sources including, *inter alia*, terrestrial, marine and other aquatic ecosystems and the ecological complexes of which they are part; this includes diversity within species, between species and of ecosystems" (United Nations, 1992, p3). The single biggest cause of biodiversity loss in Australia is the removal, fragmentation, and degradation of native vegetation (Lockwood et al., 2000). Such land-use change is also the single biggest driver of dryland salinity and rate among the largest sources of domestic greenhouse gas emissions in Australia (Lockwood et al., 2000; John et al., 2005). Nonetheless, how best to redress biodiversity loss has become much more complex than simply halting or slowing the rate of vegetative removal. While explicit economic ecosystem biodiversity valuation is uncertain and difficult, we choose to value it implicitly every day (Costanza et al., 1997), and regrettably, the implicit value is often very close to zero. Compounding the complexity of biodiversity valuation and decision-making is the lack of distinction between the public and private good, and a rigorous scientific understanding of the consequences of removal, protection, or planting of an ecosystem over time.

Economic valuation is purely anthropocentric in nature as it only considers benefits and costs relevant to human well-being. Humans benefit from natural ecosystems culturally, aesthetically, agriculturally, pharmaceutically etc., and also via the provision of such diverse services as climate regulation, soil formation, nutrient cycling, materials, fuels, quality water, etc. (Fromm, 2000). Ecosystems at the landscape level such as natural or plantation forests can yield substantial flows of economic goods and services, both before and after conversion/harvest (Balmford et al., 2002), and displacing native ecosystems with non-native ecosystems and species (e.g. the introduction of beef, wool, wheat, plantation timber production systems) can bring significant benefits to the community as whole (Bennett, 1999). These benefits ensure that there remains considerable social demand for landholders to clear or modify native vegetation for agricultural, housing, fuel, timber, plantation, or infrastructure development (Gibbons et al., 2009). On the other hand, such ecosystem modifications also result in some undesirable opportunity costs to private individuals and the general public (Bennett, 1999). At present these opportunity costs are currently ignored or undervalued in policy circles because their values are largely external to private operators (Costanza et al., 1997). Furthermore, a detrimental change in the net flow of benefits from the ecosystem (whether natural or created) eventually occurs when productive ecosystems are not managed sustainably, or the functional components are removed (Pagiola et al., 2004).

A quantification of the difference between the economic and market values of the net ecosystem goods and service flows can provide a practical means to develop a mechanism that enables outcomes that can maximize the net benefits of the ecosystem over time (Balmford et al., 2002; McHenry, 2009a). Using economic techniques to incorporate biodiversity values into decision-making processes allows a more meaningful comparison of alternative land use options of retaining, removing, modifying, or establishing ecosystems in theory (Department of the Environment and Heritage, 2005). However, in practice these economic biodiversity valuation techniques require subjective quantification and qualification of both benefits and costs (Bennett, 1999), and monetization is often difficult and irrelevant to a market-based system at the level of the private decision-maker. However, not all net ecosystem good and service flows are external to market-based decision-making, as land values are a function of both productivity and visual amenity variables (Bastian et al., 2002). The fundamental aim of this work is to refine biodiversity value theory to increase the number of the currently few examples of plantation ecosystem values that are internalized into the market price of land (for example dryland salinity, water resources, presence of nitrogen fixing organisms in soils, etc.). These examples in many regions (not all) are insufficient to facilitate first-class land use and management priorities, and new mechanisms to internalize the various biodiversity values will be required to underpin fundamental land use change from the 'bottom up' in a market economy.

## Links between Biodiversity and Ecosystem Value

The economic value of each ecosystem arises from the interdependent relationships between species, habitat components, and their organization that contribute to ecological functions, and the human welfare that can be derived from them over time (Fromm, 2000; Balmford et al., 2002). Thus the fundamental primary value of the dynamic evolutionary processes of the ecosystem's biodiversity, and the capability of the system to maintain stability gives rise to the secondary value of exported ecosystem goods and services (Fromm, 2000). For example, soil biodiversity research by Griffiths et al. (2000) explored the relationship between soil microbe biodiversity and agricultural pasture ecological function stability. The research found that while biodiversity does not confer ecological function stability directly, it does result in improved resilience and recovery from disturbance, and thus, continuous provision of exported services to maintain the pasture system over time. Whilst ecosystem secondary values rely on the primary ecosystem value, the primary value of one ecosystem also relies on the exported secondary values of other ecosystems (Fromm, 2000; European Communities, 2008). Therefore, the economic value to humans of exported services, or even the consumption of the primary ecosystem itself, can theoretically be traced back to the ecological biodiversity structures and functions, which in turn is derived from the ecological role of species as carriers of ecological functions (Fromm, 2000).

'Ecosystem biodiversity' refers to the variety of communities of organisms within particular habitats, and also the physical conditions under which they live, while 'functional biodiversity' refers to the existence of some redundancy in functional populations which underpin the capacity of ecosystems to absorb some disturbance without changing to a new equilibrium (Griffiths et al., 2000; Nunes and van den Bergh, 2001). As an example, the aforementioned pasture soil experimental results suggested that while the number of functional groups were important, the *level* of soil animal species biodiversity did not impact primary productivity as much as the *value* of species-specific, process-specific, and system-specific behavior of functional groups (Bengtsson et al., 1997; Griffiths et al., 2000). Therefore, from a human welfare perspective, economic valuations should ideally focus on the relative changes in the *value* of ecosystem benefits from land use changes rather than the *level* of change in benefits flowing from the ecosystem (Nunes and van den Bergh, 2001).

This is because the change in value of exported ecosystem goods and services has more of a direct impact on human welfare than changes in the level of goods and services.

As the ecological function gives rise to the ability of ecosystems to generate and export services, such as groundwater recharge, water nutrient removal, to generate economic value in theory only a limited number of physical and biological processes are required which vary in importance in different environmental conditions (Fromm, 2000; Nunes and van den Bergh, 2001). This line of logic at the first instance may seem to imply that plantation biodiversity is not a fundamental element for ecosystems in terms of economic or market value. This leads to the possibility of seriously considering the substitutability of species and their functional production values, although at present our level of knowledge of ecological interdependencies and species substitutability is far from perfect (Fromm, 2000; Farber et al., 2002). However, whilst the level of ecosystem biodiversity does not necessarily confer ecological function, it does result in a higher resilience, that is, an ability to recover quickly from disturbances (Tobor-Kaplon et al., 2005; Brussaard et al., 2007). This ability is the essence of the primary ecosystem value derived from biodiversity that has no direct economic or market value, yet is still valuable as a indirect contributor to maintaining the resilience of ecological functions and the provision of ecosystem goods and services over time (Bengtsson et al., 1997; Fromm, 2000).

Clearly, plantation biodiversity and ecological structures and functions will largely continue to be external to market economic decision-making if their value remains unquantified in a monetary sense, and clarifying the difference between total economic value of ecosystem biodiversity and the apparent market value to both private and policy decision-makers is essential. Unfortunately at present, even in theory, it is almost meaningless to ask: "What is the value of ecological support systems in total?", as their value to humans is theoretically infinite (Costanza et al., 1997). However, it is economically meaningful to ask "How value changes in the quantity or quality of ecosystem services may impact human welfare directly and indirectly?", to be able to represent a total economic value (Costanza et al., 1997; Farber et al., 2002). Nonetheless, this work argues that it is more practical in a market economy to also ask: "How value changes in the quantity and quality of ecosystem services from a particular area of land impacts the welfare of individual decision-makers in financial terms?" Attempting to answer this question provides a real market value, sidestepping the uncertainties of economic values, and/or enabling comparisons between the plantation ecosystem's economic and market value.

## Indirect and Direct Biodiversity Value Theory

Clarification of existing theory enables a clearer theoretical basis to integrate both positive and negative externalities of activities that influence the level of change in benefits flowing from forestry ecosystems. The identification of total economic value generated by natural assets recognizes the anthropocentric, instrumental, and utilitarian values that are gained or lost by segments of the environment that affect the welfare of at least one private individual directly. This includes the biodiversity value gained or lost by segments of the environment that affect the prices of agricultural and forestry inputs and products, in addition to the productive use of species and genetics in these industries. Separate from direct productive value, ecosystem biodiversity also has direct value for individuals in terms of aesthetics, recreation, or simply for it to exist, whether or not it is utilized now, in the future, or at all. However, the inclusion of ecological structures and function is necessary for total economic value assessments, as individual and production values of biodiversity do not recognize often indirect and complimentary relationships between humans and ecosystems (Fromm, 2000).

Indirect use values of biodiversity are associated with ecosystem infrastructure that supports economic activity (Nunes and van den Bergh, 2001). Indirect uses include functional benefits for life-support ecological functions through the provision of soil formation, climatic stability, clean air and water etc. Indirect ecological functions such as regulation of climatic processes, the hydrological cycle, processing of human induced pollution, (etc.) can be viewed as avoided health and material possession damages, and can be calculated as indirect benefits which can be quantified probabilistically using aggregated data akin to a form of insurance. The value in this case arises from the ecosystem protecting human capital, human-made capital, and natural capital against disturbances. These indirect values of ecological structure and function transcend the simple value of inputs for production (such as plantation timber) and the value of an individual ecosystem itself (Fromm, 2000). Some indirect use values manifest themselves as direct use values (Gilespie, 2000), especially in plantation and agricultural production systems, such as improved stock water quality and storm protection. Indirect use values can even include biological resources used to produce goods and services such as pharmaceuticals (Nunes and van den Bergh, 2001). In contrast, direct use values of biodiversity often refer to human uses of biodiversity in production and consumption, which can also include tourism, research, and other activities (Gilespie, 2000; Nunes and van den Bergh, 2001). A market analogy for direct and indirect use values are human-made assets for direct individual use (i.e. timber, furniture, houses) and productive assets (i.e. tools, plantations, farms), which are protected by security assets that support activities indirectly (i.e. private liability insurance and social welfare systems) (Fromm, 2000).

The reductionist approach of determining total economic value disaggregates biodiversity into more categories to calculate the total economic value as the sum of various use and non-use values with a bottom up approach (Nunes and van den Bergh, 2001; Hecht, 2005). See Equations 1 and 2. Indirect use values include vicarious values which relate to the benefits of indirect consumption through books, documentaries, and other media. Non-use values relate to benefits individuals obtain from the resource without directly or indirectly using them, and include existence values, option values, quasi-option values, and bequest values (Gilespie, 2000). Existence value is simply the benefits from knowing that certain things remain conserved and certain species and ecosystems survive (Bennett, 1999; Gilespie, 2000). Option values relate to the maintenance of the right to use a resource, without necessarily doing so, while quasi-option values refer to the benefits obtained from the opportunity to delay decisions to make the most of improved information about the resource over time. Finally, bequest values refer to the maintenance of environmental attributes for future generations (Gilespie, 2000). These non-use values have the capacity to reflect human, moral, philanthropic, or policy considerations of biodiversity protection intergenerationally (Nunes and van den Bergh, 2001). Whilst these values are all valid, they pose significant difficulty to market-based mechanisms and policymakers who may rely on balancing both development and conservation pressures.

**Equation 1. Total economic value. Source: (Hecht, 2005).**

Total economic value = use value + non-use value

Where, use value = direct use value + indirect use value, and; non-use value = existence value + option value + quasi-option value + bequest value.

**Equation 2. Total economic value simplified. Source: (Hecht, 2005).**

Total economic value = direct use value + indirect use value + existence value + option value + quasi-option value + bequest value.

Formal valuations of ecosystem goods and services provide insights into decision-making trade-offs for or against ecosystem conservation, modification, or establishment (Howarth and Farber, 2002). The aim of valuations are to clarify the value of the trade-offs between the productive benefits and environmental benefits when land is utilized for, or taken out of production (Bennett, 1999). The assessment of trade-offs require specialist knowledge of the ecosystem, and the economic decision-making regarding conservation and production to consider all potential benefits and costs generated by the natural resource (Fromm, 2000). A simple and practical example is from the perspective of administration and monitoring, where the benefits of environmental monitoring to underpin biodiversity indicators and markets should exceed the costs by the greatest absolute amount (Pannell and Glenn, 2000). Yet, an expansion of quantifying benefits and costs in terms of the economic value of the actual biodiversity goods and services are more complex, as it attempts to apply worth to ecological structure and function (Fromm, 2000). This is required to give policymakers options, and also a rationale to protect non-plantation forestry assets, as conservation planning never occurs in isolation from politics and economics (Fromm, 2000; Polasky et al., 2005).

## The Monetary Value of Biodiversity

Economic valuation of biodiversity strives to overcome the current economic causes of biodiversity loss and to ensure economic incentives are established to encourage biodiversity conservation (Emerton, 2001). Economists consider that particular choices are desirable if the benefits to the community exceed the costs from a community perspective (Gilespie, 2000). Ecological economists are involved specifically with the relationships between property rights and resource management, and model the interactions between the economy and the environment, and use new instruments of environmental policy (Martinez-Alier, 1990). These new instruments have been developed to correct many existing market failures that do not account for the costs of biodiversity or ecosystem loss (Pagiola et al., 2004). These land use planning market failures drive biodiversity and habitat loss by discounting or excluding non-market benefits in market-based plantation decision-making (Balmford et al., 2002).

Market mechanisms such as carbon prices, biodiversity credits, or premium pricing for sustainably produced goods and services, capture ecosystem values at a private level for producers to allow them to have an incentive to generate positive outcomes (Balmford et al., 2002; McHenry, 2009a, 2009b, 2010, 2011b, 2011a, 2012a). When these market-based measures have been developed and implemented well, they enable sustainably-produced goods and services to compete with conventional products that are effectively subsidized through depreciating natural ecosystems (McHenry, 2011a, 2011c). Many ecosystem services do not qualify for market trading as they are not private in nature (Farber et al., 2002). Conserving relatively intact habitats on private land alongside timber plantation and other production systems will often require compensatory mechanisms to mitigate the negative private impact (Balmford et al., 2002). This is because some of the economic value of native vegetation accrue to the broader community, while the associated costs of maintenance fall on the landholder (Gilespie, 2000). Even when compensatory mechanisms do exist, smaller incentives to landholders may be regarded as a waste of time, or even a direct insult to the private owner (Lockwood et al., 2000; McHenry, 2012b), and the high level of work involved in actively maintaining conservation areas is often underestimated by the broader community, including decision-makers.

Options amenable to landholders, such as tax reductions, or exempt status, low-interest loans, grant schemes, and other associated financial mechanisms could assist non-plantation vegetation management on private lands (Gilespie, 2000). In reality, there are a number of reasons why landholders remove, degrade, retain, improve or plant vegetation, which may or may not be related to financial benefits, or optimal for the wider society (Gilespie, 2000). Therefore, the use of both economic and informative mechanisms may have an improved chance of assisting landholders to compare their available options, while including the real value of the vegetation (both non-plantation and plantation) to the society.

The existence of land use externalities, (for example submerged ecosystems and altered river flow regimes from the construction of very large water supply dams) form of market failure, and a committed government can minimize their distortionary impact on the community and the environment (Gregory Mankiw et al., 2000; Foxon et al., 2005; Jaffe et al., 2005; McHenry, 2009a). While private businesses understandably do not invest in goods and services such as clean air and water that are often free, market mechanisms can reimburse entities for protecting the quality of goods and services while at the same time regulating unacceptable outcomes (Longo and Markandya, 2005). This has developed in the energy efficiency market with the introduction of regulatory minimum performance standards to exclude inferior appliances, while information instruments allow consumers to choose to pay a premium to obtain products exhibiting high energy and water use efficiencies. Therefore, a neat distinction cannot be made between market and regulatory measures, as all market-based measures require a regulatory and institutional setting (Diesendorf, 2007).

Establishing formally protected lands through regulation-only mechanisms may conserve habitat, but socio-economic and political constraints limit this form of ecosystem conservation (Polasky et al., 2005). The economic foundation of a decision for, or against, the protection of biodiversity requires the inclusion of all costs and benefits relating to it. However, there is an 'externality gap' between the market and economic value of biodiversity. Filling this valuation gap requires the identification of, and where possible, the monetization of the services that vegetative asset provides (Fromm, 2000). Economically biodiversity must be seen as an asset, and biodiversity conservation as an investment (Fromm, 2000; Farber et al., 2002). Neglecting conservation can be interpreted as de-investment in assets, which leads to a reduction in ecosystem service provision, which in turn leads to an economic cost (Fromm, 2000). Ensuring the continued provision of ecosystem services requires conservation of natural systems, which also in turn calls for economic valuation (Balmford et al., 2002).

Using monetary indicators for economic valuation of biodiversity enables comparisons of alternative market-based ecosystem management options, while non-economic assessments of biodiversity values do not (Nunes and van den Bergh, 2001; Pagiola et al., 2004). Monetary indicators offer the flexibility to be based on various market price valuations or even individuals willingness to pay for such services. However, with this flexibility comes the potential to derive ambiguous monetary values of biodiversity, as different valuation methodologies each have their strengths and weaknesses (Nunes and van den Bergh, 2001). While a reasonable level of flexibility is necessary for assessment methods to accommodate unusual situations, this flexibility needs to be balanced to ensure environmental considerations are not compromised (Gibbons et al., 2009). At times, ecosystem, economic, and market values are at odds with each other, as only some of the species in an ecosystem are valued due to a number of reasons (Farber et al., 2002). It is for this reason that economic biodiversity indicators and methods ought to be based on accurate biological indicators based on scientific principles (Nunes and van den Bergh, 2001).

## Market and Non-Market Biodiversity Valuation

The main obstacles to the wider application of biodiversity valuation in Australia are lack of biophysical information to support valuations, the technical accuracy of valuation techniques, and ethical concerns over valuing environmental impacts in monetary terms (Department of the Environment and Heritage, 2005). Particular ethical criticisms of economic biodiversity valuations relate to conferring of dollar figures on 'priceless' biodiverse assets, such as a non-plantation forest or river ecosystem. As always, there is a counter argument: human development decisions either implicitly or explicitly value ecosystems. Implicit valuation simply ignores biodiversity values in decision-making, and, by comparison, explicit economic valuation represents the potential biodiversity losses (Department of the Environment and Heritage, 2005). The primary purpose of economic valuation is to obtain consistent information on the costs and benefits of biodiversity conservation to inform decision-makers (Pagiola et al., 2004). This may balance the predominance of implicit valuation processes.

There are a variety of values that biodiversity can be attributed beyond individual and productive values, including security values that ensure the continued service provision from ecological functions (Fromm, 2000). Theoretically, biodiversity value can be characterized by a number of values: local versus global diversity, life diversity versus biological resources, instrumental versus intrinsic values, and so on (Nunes and van den Bergh, 2001). Biodiversity value can also be categorized in terms of an ecosystem spatially, or a habitat that is in high demand, such as areas of recreation or tourism (Nunes and van den Bergh, 2001). However, policymakers must be aware that there are unresolved issues in some valuation methods. For example, whether economic valuations at multiple levels leads to double counting of biodiversity values (McHenry, 2011a).

Market-based classifications of economic valuation include techniques such as: the human capital approach; productivity changes method; defensive expenditures;

repair/replacement expenditures; shadow projects, and; the opportunity cost method (Gilespie, 2000). In comparison, there are six major ecosystem service non-market economic valuation techniques when market valuations do not capture the social value of biodiversity: avoided cost techniques quantify the value of costs that would have occurred in the absence of certain ecosystem services; factor income techniques value the enhancement to incomes from improving ecosystem services; travel cost techniques reflect the costs people are prepared to pay to travel to enjoy ecosystem services of specific regions; hedonic pricing techniques reflect the differential prices people pay for goods that involve specific ecosystem amenities; contingent valuation techniques value ecosystem services by quantifying the differential values that people are willing to pay for hypothetical ecosystem service alternatives, and finally; replacement cost techniques use the cost of substitutes that can replace the ecosystem services (Farber et al., 2002).

The replacement cost technique is the only technique in both-market and non-market categories, as it leaves scope to sum additional techniques to derive site-specific non-market values (such as travel costs) and likely market-based values (such as plantation carbon sequestration). The replacement cost technique estimates how much it would cost to replace an environmental resource and is a promising approach to provide a substitute for an ecosystem service valuation technique (Department of the Environment and Heritage, 2005). The replacement cost technique gives value to ecosystem services by quantifying the cost of restoring or synthetically replacing it (Balmford et al., 2002). This technique does not strictly evaluate the value of biodiversity benefits, but is useful for providing an initial estimate of the resources value (Department of the Environment and Heritage, 2005). While intuitively appealing, replacement cost methods may misrepresent the 'willingness to pay' or 'willingness to accept' valuation concepts in some circumstances where social amenities are lost in the synthetic replacement (Farber et al., 2002). Nonetheless, replacement cost methods do provide an easily verifiable and practical methodological choice to reveal the lower bounds of biodiversity value to restore functional diversity and facilitate a cost-benefit analysis of land use change options.

Formal cost-benefit analyses of areas of biodiversity can be used to determine productive and individual values, although essential services may not be considered when there are significant ecosystem knowledge gaps (Fromm, 2000). For example, a robust attempt at valuing flood protection services provided by various vegetative islands (both plantation and non-plantation) must be based on complex hydrological models of topography and ecosystems, and uncertainties or errors can become considerable (Howarth and Farber, 2002). However, if the economic costs of establishing and maintaining vegetative islands on degraded land is low, then there is little practical barrier to 'over-engineering' to ensure sufficient protection as a form of insurance. If a plantation project aimed to replicate the original high-quality habitat, then this organic category of replacement cost method would have roughly comparable ecological function in terms of exported ecosystem services as the original habitat (Emerton, 2001). Replacing the original vegetation would also avoid the lost social benefits and may more accurately represent 'willingness to pay' or 'willingness to accept' values by improving, retaining, or re-establishing ecosystem services over and above flood-protection (Farber et al., 2002). However, economic valuations such as replacement cost tend to handle large-scale and long-term problems poorly, but have the potential to be suitable for looking into shorter-term and local-scale values (Pagiola et al., 2004).

## Ecosystem Service Valuation Method and Limitations

Valuation studies illuminate ecological structure and function relationships and their roles in supporting human welfare (Howarth and Farber, 2002). While economic valuation has both strengths and limitations as a decision-making tool, it is clear that information about environmental management costs and benefits are essential to ensure efficient, equitable, and sustainable outcomes. While most of the direct and indirect use values of ecosystems may be approximated quite accurately, the availability of physical data or the change in the functional ecosystem services are often limited (Pagiola et al., 2004). Variables, such as vegetation condition, percentage of vegetation types cleared in the region, and the area of any potential vegetative offset location should receive special attention during assessment (Gibbons et al., 2009).

Local historical and cultural knowledge of ecosystems and their traditional land uses is also recommended to inform biodiversity valuation studies (Dyer et al., 2008). When cultural, historical, and social systems are intimately entwined with ecosystems, the individual component values should, in theory, be a larger value than the sum, as these values are more communal and have greater interpersonal impacts than standard economic ecosystem values (Farber et al., 2002). Therefore, there is no one 'correct' method or technique to obtain ecosystem values, and there is a need for, as Farber et al. (2002) p390 describes as "conceptual pluralism, and thinking outside the box" in its development. These issues beg the question of: "How valuable are ecosystems to whom?", as ecosystem benefits can fall unequally across different groups of people, while being valuable to some and incurring costs to others (Pagiola et al., 2004).

Landscape, species, and genetic diversity that provide input into productive processes have been widely valued using the contingent valuation method (Nunes and van den Bergh, 2001). Contingent valuation asks how much a person would pay for a particular environmental outcome, or how much compensation they would be prepared for its loss (Balmford et al., 2002). The contingent valuation method is the most useful to identify and measure economic non-use values. In principal, the contingent valuation method is applicable for all biodiversity categories, except for categories that the general public is not informed about, or has little experience with, for example: ecosystem life-support function valuations. When far removed from human perception, contingent valuation becomes problematic when eliciting the economic value of ecological processes (Nunes and van den Bergh, 2001), such as the carbon fixation of trees, or respiration of soil biota. Contingent valuation is more suited to interpretations of existence and bequest values from the amount an individual would pay to know that a particular native fish exists in its natural habitat and remain so for future generations, respectively. The concern with contingent valuation is the reliability and validity of the responses (Loomis et al., 2000). These issues may be improved by including the hedonic price method, where environmental services are valued by comparing market prices of biodiversity conservation at a regional scale, such as a water body or catchment (Lockwood et al., 2000; Balmford et al., 2002).

A hybridized total economic valuation of use and non-use values could be used by utilizing government departments expenditure on specific ecosystem amenities (the hedonic method) and the additional costs that people are prepared to pay to travel to use the ecosystem (travel cost method), with the addition of a contingent valuation study of the non-use values of native vegetation that sum to add value to the construction of new habitats (replacement cost method). Using hybrid methods enables flexibility to cater for the unique circumstances of each biological system (both plantation and non-plantation), and introduces a higher level of rigor for decision-makers when choosing between various direct, indirect, or non-use alternatives, than simply the current status quo of implicit valuation.

## Conclusion

Regardless of the level of scientific rigor, high-precision, or accuracy of data utilized to underpin and verify biodiversity values, there will necessarily remain a subjective human and local element to the economic values determined. Rather than a stark trade-off between biodiversity conservation and high-value plantation commodity production, a large fraction of conservation objectives can be achieved at little economic cost with thoughtful land-use planning (Polasky et al., 2005). While economists continue to debate the validity of economic valuation methods, rightly or wrongly, they undermine the public confidence in valuation techniques (ten Kate et al., 2004). At the same time, questions of irreversibility and uncertainty raises issues for environmental valuation (Howarth and Farber, 2002). The prime reasons for the explicit valuation of biodiversity is to introduce at least some value into decision-making, and simultaneously foster a level of rigor in the analysis of the costs and benefits of various alternative options available.

This review suggests that valuations of functional biodiversity is a bridge towards market valuation of biodiversity, and new 'non-commodity' markets may close the economic and market 'externality gap' between ecosystem conservation and exploitation. However, functional biodiversity is difficult to value. Thus, the development of a hybridized 'total economic valuation' approach considering both use and non-use values of government expenditures on specific ecosystem amenities and the travel costs that visitors incur, alongside a contingent valuation of the non-use values of vegetation, and replacement cost methods for new plantations, may a be a suitable approach. Such hybrid methods enables flexibility to cater for each biological system (both plantation and non-plantation), and introduces a level of relative comparison for decision-makers that is at least an improvement on the current status quo of implicit valuation.

## References

Balmford, A., Bruner, A., Cooper, P., Costanza, R., Farber, S., Green, R. E., Jenkins, M., Jefferies, P., Jessamy, V., Madden, J., Munro, K., Myers, N., Naeem, S., Paavola, J., Raymnet, M., Rosendo, S., Roughgarden, J., Trumper, K., and Turner, R. K. (2002). Economic reasons for conserving wild nature. *Science,* 297, 950-953.

Bastian, C. T., McLeod, D. M., Germino, J., Reiners, W. A., and Blasko, B. J. (2002). Environmental amenities and agricultural land values: a hedonic model using geographical information systems data. *Ecological Economics,* 40, 337-349.

Bengtsson, J., Jones, H., and Setala, H. (1997). The value of biodiversity. *Tree,* 12(9), 334-336.

Bennett, J. (1999, 20-21 May). *Estimating the values of environmental impacts of agriculture.* Paper presented at the Country Matters Conference, Canberra.

Brussaard, L., de Ruiter, P. C., and Brown, G. G. (2007). Soil biodiversity for agricultural sustainability. *Agriculture Ecosystems and Environment,* 121(3), 233-244.

Costanza, R., d'Arge, R., de Groot, R., Farber, S., Grasso, M., Hannon, B., Limburg, K., Naeem, S., V, O. N. R., Paruelo, J., Raskin, R. G., Sutton, P., and van den Belt, M. (1997). The value of the world's ecosystem services and natural capital. *Nature,* 387, 253-260.

Department of the Environment and Heritage. (2005). *Making economic valuation work for biodiversity.* Canberra, ACT: Land and Water Australia.

Diesendorf, M. (2007). *Greenhouse solutions with sustainable energy.* Sydney: University of New South Wales Press.

Dyer, S., Grant, J., Lesack, T., and Weber, M. (2008). *Catching up: conservation and biodiversity offsets in Alberta's boreal forest.* Ottawa, Canada: Canadian Boreal Initiative.

Emerton, L. (2001). *Economic measures for biodiversity planning: an annotated bibliography of methods, experiences and cases.* Nairobi, East Africa: The World Conservation Union (IUCN).

European Communities. (2008). *The economics of ecosystems and biodiversity: an interim report.* Cambridge, United Kingdom: Banson.

Farber, S. C., Costanza, R., and Wilson, M. A. (2002). Economic and ecological concepts for valuing ecosystem services. *Ecological Economics,* 41, 375-392.

Foxon, T. J., Gross, R., Chase, A., Howes, J., Arnall, A., and Anderson, D. (2005). UK innovation systems for new and renewable energy technologies: drivers, barriers and system failures. *Energy Policy,* 33, 2123-2137.

Fromm, O. (2000). Ecological structure and functions of biodiversity as elements of its total economic value. *Environmental and Resource Economics,* 16, 303-328.

Gibbons, P., Briggs, S. V., Ayers, D., Seddon, J., Doyle, S., and Cosier, P. (2009). An operational method to assess impacts of land clearing on terrestrial biodiversity. *Ecological Indicators,* 9, 26-40.

Gilespie, R. (2000). *Economic values of the native vegetation of New South Wales.* Sydney, New South Wales: Native Vegetation Advisory Council, New South Wales.

Gregory Mankiw, N., Gans, J., King, S., and Stonecash, R. (2000). *Principals of Economics.* Sydney: Harcourt Australia.

Griffiths, B. S., Ritz, K., Bardgett, R. D., Cook, R., Christensen, S., Ekelund, F., Sorensen, S. J., Baath, E., Bloem, J., de Ruiter, P. C., Dolfing, J., and Nicolardot, B. (2000). Ecosystem response of pasture soil communities to fumigation-induced microbial diversity reductions: an examination of the biodiversity - ecosystem function relationship. *Oikos,* 90, 279-294.

Hecht, J. (2005). *Valuing the resources of Mulanje Mountain: study design*: USAID.

Howarth, R. B., and Farber, S. (2002). Accounting for the value of ecosystem services. *Ecological Economics,* 41, 421-429.

Jaffe, A. B., Newell, R. G., and Stavins, R. N. (2005). A tale of two market failures: technology and environmental policy. *Ecological Economics,* 54, 164-174.

John, M., Pannell, D., and Kingwell, R. (2005). Climate change and the economics of farm management in the face of land degradation: dryland salinity in Western Australia. *Canadian Journal of Agricultural Economics,* 53, 443-459.

Lockwood, M., Walpole, S., and Miles, C. (2000). *Economics of remnant native vegetation conservation on private property* Canberra, ACT: Land and Water Resources Research and Development Corporation.

Longo, A., and Markandya, A. (2005). *Identification of options and policy instruments for the internalisation of external costs of electricity generation.* The Fondazione Eni Enrico Mattei Nota di Lavoro Series.

Loomis, J., Kent, P., Strange, L., Fausch, K., and Covich, A. (2000). Measuring the total economic value of restoring ecosystem services in an impaired river basin: results from a contingent valuation survey. *Ecological Economics,* 33, 103-117.

Martinez-Alier, J. (1990). *Ecological economics: energy environment and society.* Oxford, England: Basil Blackwell.

McHenry, M. P. (2009a). Policy options when giving negative externalities market value: Clean energy policymaking and restructuring the Western Australian energy sector. *Energy Policy,* 37, 1423-1431.

McHenry, M. P. (2009b). Synergies between conventional soil organic carbon, farm productivity, soil sequestration and soil carbon market risk in Australia. In: E. T. Nardali (Ed.), *In No-Till Farming: Effects on Soil, Pros and Cons, and Potential.* Hauppauge, New York, USA: Nova Science. 978-1-60741-402-5.

McHenry, M. P. (2010). Carbon-based stock feed additives: a research methodology that explores ecologically delivered C biosequestration, alongside live-weights, feed-use efficiency, soil nutrient retention, and perennial fodder plantations. *Journal of the Science of Food and Agriculture,* 90, 183-187.

McHenry, M. P. (2011a). Australian carbon biosequestration and clean energy policy market co-evolution: Mechanisms, mitigation, and convergence. *Australian Forestry Journal,* 75(2) 901-913.

McHenry, M. P. (2011b). Integrating climate change mitigation and adaptation: Refining theory for a mathematical framework to quantify private and public cost-effectiveness, and carbon emissions for energy and development projects. *Renewable Energy,* 36(4), 1166-1176.

McHenry, M. P. (2011c). Soil organic carbon, biochar, and applicable research results for increasing farm productivity under Australian agricultural conditions. *Communications in Soil Science and Plant Analysis,* 42, 1187-1199.

McHenry, M. P. (2012a). Practicalities of establishing forestry carbon sequestration projects in the agricultural sector: a technical and economic analysis with implications. In: B. J. Ryan and D. E. Anderson (Eds.), *Carbon sequestration: technology, measurement technologies and environmental effects.* Hauppauge, New York, USA: Nova Science Publishers. 978-1-62081-081-7.

McHenry, M. P. (2012b). Sensitive variables for applying biochar as a fertiliser substitute and a method to sequester carbon in soils: a wheat crop scenario. In: B. J. Ryan and D. E. Anderson (Eds.), *Carbon sequestration: technology, measurement technologies and environmental effects.* Hauppauge, New York, USA: Nova Science Publishers. 978-1-62081-081-7.

Nunes, P. A. L. D., and van den Bergh, J. C. J. M. (2001). Economic valuation of biodiversity: sense or nonsense? *Ecological Economics,* 39, 203-222.

Pagiola, S., von Ritter, K., and Bishop, J. (2004). *How much is an ecosystem worth? - assessing the economic value of ecosystem conservation.* Washington, DC USA.: The International Bank for Reconstruction and Development/The World Bank.

Pannell, D. J., and Glenn, N. A. (2000). A framework for the economic evaluation and selection of sustainability indicators in agriculture. *Ecological Economics,* 33, 135-149.

Polasky, S., Nelson, E., Lonsdorf, E., Fackler, P., and Starfield, A. (2005). Conserving species in a working landscape: land use with biological and economic objectives. *Ecological Applications,* 15(4), 1387-1401.

Ten Kate, K., Bishop, J., and Bayon, R. (2004). *Biodiversity offsets: views, experience, and the business case.* Gland, Switzerland, Cambridge, United Kingdom, and London, United Kingdom: The World Conservation Union (IUCN) and Insight Investment.

Tobor-Kaplon, M. A., Bloem, J., Romkens, P. F. A. M., and de Ruiter, P. C. (2005). Functional stability of microbial communities in contaminated soils. *Oikos,* 111(1), 119-129.

United Nations. (1992). *Convention on Biological Diversity.* Rio de Janeiro, Brazil.

In: Plantations
Editor: Hai Ren

ISBN: 978-1-62808-090-2


*Chapter 2*

# MUSHROOMS AND WOODLANDS: ECOLOGICAL NEXUS

***Osarekhoe Omorefosa Osemwegie[1]*, John Aroye Okhuoya[2] and Theophilus A. Dania[2]***
[1]Biological Sciences Department, College of Science and Engineering, Landmark University, Omu Aran, Kwara State, Nigeria
[2]Department of Plant Biology and Biotechnology, Faculty of Life Sciences, University of Benin, Benin City, Edo State, Nigeria

## ABSTRACT

Mushrooms produce specialised spectra of enzymes that they use in biochemical extracellular digestion of diverse and chemically varied land organic compounds of both synthetic and natural origin.

This singular behaviour enthroned fungi especially mushrooms as a foremost ecological biodegrader of complex organic matter and recycler of interlocked elements in organic compounds.

They are characterised by sub-perennial arrays of unseen vegetative growths, their visible reproductive part remains a function of seasonality, nutrient levels and inter-specific interactions.

The extent of their interaction and role on heterogeneous and homogenous woodlands is indispensable with wide ranged benefits for agroforestry, agriculture and conservation initiatives. A conceptualization of the ecological processes expressed by mushrooms in woodland zones may underscore the synergism in the ecological complexities of the woodland ecosystems productivity.

**Keywords:** Mushrooms, woodlands, interactions, conceptualization, potential benefits

---

* Corresponding Author: Biological Sciences Department, College of Science and Engineering, Landmark University, Omu #Aran, Kwara State, Nigeria. Email: osemwegie.omorefosa@landmarkuniversity.edu.ng.

## INTRODUCTION

Mushrooms, which in various literatures are also referred to as macrofungi, toadstools, macromycetes, basidioma (sexual fruit body of basidiomycetes) or ascoma (sexual fruit body of ascomycetes) represent a biological and taxonomically distinctive group of fungi (Redhead, 1997; Labarère and Menini, 2000).

These are defined diversely as larger fungi or higher fungi of the Class Basidiomycetes or Ascomycetes and recently, Zygomycetes and non-lichenized fungi with large fruitification (fruit body). In addition, mushrooms are also fungi with typical stalk and cap configuration, fleshy or non-fleshy, fruiting body of a fungus plant which typically contains spores or spore-bearing structures visible to the naked eye (Kirk et al., 2001; Miles and Chang, 2004; Wasser, 2007).

The term mushroom is used in a restrictive sense to describe the extension of a fungus mycelium, edible toadstool or basidiomycetes (Nicholson, 1989; Adewusi et al., 1993). It also represents a polypore (non-gilled or non-lamellae mushroom), large visible fungus with medicinal values, toadstool which is inedible or poisonous, mass of interwoven hyphae, agaric (fleshy mushroom) and sporocarp (spore harbour) of a fungus rather than the mycelium (Masuka and Ryvarden, 1993; Harkonen et al., 2003).

The restrictive use by extension also includes the antiquated categorization of wood colonized obtrusive sporocarp as mushrooms and non-wood fungi as toadstool. Mushroom is described by Chang and Miles (1993) as a macrofungus with a distinctive fruit body which may be epigeous (above ground) or hypogeous (below ground) and is sufficiently large enough to be seen with the naked eyes and picked by hand. Mushrooms therefore need not be restricted to basidiomycetes or ascomycetes, fleshy or non-fleshy, edible or non-edible, medicinal or lethal, subterranean rather than epigeous or hypogeous and may grow on different organic based substrates/substrata (wood or non-wood) in diverse habitats (Bates, 2006). The term mushroom can be used interchangeably with toadstools as opposed to the antiquated categorization of the former as visible wood or edible fungal fruit-bodies and the later as poisonous or non-wood based fruit-bodies. Mushrooms may consequently include poisonous or edible, ectomycorrhizae species associated with the roots of conifers and dicotyledonous trees or saprophytic species growing on plant tissues and plant wastes or poisonous species or opportunistic parasites of tree plants (Labarère and Menini, 2000). They also exhibit varying size, colour, shape (bracket, puffballs, truffles, cup, toothed, club etc.), and texture (Figure 1) (O'Dell et al., 2004; Wasser, 2007).

The growing global consciousness and knowledge of mushroom resources and products have assisted the emergence of new areas of mycology which include but not limited to mushroom science or biology, mushroom biotechnology and mycogeography. Chang and Miles (1993) described mushroom biology as a special scientific excerpt (discipline or branch) of mycology comprising diverse aspects that include mushroom cultivation and genetics; medicinal and nutritional mushrooms; pathology, ecology and geomycology, physiology and evolutionary biology, taxonomy and toxicity of mushrooms etc.

Mushrooms are non-photosynthetic, achlorophyllous fungal organisms incapable of manufacturing their own food as do green plants. They produce a wide range of enzymes and acids that can degrade a variety of complex substrates (organic matter) and consequently have broad ecological distribution covering temperate, subtropical and tropical vegetations.

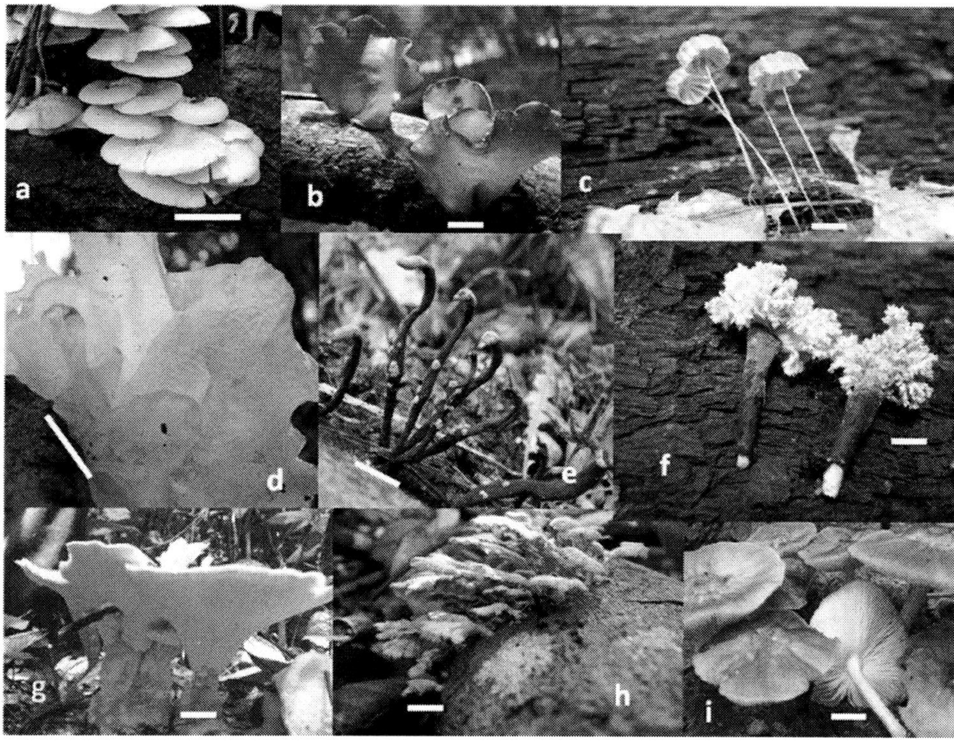


Figure 1. The diversity of mushrooms in some woodlands of Edo State, Nigeria. *Lentinus squarrosulus* (M.) Singer (a), *Auricularia auricular* Judae (b), *Marasmius graminum* (Lib.) Berk. (c), *Tremella fuciformis* Fr. (d), *Clavaria* sp. (Afro-fungus), *Xylaria hypoxylon* (L.) Grev. (f), *Bondarzewia* sp. (g), *Schizophyllum commune* Fr. (h), *Pluteus cervinus* (Schaeff. ex Fr.) Kum. (i). Bar = 0.5mm.

They exist as saprophytes, parasites of trees or as symbionts on or within the cells of the roots of higher plants (i.e. mycorrhizae) or they become an integral non-pathogenic endophytic part of a plant part (Zadrazil, 1980; Wood, 1984; Chang et al., 1993). They are also important in nature conservation, forest health, management and productivity because of their proactive ecological roles and ecosystem functions to micro- and macrofauna and flora, and mycorestoration processes (Ohga et al., 2000; Mshigeni, 2005; Stamets, 2005). The majority of symbiotic microfungi have been reported on animals, humans and plants (Kirk et al., 2001).

In return, the forest communities influence the atmosphere of mutual interactions, mushroom spore dissemination and the establishment of hitherto cryptic mycelia, strengthening genetic plasticity, and gene flow (Fries, 1981; Gregory, 1984). Mushrooms are apparently more potentially valuable to plants, humans, some animals and insects compared to other traditional thallophytes and microorganisms yet understudied as healthy ecosystem resource.

It is within these contextual premises that this paper sorts to overview the robust benevolence in the life activities of fungi in many woodland ecosystems and the conceptualization of added values to improving human and animal life especially in Africa.

# MUSHROOM INTERDEPENDENCE WITH OTHER BIOTICS

## Practice of Exploiting Edible Mushrooms from the Wild

The nutritional and the medicinal values of mushrooms are recognized in different parts of the globe with abundant reports on their nutrient contents and medicinal usage (Lelley, 1987; Arora, 1989; Quimio et al., 1990; Bhandary, 1991; Masuka and Ryvarden, 1993; Alofe et al., 1996; Kekawa, 2001; Akpaja et al., 2003; Osemwegie et al., 2006). The growing demand for mushrooms as food and/or medicine especially in highly developed countries of the world with advanced mushroom cultivation industries has boosted their commercial and foreign exchange values. It has equally improve cultivation practices in developing nations, exportation and rapid technological development that addresses improvement of yield; reduced cropping period, genetic selection of pest/pathogen resistant variety and genetic engineering for improved valued product yield (Oei, 1991; Chang and Miles, 1993; Mshigeni et al., 2003; Miles and Chang, 2004). In many developing countries of Africa, mushrooms are still being sourced from the wild rather than from approved cultivation cottages, farms or certified retail markets. This practice predispose mushroom hunters (gatherers) to various degrees of danger or worst still life threatening conditions from attack by wild animals and insect. Although, this have been the ancient recreational practice of African, it is the direct result of poor access to information, training and technology, ignorance of the possibility of mushroom cultivation, lack of access to affordable mushrooming spawns, dis-interest by veritable farmers, low numbers and slow developing mushroom cultivation industries and lack of funds *inter alia*. Over-reliance on mushrooms from the wild or woodland forests in many African nations slowly depresses wild mushroom diversity. This consequently has feedback effects of forest fragmentation and gap, deforestation (including firewoodcollection) and other general anthropogenic disturbances. It may also expose consumers to the risk of mistakenly gathering poisonous with edible mushrooms leading to mycetisma or mycotoxicosis (Quimio et al., 1990; Oei, 1991; Akpaja et al., 2003; Osemwegie et al., 2006). This may possibly have a long term negative consequence on the forest or forest-floor ecosystem balance and pressured forest productivity. The domestication of many African edible mushrooms is slow and developing due to subjective reasons of socio-infrastructural dysfunction, lack of political will power and poverty.

The indirect effects of tree loss due to deforestation activities, wood fuel (firewood and charcoal) gathering, animal grazing by cattle rearing nomads and the practice of bush burning on mushrooms diversity or the distribution, habitat selection and evolution of mushrooms cannot also be overlooked. Though, empirical data are dearth on the impact of each of these on the mushroom species richness, diversity and delivery of ecosystem function, overall forest health and productivity, and climate for most African countries (Egli et al., 2006).

Alabi (1991), Osemwegie et al. (2006) and Idu et al. (2007) reported low incidents of mushroom poisoning and death from mycophagy of wild edible mushrooms in Nigeria which may be collected for commerce, poverty alleviation or food subsistence. This may be due to factors such as lack of documented evidences in orthodox medical records, unpopular use of modern health care centres by most rural communities, poor befuddled diagnostic and non-reporting of incidences due to mushrooms poisoning or death. Mushrooms are widely reported in scientific literature to be good sources of food, tonic and, in some cases medicine since prehistoric times (Chang, 1980; Alofe, 1991). The nutritive nature, cultural knowledge and values that are handed down generations were however more recent (Chang and Miles, 1993; Miles and Chang, 1997). Vegetations in Africa have been reported as hotspots of mushroom diversity and new taxa, mapping of identified mushrooms worldwide remained challenging to researchers while documentations on mycogeography are dearth. Mueller et al. (2007) concurred with the hypothesis that mushroom mapping challenge is derived from the variation between the location of mushroom fruit bodies and the mycelia activity as well as the distribution of mycelia. Mushrooms contain 20-45% of protein (dry matter) and rich in all essential amino acid whose quality competes favourably with those of plant and animal origin (Lelley, 1987). In addition, they also possess polymeric carbohydrate diverse biochemical characterization, origin, and various low molecular weight carbon compounds that include glucose, fructose, galactose and threalose; minerals nutrients notable amongst which are potassium, phosphorus and iron. They are also very rich in crude fibre and vitamins particularly thiamine (B1), riboflavin (B2), panthotenic acid (B3), ascorbic acid (C) and biotin (H) (Labarère and Menini, 2000).

## Other Explorative Uses

In developed countries of the world however, mushrooms are also easily and readily processed, dried, pickled or canned for storage until ready for shipment to end users/consumers. Mushrooms growing popularity is acknowledged in literatures to enhance both foreign and local commerce, and agriculture (e.g. animal husbandry, crop and tree farming, fertilization of agricultural soils and biological control of pathogens and pests etc.). It is also fundamental in the bioconversion of solid wastes of industrial, domestic and agricultural origin; biotechnology such as bioremediation or mycorestoration of arable lands contaminated by either heavy metals or agrochemical products e.g. pesticides and herbicides or petroleum hydrocarbon and other effluents of diverse origins (Onianwa, 1995; Okeke et al., 1996; Ochiel et al., 1997; Isikhuemhen et al., 2003; Anoliefo et al., 2002; Wasser, 2007). Furthermore, mushrooms are also variably recognized in industries as sources of amino acids, antibiotics, enzymes, organic acids, food, beverages, hormones (e.g. abscisic acid, zymosterol) and natural products that are alternative substitute to synthetic chemicals in biopulping. This is in addition to the wide range of new mushroom products being explored in many aspects of human lives (Agu, et al., 1993; Kirk *et al.*, 1993; Dreyfuss and Chapela, 1994; Bucher et al., 2004; Mshigeni, 2005). They are equally valuable in folk medicine practice in Africa, Asia and South America despite the paucity of information on the patency of folk knowledge of medicinal mushrooms, mushroom genetic resources and biodiversity data e.g. species composition, richness and diversity (Alabi, 1991; Ryvarden et al., 1994; Chang and Mshigeni, 2001; Akpaja *et al.*, 2003; Osemwegie et al., 2006; Idu et al., 2006). Little is also documented on mushroom biogeography, aboriginal (native, endemic or indigene) and introduced macrofungal species in different locations around the globe contrary to reports of their earlier discovery relative to other group of fungi (Miles and Chang, 1997; Mueller et al., 2007). The holistic summary of the benefits of mushrooms to man including their ecological role in forest ecosystem stability, development and community functions cannot be over-emphasized (Gilbertson and Bigelow, 1998; Read and Perez-Moreno, 2003). These potentials remain untapped in Nigeria and some other African countries.

Mushrooms which perhaps are valued for their edibility, medicinal uses, diverse domestic and ecological relevance has encouraged global research interests (Jain, 2000; Labarère and Menini, 2000; Stamets, 2000; Kirk et al., 2001; Miles and Chang, 2004; Mshigeni, 2005). It has consequently become necessary to join in the global initiatives at understanding our indigenous mushroom resources, identifying and preserving such resource pools by minimizing threats to their diversity and ecological functions.

## Forest Community and Litterfall

The forest ecosystems (woodlands) are complex with prolific reports on their community structure, functions and composition with respect to animals, insects and leafy (herbaceous) plants ecography compared to macrofungi (Waring and Schlesinger, 1985). Fungi and especially mushrooms have hitherto been recognised as an integral but fundamental part of the forest community and plantations, farms, gardens and other places with high deposits of organic matter. Shigeki et al. (1994) and Takashi (2007) enunciated the role of fungi and mushrooms in woodland ecosystems or forest's mineral cycles and the importance of lignin as a regulating factor in the decomposition of litter. Where it not for the unceasing decay activities of fungi and more especially mushrooms, humans and other living organisms would have become suffused or suffocated in dirt or drowned neck-deep in leaves. Consequently, fungi and mushrooms are also affected by a huge range of interconnected ecosystem activities such as nutrient acquisition, competition for limited space, decomposition and litterfall dynamics, and biogeochemical cycles (Coûteaux et al., 1995; Sala et al., 2000; Kauserud et al., 2008). The heterogeneous and the homogeneous forests of the world therefore suffice as the hottest spot of mushroom diversity (Myers et al., 2000)

According to Simmons (2003) litterfall is relevant to the movement of various organic and inorganic matter through woodland ecosystems especially those characterised by a rich expanse of heterogeneous trees and ecosystem services. Proctor (1983) and Dantas and Phillipson (1989) observed that litterfall is important in the estimation of primary productivity, stand vitality, indices of seasonal phenomena related to plant phenology, energy and nutrient fluxes, and as bioindicators of ecosystem health. The word litter is reported by McIntosh (1964) in the concise oxford dictionary, and Eagle and Hawkins (1975) in the oxford illustrated dictionary of English to mean (i) rushes, straw and other materials used in making animal beddings; (ii) straw and dung for farmyard; (iii) state of untidiness or disorderly accumulation of papers or make place untidy, scatter or leave lying. Ecologically, litter referred to a layer of dead plant material or any material especially of organic origin lying on the surface of the soil such as shed animal skin, plant parts or organs (Proctor, 1983; Simmons, 2003). This material does not however include standing dead matter such as tree stumps, dead trees and felled tree trunks which render the aforementioned definitions contextually unsatisfactory to an ecologist concerned with the functions of an ecosystem. Furthermore, Proctor (1983) defined litter as dead or decaying organic matter whose source may be from above or below ground plant parts while Maguire (1994) and Mudrick et al. (1994) remarked that it represents a major biological pathway for essential elements transfer from vegetation to soil and vice versa. Litterfall is therefore defined as the organic debris or litter falling from the above ground parts of a plant onto the forest or plantation floor (Onyibe, 1990). In the same vein, Proctor et al., (1983) and Clark et al. (2001) described it as the pathway for the transfer of organic and chemical elements from vegetation to the soil surface in forest ecosystem. In contrast to this, Simmons (2003) defined litterfall as the constant rain of organic debris on the forest floor while also describing it in a functional term as the transfer of organic matter (carbon, energy and nutrients) from the tree canopy to the forest floor. Consequently, the characteristic components of litterfall will include leaves, buds, twigs, flowers, fruits, seeds, glumes and coarse woods of not more than 2cm diameter of which only the leaf litter has been extensively studied with work on all other components dearth (Vitousek, 1984; Proctor, 1984; Dantas and Phillipson, 1989; Simmons, 2003). The litterfall therefore constitutes over 65% of the woodland degradable biomass while organic biomolecules of animal origin and standing dead or felled trees composed the remaining 25%. This reaffirms the irreplaceable role of forest litters as the conduit in the cycling of elements valuable for tree nourishment and soil *tilt* (English term that describes the structure and quality of a soil), growth and healthy performance of the forest ecosystem. Fungi along with bacteria rely mostly on the litters as their primary food resource and stimulus in the ceaseless process of decomposition and habitat selectivity. This natural process is harnessed in organic farming to for composting, humus (humic substances) formation and other geomycological mineralization processes.

Literatures are equally numerous on litterfall estimates either relative to their rate of accumulation or disappearance or decomposition or nutrient content in various woodland stands across the world spreading through both temperate and tropical climates (Aerts, 1997). Vitousek (1982), Simmons (2003) and Vallinga (2004) reported that the ratio of leaf fall to litter accumulation is higher in the tropics and low at higher latitudes. This acknowledges the fundamental and simultaneous effect of climate and edaphic variables on dead forest biomass decomposition activity as well as the metabolic diversity and activities of microbiotic degraders (Weedon et al., 2009). Woodland litters have also been widely studied in relation to fauna and flora diversity, ecological performance and overall forest productivity (Dantas and Phillipson, 1989; Molofsky and Augspurger, 1992; Finotti et al., 2003). Reports are however scanty on the interaction between litterfall, mushrooms and climate relative to elemental extraction, accumulation and translocation, and how this interaction affects mushroom metabolic selectivity and diversity structure in woodland ecosystems across temperate and tropical latitudes. The recognition of the forest as the largest carbon pool and the attendant understanding of the variations in the physical chemical and structural properties of litter elements are consequently fundamental to estimating decomposition rates and niche selectivity **in** any woodland ecosystem. Invariable, more researches to uncover the total intricate workings and ecological connectivity between mushroom eco-processes (e.g. mineralization, comminution, mobilization, utilization by metabolism or bioaccumulation and other woodland ecosystem variables of climate, especially moisture and temperature) and tree diversity or trait structure are still ongoing.

Information and studies on the eco-diversity of mushrooms limited in scope, scanty and failed provide true representation of global diversity, nature, heritage and resources. Therefore, it is important to connect with a global initiative to document, create inventories and identify more mushroom genetic resource. The knowledge may just be important in (i) identifying species vulnerable to extinction, estimating extinction rates and establishing sustainable conservation strategies or basis for preserving mushroom hotspot areas, (ii) estimating, monitoring and predicting mushroom diversity loss and extinction rates, (iii) contribute to global inventory of mushrooms with the discovery of previously cryptic (undescribed) taxa, habit and habitat (and niche), (iv) discover more ethnomycological knowledge and practices especially in other remote areas of Africa and Asia, this may be as variable as there are cultures, (v) boosting understanding of mushroom seasonality and other ecological behaviours that affect the provision of healthy ecosystem functions, goods and services, (vi) identifying and separating resident (indigent) and immigrant (introduced) mushroom resources and, (vii) generating data that could select mushrooms as very sensitive bioindicator of climatic change. Mushrooms from various empirical studies are incontrovertibly a major nexus between plants, their nutrient availability and nutrient resource utilization through the use of innumerable hydrolysing enzymes in natural elemental cycling phenomena.

## MUSHROOMS VERSUS TREES

Mushrooms and plants especially trees have *ab initio* enjoyed a ceaseless relationship that could be dated back to many millennia long before the recognition of mushrooms by humans. The extent and the nature of this relationship are constantly evolving with the environment and may not be fully understood by modern science. However, mushroom diversity estimates by Hawksworth (1991 and 2001) that put global fungi diversity at 1.5 million was extrapolated from native plant species diversity, reemphasizing the inherent mutuality of the two groups of organism. Subsequent attempts at diversity estimate of terrestrial fungi including mushrooms could not also be divorced from trees/plants diversity (Manoharachary et al., 2005; Crous et al., 2006; Chaverri and Vilchez, 2006; Wasser, 2007). Mueller et al. (2006) and Hammond (1992) reaffirms this hypothesis by identifying that the ratio of plant-fungus is fundamental to arriving at the actual global estimate of mushrooms. Estimating mushroom diversity in Africa is constrained because the forest and trees were badly documented, confounded by latitude and gradient coupled with dearth of grants and expert mycologists especially mushroom taxonomists (Gryzenhout et al., 2012). These coupled with the emerging controversies on fungal pleomorphism could disconfirm existing modest assumption of global fungal estimate and contributed directly or indirectly to the poor fungal biodiversity data from Africa. Therefore, it is logical to hypothesise that the architectural end result of any woodland ecosystem across the globe is hitherto predetermined by a plethora of complex interactive networks. We may also affirm that the woodland ecosystem provides more resources, niche opportunities and surfaces for fungi to thrive due to its profound carbon and mineral storage capacity. In view of this, it is expected that global fungal biodiversity is to be more than what is estimated by Hawksworth (1991) and Wasser (2007). Consequently, Labarère and Menini (2000) concluded that the knowledge of wild mushroom species in the world over is poor and challenging researchers into undertaking more studies on the biodiversity of fungi especially those that are macroscopic and less cryptic. This is to improve the existing record on the 7% (about 100,000) fungi and 10% (14,000) mushrooms estimated by Wasser (2007), as species already described globally.

The woodland patches around the globe represent large pools of carbon which are mobilized by a conglomerate of microorganisms from litterfall and dead coarse woods in a phenomenal biogeochemical process that stabilizes the ecosystem. Detritus food web and forest floor decomposition dynamics is driven by fungi activities that provide the impetus for detrital ecological succession involving bacteria, actinomycetes and other soil animalcules. The role of fungi and mushroom in the maintenance of the forest (woodland) ecosystem was linked to factors such as feeding habit.

This is exclusively by extracellular digestion of dead woodland biomass using biochemical enzymes and acids, food resource distribution and microenvironmental conditions. Studies have reported the dominance of ligninolytic, hemicellulolytic and cellulolytic mushrooms (Figure 2) as the main decomposers of recalcitrant substances, xenophobic and forest floor litter in many woodland ecosystems (Lynch and Thorn, 2006). Consequently, this causes the release of renewable nutrient resource for use by the trees. Mushrooms also influenced the long-term speciation, natural selection of resistant trees and the slow disappearance of susceptible ones to disease causing attributes. Although, mushrooms have only been associated with narrow disease causing potentials in plants compared to their microfungi counterpart, they are still capable of depressing net productivity of any poorly managed forestry and agroforestry.

The mechanism of competition by mushrooms for limited forest floor resources and space with other microorganisms and/or detritus feeders may have stimulated the evolution of acquired mutualistic behaviour with trees, consequently tilting the complex ecological balance of ecosystem functions, metabolic diversity of decomposers and community stability of the woodland ecosystem. The ecological role of fungi in the recycling of carbon, nitrogen, oxygen, phosphorus, water and other minerals in woodland ecosystems is pivotal in determining the net productivity, regeneration capacity and plant species composition of any vegetation.

Humans have for many generations relied on woodland ecosystems for diverse sustainable non-wood resources and products including edible and medicinal mushrooms. They have contributed further to the extinction of trees and invariable loss of mushrooms diversity due to geometric population expansion pressure and activity. This ecosystem is incontrovertibly the largest repository (pool) of mushrooms across the globe which also reemphasises their significance to the sustenance of healthy woodland vegetations.

Limited trees in diverse woodland have been reported to evolved root-pathogen defensive mechanisms and mobilize nutrients more efficiently through the assistance of macrofungi inhabiting their root system (mycorrhizae). The fungus help trap and/or manufacture nitrate based nutrients for the host plant while receiving energy-rich carbohydrate resource in return.

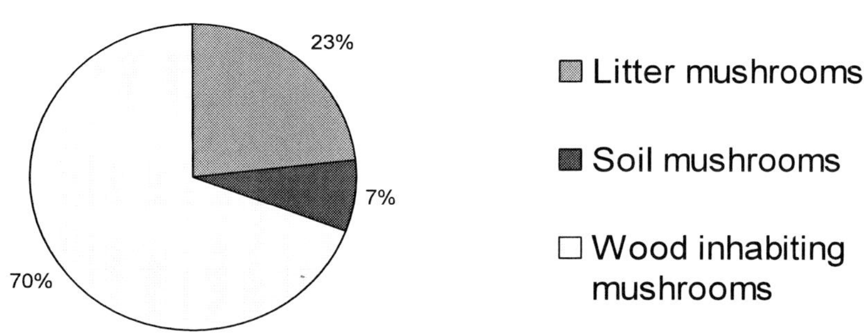


Figure 2. The distribution of different arbitrary group of woodland fungi from a study of forests and agroforest survey in Edo State, Nigeria (Osemwegie, 2008).

This is in addition to their role in the fighting off invasion of the host by other potentially dangerous soil associated pests and pathogens through the expression of allelopathic characteristic, superior feeding vigour, and uncanny antagonism. Knowledge has also emerged on many other fungi of diverse nature that are located within the host plant tissues called endophytes. They co-evolved with many species of trees in a symbiotic relationship with the mechanism of symbiosis currently not well understood but hypothesized as asymptomatic. Empirical work of Vega et al. (2008) showed that endophytes may benefit the host plants by preventing pathogenic organisms from colonizing it, creating a "barrier effect" where it outcompete the invading pathogens or produce chemicals inimical to the growth of competitors. Fungi that are categorized as endophytes also contribute positively to the growth and transpiration rates of the host coupled with protection against drought, heat and overgrazing due to localised toxic content (Gao et al, 2010). Majority of soil fungi are efficient in nutrient mineralization, bioweathering and $CO_2$ evolution while also regulating the turn-over of carbon and nitrogen.

Mushrooms and other non-mushroom forming fungi are parasitic and often looked upon as evil while expressing their natural roles in maintaining global equilibrium through processes that create new habitat and drive evolution. Based on this, humans may need better understanding of parasitic fungi, the management of ecosystem economy and their contiguous ecological significance to mount restraint against *fungal annihilation*. The constant use of synthetic chemical fungicide to kill parasitic members of fungi is recently identified as having negative consequences on the evolution of economically valuable endophytes that could benefit humans especially in the area of medicine. It is also conceptual that some mushrooms are pleomorphic in nature and the sexual teleomorphic mushroom phase may have a parasitic anamorphic representative vulnerable to chemical attacks by humans. Although the effect of constant human attacks of parasitic anamorphs on their corresponding mushroom teleomrphs is not fully understood but may result in the decimation of mushroom diversity and eventual extinction of hyper-susceptible species, loss of potentially valuable fungal resources and productivity of vegetations (woodland ecosystems). Parasitic fungi may however attain equilibrium with their host in time and space, transmuting into potentially useful endophytes. The persistent chemical attacks on fungal parasites predates and drives the evolution of resistant species, inject toxin or residue into ecosystem food chain that may become biological magnified from one trophic level to another, and the environment with potentially dangerous consequences to humans (Kaewchai et al., 2009). The voracious nature of fungal nutrition results in the subterranean spread for miles of mycelium foraging for carbon and nitrogen, and creating a compact network capable of protecting topsoil leaching, *in situ* mycofiltration as well as adsorption of carbon/nitrogen, bacterial, insect and nematode elements from runoff, mobilization of usable nutrient elements valuable to the health of plant vegetation and inclined interaction with earth metals. Mushrooms therefore have the selective advantage in wastewater, effluent and sewage treatment *in vitro* studies over bacteria and other soil organisms. Furthermore, their broad ecological functions in the delivery of ecosystem's goods and services is earnestly harnessed for the improvement of humans livelihood (Figure 3) in the areas of agriculture (bioferfilizers, biocontrol), environment (biodegradation, bioremediation), industries (brewing, pharmaceutic, nutrition, biopulping, biogas, biofuel, dye extraction) and economy (mushroom farming, food security, mushroom retailing/commerce). Therefore, the question is whether mushrooms are worth preserving despite their unobtrusive nature and if there is enough political will to form and implement policies that can force preservation agenda in line with the global biodiversity treaty of the 1992's Rio de Janeiro Biodiversity Summit. It also supported the conceptualization that fungi activity structures woodland ecosystem community and also made penetration of such ecosystem by humans for gathering of wood and non-wood products including game hunting possible. These phenomenal human activities also benefit the mushrooms in spore dispersal and dissemination. Mushrooms liberalise the premier nutrient resource pool of many woodland ecosystem and generate the force of inertia that drive the establishment of secondary succession as well as sustained climax vegetation. Otherwise, forest lives would have been suffocated by a growing heap of undegrading mass of organic wastes. This would equally further have a negative feedback effect on both domestic and industrial waste management processes; biodiversity and human lives in the long-term emphasizing the inseparable link between mushrooms and woodland vegetations.

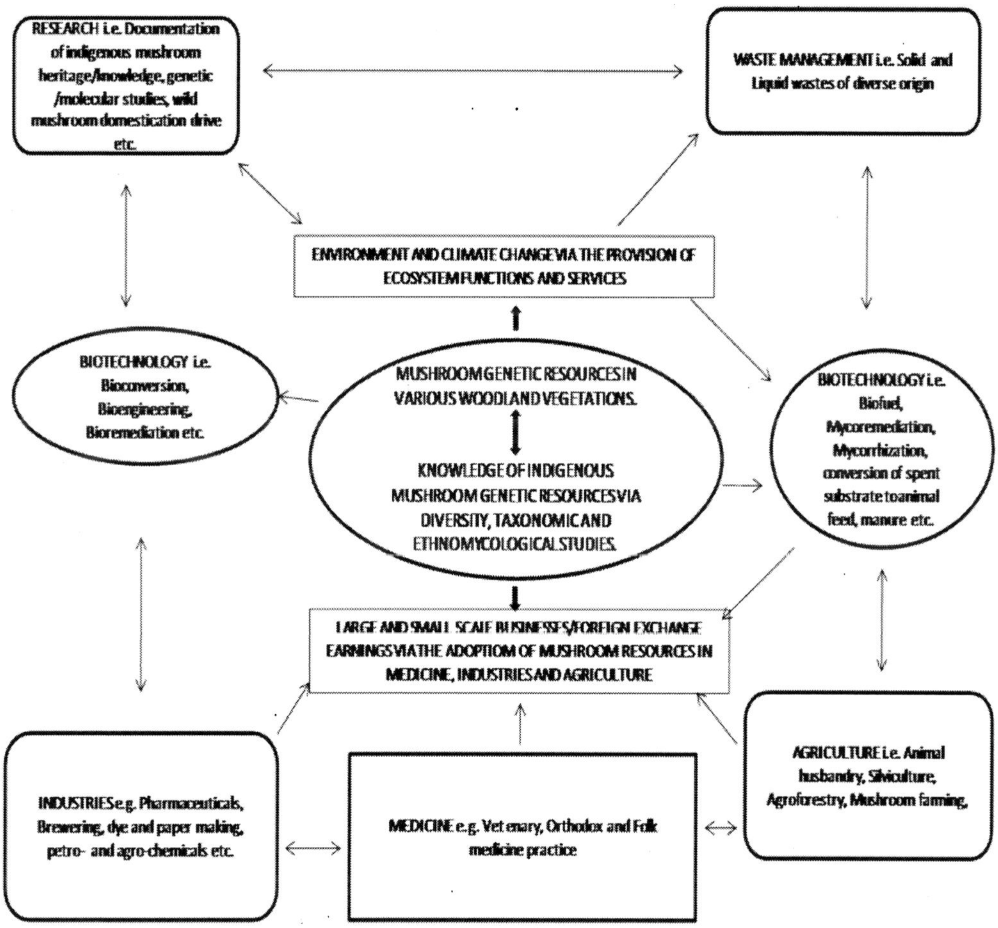


Figure 3. Model illustrating woodlands as mushroom hotspots and multidimensional benefits of mushrooms to humans.

In recent emerging studies on soil microbial decomposition, plant and mushroom biodiversity, forest management practices, regeneration and forest structure across the world emphasized the obligate roles of mushrooms, a subgroup of the fungi race, in the delivery of inherent ecological goods and services that support woodland ecosystems. The multidimensional services comprised the promotion of optimal nutrient uptake, pathogens suppression as well as their inocula through superior competitive capacity for food and niche. In addition, mushrooms contribute to the regulation and mineralization of nutrients by biodegrading ecosystem wastes amongst others and natural mycofiltration processes that remove elements from topsoils capable of potentially suppressing the healthy development of plant vegetations. The dimension of mushrooms ecological versatility underscores their connection to the woodland ecosystems. This study has provided insights into how tree or plant diversity loss results in a corresponding loss of mushroom diversity.

## DISCUSSION

Fungi, more especially mushrooms are proven to be crucially important in the improvement of humans and animals through their mysterious capacity to deliver efficiently ecosystem goods and services that improves forest health, productivity and products. Deregulated exploitation of diverse woodlands for edible and medicinal mushrooms would have unconsciously but logically driven some mushrooms to extinction and causes selective evolutionary changes in others through speciation. Although, a large part of the local populace provided a well-articulated explanation that bothers on the priority to feed families and improve livelihood while misunderstanding the paired capacity between utilization to quench food need or poverty and conservation (Boa et al., 2004) The negative consequences of wild mushroom exploitation for human purposes are more popular in literatures than the positive effects that are slower in manifesting. This is confined in the dynamism of evolution of trees, mushrooms and detritus animals dependent on mushroom feed. The transmutation of the naturally occurring phenomena of fungi ecosystem activities for the benefit of humans is still evolving globally as some of the activities may yet be cryptic or artificially irreproducible. This lends credence to the fact that the importance of fungi contrasted widely to their extant study capacity and it is by no means a happenstance that African forests abound with explorative opportunities in mycology with diverse species of new fungal taxa still cryptic (Gryzenhout et al., 2012). Fungi are capable of serving as renewable raw materials for industries dealing in mycopestides i.e. mycofungicides, mycoherbicides, mycoinsecticides etc., paper (biopulping), dye, wasteland development, fertilizers (mycofertilizers), animal feed (single cell protein, probiotics, additive etc), waste management, medicine, drugs and hormones etc. but currently under explored in Africa (Figure 3).

Human activities, especially deforestation, negatively affect the mushroom population and community composition leading to loss of diversity, decelerated rate of biogeochemical recycling, depleted potential habitat and ultimately climate change. Mushrooms are also suitable bio-indicators of woodland ecosystem stress and disturbances parameters resulting from forest management practices or a healthy forest. It is therefore pertinent to initiate a sustainable regime of forest conservation and preservation practices especially in African that would preserve the continent's vegetations and biodiversity heritage. It is also essential to challenge researchers and ecologist in exploring cheap, more environmentally-friendly alternatives to chemical eradication of economic parasites and/or protecting fungi from activities that facilitate their decimation.

Suffice to say that low to zero parasite population could result in life-threatening ecosystem consequences that may compromise the development of immunity or loss of host suppression of pathogen activity and limit the evolution of symbiotic species. This perception obtains from the theory that suggested that symbiotic relationships evolved from a long line of parasitism that culminated in host-parasite equilibrium, a point where the parasite becomes powerless in causing host infection.

## Conclusion

Mushrooms have remained an invaluable but obtrusive part of fungi currently evolving an identity as a scientific discipline – *Mushroom Science* and gaining popularity as viable resources for the production of many novel products. The bio-prospecting, characterization and bio-exploration of this group of fungi have suffered great neglect in many African countries and other developing nations of the world. This may have been precipitated largely by ignorance of the inherent potential abound in mushrooms, complete lack of adequate capital, infrastructural and human capacities, non-availability of sponsorship sources (non-governmental and governmental), uncertainty of profit generation and lack of preferential for mushrooms and mycology. It is therefore important to set out measures that help moderate and minimize woodlands encroachment by humans for recreational and non-recreational (logging, farming, construction etc) activities. This can be achieved by introducing conservation laws, red-tapping species at risk of extinction, popularizing domestication initiatives and values of mushrooms to forest health and management through government agencies, and developing through many tertiary Institutions the necessary capacity to sustain as well as improve research with further applications of mushrooms and other fungi in improving livelihood. Furthermore, connecting to global initiative on forest conservation and international collaboration that is based on instrumentation, training, exchanges, funding and resource materials (books, spawns, journals) can also promote bio-exploration of mushrooms in some African countries.

## References

Alabi, R.O. (1991). Mycology and Nigerian culture: past, present and future. In: *Proceedings of 1st conference on African Mycology*, Mauritius, June 10-15 1990. pp 43-52.

Adewusi, S.R.A., Alofe, F.U., Odeyemi, O., Afolabi, O.A and Oke, O.L. (1993). Studies on some edible wild mushrooms from Nigeria 1: nutritional, teratogenic and toxic considerations. *Plant Foods and Human Nutrition* 43:115-121.

Aerts, R. (1997). Climate, leaf litter Chemistry and leaf litter decomposition in terrestrial ecosystems: triangular relationship. *Oiko* 79:439-449.

Akpaja, E.O., Isikhuemhen, S.O and Okhuoya, J.A. (2003) Ethnomycology and usage of edible and medicinal mushrooms among the Igbo people of Nigeria. *International Journal of Medicinal Mushrooms* 5:313-319.

Alofe, F.V., Odeyemi, O and Oke, O.L. (1996). Three edible wild mushrooms from Nigeria: their proximate and mineral composition. *Plant Foods Human Nutrition* 49(1):63-73.

Anoliefo,G.O., Isikhuemhen,S.O and Ohimain,E. (2002). Phytoassessment of soil polluted with Farcados blend crude oil exposed to a white rot fungi; *Pleurotus tuberregium* (Fr.) Sing. In: Proceedings of the 2002 national conference on environmental science and technology. Greensboror, North-Carolina, USA. September 8-10. pp43-57.

Arora, D. (1989). Ethnobotany and plant domestication: Global perspective. In: Mehods and approaches in ethnobotany. K. S. Jain (ed.). Society of ethnobotany. Lucknow. pp 49-57.

Bates,S.T. (2006). Preliminary checklist of Arizona macrofungi. *Canotia* 2:47-78.

Bhandary, H.R. (1991). Some edible and medicinal mushrooms from Nepal, *Network* (Newsletter of the national committee for APINMAP/S CAMAP, Nepal) 1: 1-2.

Boa, E (2011). From Chipho to Msika: An introduction to mushrooms, trees and forests, In: Mushrooms in forests and woodlands: resource management, values and local livelihoods, Anthony B Cunningham and Xuefei Yang (eds.) Earthscan Ltd, Washington, DC, USA. 217 pp.

Bucher,V.V.C., Pointing,S.B., Hyde, K.D and Reddy,C.A. (2004). Production of wood decay enzymes, loss of mass and lignin solubilization in wood by diverse tropical fresh water fungi. *Microbial Ecology* 43:331-337.

Chang, S.T and Miles P. G. (1993). Mushrooms: trend in production and technological development. *UNIDO-Genetic Engineering and Biotechnological Monitor* 41-42:73-84.

Chaverri, P and Vilchez, B. (2006). Hypocrealean (Hypocreales, Ascomycota) fungal diversity in different stages of tropical succession in Costa Rica. *Biotropica* 38:531-543.

Clark, D.A, Brown,D.W., Kicklighter,J.Q., Chambers,J.R., Thomlinson and Ni, J. (2001). Measuring net primary production in forests: concepts and field methods. *Ecol. Appl.* 11:356-370.

Couteaux, M.M., Bottner, P and Berg, B. (1995). Litter decomposition, climate and litter quality. *Tree* 10:63-66.

Crous, W.P., Rong, H.I., Wood, A., Lee, S., Glen, H., Botha, W., Slippers, B., Beer, Z.W., Wingfield, J.M and Hawksworth, L.D (2006). How many species of Fungi are there at the tip of Africa?. *Studies in Mycology* 55:13-33.

Dantas, M and Phillipson, J. (1989). Litterfall and litter nutrient content on primary and secondary Amasonian 'terra firme' rainforest. *Journal of Tropical Ecology* 5: 27-36.

Dreyfuss, M.M and Chapela, I.H. (1994). Potential of fungi in the discovery of novel, low-molecular weight pharmaceuticals. In: The discovery of natural products with therapeutic potential. V.P. Gullo (ed.) Butterworth-Heinemann, London. pp 49-80.

Eagle,D and Hawkins,J. (1975). *The Oxford illustrated dictionary*. Oxford University Press, USA. 998 p.

Egli,S., Peter,M., Buser,C., Stahel,W and Ayer,F. (2006). Mushroom picking does not impair future harvests-results of a long-term study in Switzerland. *Biological Conservation* 129:271-276.

Finotti,,R., Freita,R.S., Cerqueira,R and Vieira,M.V. (2003). A method to determine the minimum number of litter traps in litterfall studies. *Biotropica* 35: 419-421.

Gao, FK, Dai,CC and Liu,XZ. (2010). Mechanisms of fungal endophytes in plant protection against pathogens. African Journal of Microbiology Research 4(13): 1346 – 1351.

Gilbertson, R.L and Bigelow, D.M. (1998). Annotated checklist of wood rotting Basidiomycetes of the sky Islands in Southeastern Arizona. *Journal of the Arizona-Nevada Academy of Science* 31:13-36.

Gregory P.H (1984). The fungal mycelium: an historical perspective. *Transactions of the British Mycological Society* 82: 1-11.

Gryzenhout,M., Jefwa,J.M and Yorou,N.S. (2012). The status of mycology in Africa: a document to promote awareness. IMA Fiungus 3(1): 99-102.

Hammond, P.M. (1992). Species inventory. In: Global Biodiversity: status of the earth's living resources B. Groombridge (ed.). Chapman and Hall, London. pp 17-39.

Harkonen, M., Saarimaki, T., Nwasumbi, L. and Niemela, T. (1993). Collection of the Tanzanian mushroom heritage as a form of development of cooperation between the University of Helsinki and Dares Salam. *Aquilo Serie Botanica* 31: 99-105.

Hawksworth, D.L. (2001). The magnitude of fungal diversity: the 1.5 million species estimate revisited. *Mycological Research* 105:1422-1432.

Hawksworth, D.L (1991). The fungal dimension of biodiversity; magnitude, significance and conservation. *Mycological Research* 95:641-655.

Idu, M., Osemwegie, O.O and Akinnibosun, H.A. (2006). Floristic diversity of Okomu Forest Reserve in Nigeria. *Phytotaxonomy* 6:14-22.

Isikhuemhen,S.O, Anoliefo,G.O and Oghale,O. (2003). Bioremediation of crude oil polluted soil by the white rot fungus *Pleurotus tuber-regium* (Fr.) Sing. *Environmental Sciences and Pollution Reseach* 10:108-112.

Jain, S.K. (2000). Human aspects of plant diversity. *Economic Botany* 54:459-470.

Kaewchai,S, Soytong,K and Hyde,K.D. (2009). Mycofungicides and fungal biofertilizers. *Fungal Diversity* 38: 25- 50.

Kauserud,H., Stige,C.L., Vik,O.J., Okland,H.R., Hoiland, K. and Stenseth,C.N. (2008). Mushroom fruiting and climate change. *Proceedings of the National Academy of Sciences* 105:3811 – 3814.

Kekawa, T. (2001). Beneficial effects of mushrooms, edible and medicinal, on health care. *National Journal of Medicinal Mushroom* 4:76-79.

Kirk, P.M., Cannon, P., David, J.C and Stalpers, J.A. (2001). Ainsworth and Bisby's dictionary of the Fungi. 9th edition, CABI Bioscience, UK centre. 624p.

Labarere, J and Menini, U. (2000). Collection, Characterization, Conservation and utilization of mushrooms, germplasm resources in Africa. In: The proceedings of the First International Congress for the Characterization, Conservation, Evaluation and Utilization of Mushroom Genetic Resources for Food and Agriculture. FAO Bordeaux/France 9-13, March 1998.

Lelley, J. (1987). Edible mushrooms as a weapon against starvation. *Mushroom Journal of Tropics* 7:135-140.

Lynch, J. D. M. and Thorn, G. R. (2006). Diversity of basidiomycetes in Michigan Agricultural soils. *Applied and Environmental Microbiology* 72: 7050 – 7056.

Maguire, D.A. (1994). Branch mortality and potential litter Fall from Douglas Fir trees on stands of Varying density. *Forest Ecology Management* 70:41-53.

Masuka,A.J and Ryvarden, L. (1993) Aphyllophorales (basidiomycetes) of Zimbabwe. *The African Journal of Mycology and Biotechnology* 1: 24-32.

McIntosh, E. (1964). *The concise oxford dictionary of current English.* Clarendon Press. Oxford. 1558 p.

Miles, P.G and Chang, S.T. (2004). *Mushrooms: cultivation, nutritional value, medicinal effect and environmental impact*. 2nd edition. CRC Press. 480p.

Miles, P. G and Chang, S. T. (2004). *Mushrooms: cultivation, nutritional value, medicinal effect and environmental impact*. 2nd edition. CRC Press. 480p.

Molofsky, J and Augspurger, K.C. (1992). The effect of leaf-litter on early seedling establishment in a tropical Forest. *Ecology* 73: 68-77.

Manoharachary, C., Sridhar, K., Singh, R., Adholeya, A., Suryanarayanan, T.S, Rawat, S and Johri, B.N. (2005). Fungal biodiversity: distribution, conservation and prospecting of fungi from India. *Current Science* 89:58-70.

Mshigeni, E.K. (2005). Unveiling Africa's treasures of medicinal mushrooms: contributions of Professor Shu-Ting Chang. *International Journal of Medicinal Mushrooms* 7:23-28.

Mshigeni,E.K. 2003. Africa's mushrooms: a neglected bioresource whose time has come. *Discovery and Innovation* 15: 121-124.

Mshigeni,E.K., Chang,S.T and Gwanama,C. (2003). Surprises scientific charm, socioeconomic potential, and possibilities in the mushroom world. *Discovery and Innovation* 15: 1-2.

Mueller, M.G., Schmit, P.J., Leacock, R.P., Buyck, B., Cifuentes, J., Desjardin, E.D., Halling, E.R., Hjortstam, K., Hurriaga, T., Larsson, H. K., Lodge, J. D., Way, W. T, Minter D., Rajchenberg, M., Redhead, A.S., Ryvarden, Trappe, M.J, Watling, R and Wu, Q. (2007). Global diversity and distribution of macrofungi. *Biodiversity and Conservation* 16: 37 - 48.

Mudrick, D.A., Hoosein, M., Hicks, R.R., Jr and Townsend E.C (1994). Decomposition of leaf litter in an Appalachain Forest: effects of leaf species, aspects, slope position and time. *Forest Ecology Management* 68:231-250.

Myers,N., Mittermeier,G.C., Da Fonseca,B.A.G. and Kent, J. (2000). Biodiversity hotspots for conservation priorities. *Nature* 403(24): 853 – 858.

O'Dell,T.E., Lodge,D.J. and Mueller,G.M. (2004). Approaches to sampling macrofungi. In: Biodiversity of fungi: inventory and monitoring methods. Mueller,G.M., Bills,G.F. and Foster,M.S. (eds). Elsevier Academic Press, San Diego. pp163-168.

Ohga, S., Cho, N.S., Thurston, C.F and Wood, D.A (2000). Transciptional regulation of Laccase and Cellulase in relation to Fruit body formation on the mycelium of *Lentinula erodes* on a sawdust-based substrate *Mycoscience* 41:149-153.

Osemwegie,O.O., Eriyaremu, G.E and Abdulmalik, J. (2006). A survey of macrofungi in Edo/Delta region of Nigeria, their morphology and uses. *Global Journal of Pure and Applied Sciences* 12:149-157.

Oei, P. (1991). Some aspects of mushroom cultivation in developing countries *Mushroom. Science* 13:777-780.

Onianwa, P.C. (1995). Petroleum hydrocarbon Pollution of urban topsoil in Ibadan City, Nigeria. *Environment International* 21: 341-343.

Ochiel., G.S, Evans, H.C and Eilenberg, J. (1997). *Cordycepioideus,* a pathogen of Termites in Kenya. *Mycologist* 11:7-9.

Okeke, B. C., Smith, J.E., Paterson, A and Watson-Craik, I.A. (1996). Influence of environmental parameters on Pentachlorophenol biotransformation in soil by *Lentinula edodes* and*Phanerochaete chrysosporium*. *Applied Microbiology and Biotechnology* 45: 263-266.

Onyibe H.J (1990). Ecology of the weed flora and litter fall in a mature rubber plantation in Bendel State, Nigeria. Ph.D thesis. Department of Botany, University of Benin.

Osemwegie, O. O. (2008). Ecodiversity study of mushrooms in a rubber plantation and lowland forest in Iyanomo, Benin City, Edo State. Ph.D Thesis. University of Benin, Benin City, Edo State, Nigeria.

Quimio, T.H, Chang, S.T, Royse, D.J and Memni, U.G. (1990). Technical guidelines for mushroom growing in the tropics. F.A.O plant production and protection paper, 106. Food and agriculture organization of the United Nations Roma.

Proctor, J. (1983). Tropical Forest litter fall. In: British Ecological Society Rain Forest Symposium. Sutton, S., Chadwick, A and Whitmore, T.C (eds). Blackwell Scientific Publications, Oxford. 566 p.

Proctor, J., Anderson,M., Fogden,S.C.L and Vallack,H.W. (1983). Ecological studies in four contrasting lowland rain forests in Gunung Mulu National Park, Sarawak II. Litterfall, litter standing crop and preliminary observations on herbivory. *Journal of Ecology* 71: 261-283.

Read D.J and Perez-Moreno J. (2003). Mycorrhizas and nutrient cycling in ecosystems. A journey towards Relevance?. *New Phytologist* 157:475-92.

Redhead. S.A (1997). Standardized inventory methodologies for components of British Colombia's biodiversity: Microfungi Resources. Inventory Committee, Vancouver. pp 1-59.

Sala,E.O., Chapin III,S.F., Armesto,J.J., Berlow,E., Bloomfield,J., Dirzo,R.., Sanwald-Huber,E., Huenneke,F.L., Jackson,B.R., Kinzig,A., Leemans,R., Lodge,M.D., Mooney,A.H., Oesterheld,M., Poff,L.N., Sykes,T.M., Walker,H.B., Walker,M. and Wall,H.D. (2000). Global biodiversity scenerios for the year 2100. *Science's Compass* 287: 1770 – 1773.

Shigeki, I., Shigeki, S and Osamu, Y. (1994). Analysis of mushroom diversity in successional young Forests and equilibrium evergreen broad-leaved forests. *Mycoscience* 35:1-14.

Simmons, A.J. (2003). Litter fall: A laboratory exercise in ecology. *Watershed Methods Manual.* pp1-9.

Stamets,P. (2005). *Mycelium running: how mushrooms can help save the world.* Ten Speed Press, Berkeley, CA. 339 p.

Takashi, O. (2007). Ecology of ligninolytic fungi associated with leaf litter decomposition. *Ecological Research* 22: 955-974.

Vega FE, Posada F, Aime MC, Pava-Ripoll M, Infante F, Rehner SA (2008). Entomopathogenic fungal endophytes. *Biol. Control*, 46: 72-82.

Vitousek, P.M. (1984). Litter fall, nutrient cycling and nutrient limitation in trophical forests. *Ecology* 65:285-298.

Wasser, P. S. (2007). A book review: Mycelium running: How mushrooms can help save the world. *Herbalgram* 76:50-57.

Waring, R.H and Schlesinger, W. H. (1985). *Forest ecosystems: concepts and management.* Academic Press, Orlando. 340p.

Weedon,J.T., Cornell,W.K., Cornelissen,J.H.C., Zanne,A.E., Wirth,C and Coomes,A. (2009). Global meta-analysis of wood decomposition rates: a role for trait variation among tree species? *Ecology Letters* 12: 45 – 56.

Wood, D.A (1984). Microbial processes in mushroom cultivation: a large scale solid substrate fermentation. *Journal of Chemical Technology and Biotechnology* 34B: 232-240.

Zadrazil., F. (1980). Conversion of different plant wastes into feed by Basidiomycetes. *European Journal of Applied Microbiology and Biotechnology* 9: 243-248.

In: Plantations
Editor: Hai Ren

ISBN: 978-1-62808-090-2


*Chapter 3*

# THE USE OF FOREST PLANTATIONS IN THE SEMIARID SAHEL REGIONS: IMPACTS ON THE ABUNDANCE AND DIVERSITY OF SOIL LEGUME-NODULATING RHIZOBIA AND ARBUSCULAR MYCORRHIZAL FUNGAL COMMUNITIES

***Godar Sene[1,2*], Mansour Thiao[1,3], Anicet Manga[4], Seynabou Sene[1,3], Damase Khasa[5], Aboubacry Kane[1,3], Mame Samba Mbaye[3], Ramatoulaye Samba-Mbaye[1,3] and Samba Ndao Sylla[1,3]***

[1]Laboratoire Commun de Microbiologie LCM IRD/ISRA/UCAD, Dakar-Sénégal
[2]Institut Sénégalais de Recherches Agricoles, Bel-Air Dakar-Sénégal
[3]Département de Biologie Végétale, Faculté des sciences et Techniques, Université Cheikh Anta Diop, Dakar-Fann, Sénégal.
[4]UFR S2ATA, Section Production Végétale et Agronomie, Université Gaston Berger, Saint Louis-Sénégal
[5]Centre d'Étude de la Forêt, Chaire de Recherche du Canada en Génomique Forestière et Environnementale et Institut de Biologie Intégrative et des Systèmes, Université Laval, Québec, Québec, Canada

## ABSTRACT

Several fast-growing and multipurpose tree species such as exotic and valuable native species have been widely used in West Africa. These man-made forest plantations usually focused upon the trees and are defined mainly in relation to their capacity to produce timber and prevent catastrophic events such as damage by wind. In recent years,

* Corresponding author: E-mail: senegodar@yahoo.fr/senegodar@gmail.com Tel.: (+221) 77 515 82 63; Fax: (+221) 33 849 33 02.

however, there has been a growing awareness amongst plant ecologists and soil microbial ecologists that understanding the connectivity between their study organisms is of utmost importance. The interactions between plants and soil microorganisms are particularly important because plants represent the main pathway through which carbon, the element that severely limits microbial growth, enters into soil. From a reciprocal viewpoint, microbial associations have been pointed as an important strategy to guarantee plant survival under semiarid conditions. However, there are several recent studies that have been carried out on the devastating ecological impact resulting from anthropogenic dispersal of exotic plants. They suggest that exotic tree and shrub plantations could interact with soil microbial communities and disrupt mutualistic associations between the existing ecological associations within native communities, leading to soil fertility depletion.

Given the growing body of empirical evidence of the importance of these perennial plants, it is becoming increasingly important to understand the impacts of such dispersal of man-made forestry on soil biological properties. In this chapter, we will expose and discuss some of the relevant research work that has been implemented in Sub-saharian ecosystems with special emphasis on studies that have dealt with tree plantation impacts on soil microbial communities. Particularly, we discussed here the influences of 28-years-old tree plantations of tropical, subtropical, and exotic tree species on the soil legume-nodulating rhizobia and arbuscular mycorrhizal fungal communities.

**Keywords:** Degradation of Sahelian Ecosystems; Reforestation Systems; Ecological Impacts; Soil Fertility; Rhizobia; Mycorrhizae

## 1. Introduction

Degradation of forest lands due to agricultural expansion, overgrazing, fire, and tree cutting is a serious environmental concern in many Sahelian countries (Donfack et al., 1995; Diouf et al., 2002; Tappan et al., 2004; Bai et al., 2008; Vincke et al., 2010). This leads to losses of biodiversity, a decline in soil fertility, and a deterioration of soil physical and biological properties (Tiffen and Mortimor, 2002; Mungai et al., 2005; Guadarrama et al., 2008; Curlevski et al., 2010; Stürmer and Siqueira, 2011). In such a degraded land, the regeneration with tree forestry upstream is threatened, suggesting urgent needs of ecological rehabilitation through which soil quality could be improved to support biological productivity. There is an increasing evidence that forest plantations can play a key role in ecosystem rehabilitation or restoration (Donfack et al., 1995; Fortin et al., 2008; Vincke et al., 2010), and many studies have supported that the presence of the tree species provide a number of ecological advantages through increased soil organic matter content, biodiversity conservation, and improved soil microbial activity, and nutrient cycling rates (Peichl et al., 2006; Hobbie et al., 2006, 2007; Rivest et al., 2009). Hence diverse exotic and indigenous tree species, economically important, with edible and highly nutritional crops for both human and animal consumption have been widely used in the West African area (Ingleby et al., 1997; Diagne et al., 2001; Douf et al., 2002; Duponnois et al., 2005). However, the models and ideas guiding these applied fields have mostly come from sources other than restoration ecology and conservation of soil biology (Kisa et al., 2007; Faye et al., 2009; Bilgo et al., 2012).

Man-made forest plantations in these areas usually focused upon the tree species, and are defined mainly in relation to their capacity to produce timber and prevent catastrophic events such as damage by wind (Diouf et al., 2002). Such concepts of management practices, which did not consider the other resources existing in the ecosystem, lack an ecological basis (Fortin et al., 2008). It has been demonstrated that aboveground and belowground communities are intimately linked (Bever et al., 1997; van der Heijden et al., 1998, 2006, 2008) and these linkages greatly affect ecosystem properties (Xingjun et al., 2005; Sanon et al., 2009). The changes in plant communities will therefore exert a great impact on the belowground microbial communities (Sene et al., 2012a, 2012b, 2013). However, available data on the impact of vegetation changes (tree plantations) on the soil microbial communities are rather scattered, and there are still many gaps in our knowledge. The subject has been ably studied by Akpo et al. (2003), who draw attention to the role of the tree plantations on the belowground flora and soil chemical conditions. Soil microbes play key roles in ecosystems and mediate many ecological processes that are central to ecosystem functioning, including nutrient acquisition (Dommergues et al., 1999; Sene et al., 2010), nitrogen cycling (Fortin et al., 2008), carbon cycling (van der Heijden et al., 2008), soil formation (Rillig and Mummey, 2006), decomposition processes (Peichl et al., 2006; Hobbie et al., 2006, 2007), and the regulation and maintenance of plant biodiversity (van der Heijden et al., 1998, 2008). Further, biotic and environmental factors drive the activity, structure, and diversity of soil microbial communities, which are controlled by many factors including plant species (van der Heijden et al., 1998, 2008). In this chapter, we will expose and discuss some of the relevant research work that has been implemented in Sahelian ecosystems with special emphasis on studies that have dealt with tree plantation impacts on soil microbial communities. Particularly discussed are the influences of exotic and indigenous tree species on the soil symbiotic microorganisms (legume-nodulating rhizobia and arbuscular mycorrhizal fungal communities).

## 2. Soil Microbial Communities: Biological Means of Improving Nutrient Uptake in Trees

Soil and water are the basic natural resources essential for ecosystem functioning and soil is a non renewable source upon which the mankind depends for his survival. From a biological perspective, soil constitutes a diverse ecosystem in which plant roots and microorganisms coexist and interact for nutrient uptakes (Dommergues et al., 1999; Smith and Read, 1997, 2008). Microbial communities in the soil are composed of bacteria, actinomycetes, fungi, protozoa, and nematodes. Plants interact with guilds of these beneficial microorganisms like $N_2$-fixing bacteria (Kahindi et al., 1997; Dommergues et al., 1999) and mycorrhizal fungi (Smith and Read, 1997, 2008), stimulating their productivity by supplying limiting nutrients (Bala and Giller, 2006; van der Heijden et al., 2008; Schnitzer et al., 2011).

Among the plant-microbe interactions, the legume-rhizobia symbiosis that converts nitrogen gas ($N_2$) into ammonia is perhaps the best studied (Kahindi et al., 1997). The fixation of $N_2$ makes up to approximately 80 % (by volume) of the air that we breathe (Dommergues et al., 1999). The nitrogen gas cannot be used directly by plants or animals to synthesize essential biological nitrogen-containing compounds such as amino acids and nucleotides

(Kahindi et al., 1997; Dommergues et al., 1999). First, it has to be converted into ammonia and, among the wide range of bacteria that could fix $N_2$, legume-nodulating rhizobia (LNR) are of the most important for their potential in maintaining soil fertility (Graham, 2008; Noel, 2009; Lindström et al., 2010).

Mycorrhizal fungi are soil-dwelling microorganisms that form mutualistic relationships with over 80% of all vascular plants and affect plant fitness and competitive interactions (Johnson et al., 1997). They are commonly known not only to assist host plants with phosphorus uptake (Smith and Read, 2008), but may also to provide other benefits including protection from pathogens (Cardoso and Kuyper, 2006), assisting with the uptake of water and other nutrients such as nitrogen and copper (Smith and Read, 2008; Sene et al., 2010). Hyphae of arbuscular mycorrhizal fungi (AMF) also play a role in the formation and structural stability of soil aggregates (Rillig and Mummey, 2006), and contribute to the composition of plant community structures (van der Heijden et al., 2006). In return, LNR and AMF receive photosynthetic products from the host plant (Dommergues et al., 1999; Smith and Read, 2008). Various studies suggested that a specific interaction that influences both the nodulation and mycorrhizal colonization processes occurs between the LNR and AMF in legumes (Kahindi et al., 1997; Duponnois et al., 2005; van der Heijden et al., 2008; Sene et al., 2010). It is well documented that growth and productivity of the legumes were dependent on the combination of LNR and AMF, revealing that positive interactions between compatible symbionts could significantly increase growth and productivity. Container experiments with peanuts (*Arachis hypogaea* L.) demonstrated that under open air environmental conditions, biomass production and pod yield in dual inoculated plants with bradyrhizobia and mycorrhiza is generally greater than in non-inoculated plants, and more nodules were formed (Sene et al., 2010). Numerous other papers have also reported that both the nitrogen-fixing bacteria and the mycorrhizal fungi play important roles in nutrient relationships in natural forest stands (van der Heijden et al., 1998, 2006, 2008; Fortin et al., 2008; Noel, 2009) and undoubtedly they will play significant roles in man-made forest systems. The LNR as well as the AMF are thought to play an important role in the establishment and succession of plant communities (Cardoso and Kuyper, 2006; van der Heijden et al., 1998, 2008; Oehl et al., 2010; Schnitzer et al., 2011). The natural indigenous population of these soil symbiotic microorganisms, if available, could therefore facilitate ecosystem resilience when faced with environmental disturbances. However, benefits of the tree plant-microbe symbiosis, whether the food, fodder, soil fertility, or forestry, will depend on the management of natural biodiversity of both the plant host and the symbiotic microorganisms present in the soils. Only when we better understand the factors regulating this biodiversity we will be able to conserve it for the future and exploit it fully in Sahelian ecosystems. In this chapter, we outline the ecological role of man-made forestry on the soil symbiotic microorganisms in the semi-arid Sahel region.

## 3. Influence of the Man-Made Tree Species on the Indigenous Microbial Communities

The main prerequisites for choosing plant material is to select species which are adapted to the biotic (pests) and abiotic (climate, soil) environment of the specific site (Larsen, 1995; Diouf et al., 2002). Local species and populations usually ensure a certain degree of adaptedness, since the population genetic structure reflects the fluctuations in local environmental forcing functions (Forrester et al., 2006; Lemma et al., 2007). A recent study by Ndoye et al. (2012) in two contrasting rainfall sites of Senegal has indicated positive effects of *Acacia senegal* trees on soil mycorrhizal potential and enzyme activities. In contrast, there are some examples which have demonstrated the dangers of using unadapted plant species (Xingjun et al., 2005; Silva et al., 2007; Jordan et al., 2008; Mascaro et al., 2008). To date, relatively few studies have investigated in this aspect in the semi-arid Sahel region, and they were only focused on exotic plant species. The main conclusions drawn from these studies were that the exotic tree species being propagated in this area apparently have a major influence on the soil belowground microbial communities and soil physic-chemicals (Kisa et al., 2007; Remigi et al., 2008; Faye et al., 2009). For instance, in a study in Senegal at Ngane, Remigi et al. (2008) showed that *Acacia holosericea*, an exotic tree species, induced strong modifications in soil microbial functionalities (assessed by measuring the patterns of *in situ* catabolic potential of microbial communities) and reduced soil resistance in response to increasing stress or disturbance (salinity, temperature, and freeze-thaw and wet-dry cycles). The authors also indicated that *A. holosericea* strongly modified the structure of AMF communities. Such a trend of ecological effects of exotic tree plant introduction was further confirmed in a study by Bilgo et al. (2012), showing strong modifications in soil microbial functions. A recent study from Faye et al. (2009) also highlighted that *A. holosericea* plant species could modify the structure of *Bradyrhizobium* populations and their effectiveness on *Faidherbia albida* (Del.) a. Chev. (a native endemic Sahelian *Acacia* species) growth. It is worth noting that such a modification could contribute limiting the natural regeneration of *F. albida*. Indeed, it has been previously reported that some legume species require LNR to successfully coexist with other plants in natural and also man-made communities (van der Heijden et al., 2006, 2008). Therefore, disturbance of native LNR communities could alter plant competitive interactions, more particularly, decrease the capacity of legumes to out-compete other plants (Faye et al., 2009).

A study in Burkina Faso at Gampela by Kisa et al. (2007) has also reported that the exotic tree species *Eucalyptus camaldulensis* significantly modified the soil bacterial community. The authors showed that both community structure (assessed by denaturing gradient gel electrophoresis profiles) and function (assessed by substrate-induced respiration responses including soil catabolic evenness) were significantly affected after 12 months of culture. They additionally showed that the changes in the bacterial structure and function were accompanied by disturbances in the composition of herbaceous plant species layer.

# 4. Influence of the Man-Made Tree Species on the Indigenous Soil Symbiotic Microorganisms: Case Studies from the Forest Reserve of Bandia, Senegal

## 4.1. Site Description

To test the hypothesizes related to the ecological impacts of the man-made forest establishment on soils symbiotic microorganisms, Sene et al. (2012a, 2012b, 2013) conducted experiments in different habitats, i.e. planted forest reserve zones and their surrounding and deforested lands at Bandia (14°30' N, 17°0' W) in coastal Senegal (10 km from the Atlantic Ocean).

**Table 1. Summary of plant species composition for each vegetation patch: the deforested and man-made forest systems (adapted from Sene et al., 2012a)**

| | Deforested land | | | | Man-made forest reserve land | | | |
|---|---|---|---|---|---|---|---|---|
| Families | Genera (number) | Relative Frequency (%) | Species (number) | Relative Frequency (%) | Genera (number) | Relative Frequency (%) | Species (number) | Relative Frequency (%) |
| Acanthaceae | 0 | 0 | 0 | 0 | 1 | 1.5 | 1 | 1.1 |
| Aizoaceae | 1 | 2 | 1 | 1.8 | 0 | 0 | 0 | 0 |
| Amaranthaceae | 4 | 8.2 | 5 | 8.9 | 5 | 7.2 | 3 | 3.4 |
| Ampelidaceae | 0 | 0 | 0 | 0 | 1 | 1.4 | 1 | 1.1 |
| Anacardiaceae | 0 | 0 | 0 | 0 | 1 | 1.4 | 1 | 1.1 |
| Anthericaceae | 0 | 0 | 0 | 0 | 1 | 1.4 | 1 | 1.1 |
| Araceae | 1 | 2 | 1 | 1.8 | 1 | 1.4 | 1 | 1.1 |
| Apocynaceae | 1 | 2 | 1 | 1.8 | 1 | 1.4 | 1 | 1.1 |
| Asteraceae | 2 | 4.1 | 2 | 3.6 | 3 | 4.3 | 3 | 3.4 |
| Balanitaceae | 0 | 0 | 0 | 0 | 1 | 1.4 | 1 | 1.1 |
| Capparaceae | 2 | 4.1 | 2 | 3.6 | 3 | 4.3 | 3 | 3.4 |
| Caryophyllaceae | 1 | 2 | 1 | 1.8 | 0 | 0 | 0 | 0 |
| Combretaceae | 1 | 2 | 1 | 1.8 | 1 | 1.4 | 2 | 2.2 |
| Commelicaceae | 1 | 2 | 1 | 1.8 | 1 | 1.4 | 2 | 2.2 |
| Convolvulaceae | 3 | 6.1 | 5 | 8.9 | 2 | 2.9 | 5 | 5.6 |
| Cucurbitaceae | 3 | 6.1 | 3 | 5.4 | 2 | 2.9 | 2 | 2.2 |
| Cyperaceae | 1 | 2 | 2 | 3.6 | 4 | 5.8 | 7 | 7.9 |
| Euphorbiaceae | 3 | 6.1 | 4 | 7.1 | 3 | 4.3 | 3 | 3.4 |
| Fabaceae | 9 | 18.4 | 10 | 17.9 | 11 | 15.9 | 16 | 18 |
| Lythraceae | 0 | 0 | 0 | 0 | 1 | 1.4 | 1 | 1.1 |
| Malvaceae | 2 | 4.1 | 2 | 3.6 | 7 | 10.1 | 7 | 7.9 |
| Meliaceae | 1 | 2 | 1 | 1.8 | 1 | 1.4 | 1 | 1.1 |
| Myrtaceae | 0 | 0 | 0 | 0 | 1 | 1.4 | 2 | 2.2 |
| Onagraceae | 0 | 0 | 0 | 0 | 1 | 1.4 | 1 | 1.1 |
| Poaceae | 9 | 18.4 | 10 | 17.9 | 10 | 14.5 | 15 | 16.8 |
| Rhamnaceae | 1 | 2 | 1 | 1.8 | 1 | 1.4 | 1 | 1.1 |
| Rubiaceae | 1 | 2 | 1 | 1.8 | 3 | 4.3 | 5 | 5.6 |
| Sapindaceae | 1 | 2 | 1 | 1.8 | 0 | 0 | 0 | 0 |
| Solanaceae | 1 | 2 | 1 | 1.8 | 1 | 1.4 | 2 | 2.2 |
| Ulmaceae | 0 | 0 | 0 | 0 | 1 | 1.4 | 1 | 1.1 |
| Total | 49 | 100 | 56 | 100 | 69 | 100 | 89 | 100 |

The climate is typical of semiarid lands of West Africa with mean temperature ranging from 25°C to 35°C, and an annual rainfall ranging from 400 to 700 mm, occurring mostly during a single short rainy season lasting for 2-3 months. Mineral compositions of the soils overly solid sand-clay base materials (Sene et al., 2012a). An inventory of plant species was carried out, and the summary of plant compositions for each vegetation patch is presented in Table 1.

Briefly, the man-made forest land is a composed of 89 plant species including 15 dominant arboreal plant species, 2 shrub species, and 72 herbaceous plant species. Trees are planted in monospecific stands of about 28 years and comprised 9 indigenous species: *A. nilotica* (L.) Willd. ex Del., *A. senegal* (L.) Willd., *A. seyal* Del., *A. tortilis* (Forsk.) Hayne, *Celtis integrifolia* Lam., *Grewia bicolor* Roth, *Sclerocarya birea* Hochst., *Sterculia setigera* Del., and *Zizyphus mauritiana* Lam.; and 7 exotic species: *Azadirachta indica* A. Juss., *E. camaldulensis* Dehnh., *E. microtheca* F. Muell., *Cassia siamea* Lam., *Gliricidia sepium* (Jacq.) Kunth, *Hardwikia binata* Roxb., and *Prosopis juliflora* (Sw.) DC. These species are commonly planted throughout the Sahel and were growing on the same soil type as at Bandia. They were established in 1980 with a spacing of 4.5 x 4.5 m. The shrub plant community is dominated by *Bossia senegalensis*. The common herbaceous plants at the time of sampling were: *Mitracarpus villosus* Cham. and Schltdl., *Acalypha crenata* Hochst. ex A Rich., *Achyranthes aspera* L., *Cucumis melo* L., *Rhynchosia minima* (L.) DC., *Stylochyton hypogaeus* Lepr., *Commelina benghalensis* L., and *A. aspera*; *M. villosus* have the more important patches beneath the tree plantations.

In the deforested land, all woody plants were harvested about 30 years ago, resulting in the disturbed land vegetation cover being dominated by shrubs and annual plant species. This land is composed of 56 species including *A. seyal* and *A. nilotica* trees, which are isolated in individuals, Combretaceae shrubs and annual plant species. The most dominant herbaceous species at the period of sampling were: *Cassia obtusifolia* L., *Brachiaria racemosa* (L.) Stapf, *Mitracarpus villosus* Cham. and Schltdl., *Digitaria ciliaris* (Retz.) Koeler, *Cyperus rotundus* L., *Fimbristylis exilis* Roem. and Schult., *Enteropogon prieurii* (Kunth) Cl., and *Cyperus esculentus* L.; *Cassia obtusifolia* has the more important patch.

## 4.2. Influence of the Man-Made Tree Species on the Indigenous LNR Populations

Many studies have supported that tree plants provide a number of ecological advantages through increasing soil organic matter content, biodiversity conservation, and improving soil microbial activity and nutrient cycling rates (Munguai et al., 2005; Forrester et al., 2006; Peichl et al., 2006; Hobbie et al., 2006, 2007). Studies conducted at the Bandia ecosystem by Sene et al. (2012a, 2013) have shown that significant variability exists in the LNR populations estimated in different exotic and native tree plantations established in this area. The authors showed close positive correlations between the sizes of LNR populations estimated both in the dry and rainy seasons and the presence of legume tree hosts (Sene et al., 2013). More importantly, they indicated significant increases in *Rhizobium* spp. (fast growers LNR) population densities in response to planting with Sahelian acacia species, and a high number of genotypes were fitted to these tree plantations, which undoubtedly suggest that enrichment of soil *Rhizobium* spp. populations is host-specific.

This result is consistent with the known influence of host legume-nodulating species on soil LNR. In fact, it is recognized that Sahelian acacia species may be nodulated by strains of *Rhizobium* spp. and/or by strains of *Bradyrhizobium* spp. (slow growers LNR) (Dreyfus et al., 1988; Bala and Giller, 2006; Diouf et al., 2007). More interestingly, they develop extensive root systems near the soil surface (Gueye M. pers. com.), and usually have a large number of natural nodules during the rainy period (Diouf D. pers. com.). This could naturally facilitate the LNR multiplication and diversification through rhizosphere effects and senescence of nodules as well as provision of organic nitrogen, which could interact with soil organic carbon to enhance soil structural stability (Andrade et al., 2002; Wolde-Meskel et al., 2005; Mwenda et al., 2011). This recalls a study by Lawson et al. (1987), showing large differences in *Rhizobium leguminosarum* bv. trifolii population, which was caused by the presence of homologous legume hosts.

By contrast, Sene et al. (2012a) have further indicated that infrequent or delayed nodulation still occurred in their different trap experiments for soil samples from the deforested lands. This suggests that a low number of LNR reside in these soils (Sene et al., 2012a), and this is particularly worrisome for the LNR belonging to the fast grower groups (*Rhizobium* spp.). Moreover, the authors have indicated that the 16S rRNA sequences analyses have shown reduced diversity of sub-clusters. This pattern could be simply attributed to the scarcity of the tree plants, and the less diverse tree species available in this land use. Because of the limited energy and nutrient sources, only the LNR that could thrive with the few available tree plants could be present, and at low numbers. The authors also showed a low number of LNR populations in plantations where the tree plant does not form symbiotic association with the LNR (Sene et al., 2013), possibly attributable to the lack of homologous hosts, and/or a lower level of saprotrophic adaptation to the environmental conditions beneath these tree plantations. Compared with these studies, the legume-related effects as early observed by Woomer et al. (1988), and recently confirmed by Faye et al. (2009), appear to be an important way to maintain high levels of LNR populations.

### *The Linkages between Plant Species Composition and Soil Microorganisms: What about LNR within Man-Made Tree Plantations?*

Because of the direct ecological linkages existing between plant species composition and soil symbiotic microorganisms (Bever et al., 1997; van der Heijden et al., 1998; 2006), most experiments suggest that disturbance or change in the environment that affect aboveground plants will also affect the soil microsymbionts (Woomer et al., 1988; Kisa et al., 2007; Silva et al., 2007; Remigi et al., 2008; Faye et al., 2009; Bilgo et al., 2012). However, in the case of the LNR as affected by man-made forestry, details of the results may depend on both the composition of natural LNR population and symbiotic characteristics of the present host legumes as reported by Sene et al. (2013). These authors have first revealed that the development of herbaceous vegetation was drastically inhibited in plantations of *G. sepium* and *P. juliflora* exotic species (Table 2), possibly attributable to allopathic and/or competition effects. As it appears to be mainly the case for the AMF (Kisa et al., 2007; Silva et al., 2007; Sene et al., 2012b), the observation mentioned for the herb vegetation compositions apply to the LNR as well (Bever et al., 1997; van der Heijden et al., 2006, 2008). However, Sene et al. (2013) have indicated a high genetic diversity of LNR for *P. juliflora* tree plantation, which did not support the hypothesis that above- and below-ground diversity could be causally related.

Since they have further shown with the cross-nodulation tests that a large spectrum exists for *P. juliflora* tree species, this indicates an extant diversity of compatible LNR across the study area.

The authors have also shown a large spectrum for nodulation for *A. seyal* tree species and accordingly, this tree plantation harbored a high genetic diversity of soil LNR. *A. seyal* tree species is native to a broad belt across Africa between the Sahelian and moist forest zones, from Senegal in the west, to Sudan and Kenya in the east, suggesting adaptation to the LNR from this area. Likewise, the exotic *P. juliflora* was first introduced in Senegal, Africa, in 1822 (Diouf et al., 2002).

**Table 2. Richness of shrub and herb species, diversity indexes [richness (R), diversity (H') and evenness (1-D)] of the IGS-RFLP genotypes, and of the spore morphotypes recorded in each tree plantation (adapted from Sene et al., 2012b, 2013)**

| Tree species | Vegetation | | Diversity index of LNR | | | Spore diversity | | |
|---|---|---|---|---|---|---|---|---|
| | Richness of shrub species | Richness of herb species | *R* | *H'* | 1-D | *R* | *H'* | 1 – *D* |
| Indigenous | | | | | | | | |
| *A. nilotica* | 4 | 20 | 7 | 1.531 | 0.899 | 3 | 0.611 | 0.430 |
| *A. seyal* | 4 | 17 | 15 | 2.963 | 0.962 | 3 | 0.530 | 0.405 |
| *A. senegal* | 3 | 24 | 10 | 1.957 | 0.923 | 3 | 0.603 | 0.428 |
| *A. tortilis* | 5 | 14 | 9 | 1.643 | 0.913 | 2 | 0.491 | 0.313 |
| *C. integrifolia* | 4 | 37 | 5 | 1.228 | 0.754 | 4 | 1.04 | 0.799 |
| *G. bicolor* | 3 | 19 | 3 | 0.747 | 0.585 | 2 | 0.488 | 0.307 |
| *S. birrea* | 2 | 19 | 6 | 1.219 | 0.893 | 4 | 0.663 | 0.503 |
| *S. setigera* | 3 | 17 | 6 | 1.283 | 0.831 | 4 | 0.709 | 0.580 |
| *Z. mauritiana* | 4 | 27 | 3 | 0.786 | 0.532 | 2 | 0.489 | 0.352 |
| Exotic | | | | | | | | |
| *A. indica* | 2 | 15 | 2 | 0.455 | 0.336 | 4 | 0.880 | 0.575 |
| *E. camaldulensis* | 2 | 12 | 5 | 1.095 | 0.787 | 3 | 0.679 | 0.442 |
| *E. microtheca* | 4 | 29 | 6 | 1.382 | 0.893 | 2 | 0.386 | 0.228 |
| *C. siamea* | 4 | 15 | 6 | 1.252 | 0.849 | 2 | 0.410 | 0.324 |
| *G. sepium* | 3 | 8 | 3 | 0.841 | 0.548 | 3 | 0.359 | 0.264 |
| *H. binata* | 4 | 14 | 4 | 0.987 | 0.723 | 4 | 0.747 | 0.506 |
| *P. juliflora* | 3 | 8 | 9 | 1.943 | 0.911 | 3 | 0.669 | 0.441 |

*R* Margalef (richness), *H'* Shannon (diversity), *1-D* Pielou (evenness) indexes.

According to the authors, this old introduction may contribute to evolutionary adaptations between *P. juliflora* and native LNR strains, favouring symbiotic relationships that could explain the occurrence of compatible LNR with *Prosopis* species, highlighting the high genetic diversity of LNR recorded in *P. juliflora* plantations. It has been reported, however, that relative permissiveness of exotic plant species to the diversity of LNR may not guarantee infectiveness and effectiveness in $N_2$-fixation (Faye et al., 2009). Thus, further work is needed to understand the functional significance of the genetic diversity pattern in *P. juliflora* plantations and its impact on soil quality and vegetation structure.

In contrast, Sene et al. (2013) have indicated a restricted host range for the newly introduced *G. sepium* species, suggesting that this exotic legume species may have low compatibility with the native LNR. Moreover, the authors showed that *G. sepium* tree plantation harbored less genetic diversity of LNR. There is a large body of investigations which has supported the idea that the introduction of a leguminous tree plant is capable of promoting the selection of particular LNR genotypes (Xingjun et al., 2005; Faye et al., 2009), by penalizing those fail to fix $N_2$ (Kiers et al., 2003) and thus, reducing the diversity of soil LNR. However, in the case of the *G. sepium* plantation, Sene et al. (2013) have indicated that both the size and genetic diversity of soil LNR are still low; and this suggests that other factors to which the LNR populations are most sensitive could also be considered. One can therefore suspect release of toxins from the tree root exudation and/or leaf litters decomposition, which usually occur in the ecotone of exotic plant species (Kisa et al., 2007; Remigi et al., 2008; Faye et al., 2009; Sanon et al., 2009). These compounds, when accumulated into soil, could not only decrease the belowground herb vegetation, but also act as allelochemicals on LNR growth and survival.

## 4.3. Influence of the Man-Made Tree Species on the Diversity of Indigenous Mycorrhizal Fungi

Propagules of AMF in the soil normally take the form of spores, root fragments or hyphal networks and are sensitive to slight soil modifications by any degrading agent (Smith and Read, 1997; Sanon et al., 2009; Oehl et al., 2010). Since it has been recognized that land degradation can decrease AMF taxa richness (Duponnois et al., 2001), it could be expected to observe smaller numbers of AMF taxa in a deforested land in comparison with a reforested zone.

However, Sene et al. (2012b) have reported that the deforested land ecosystem at Bandia has preserved higher mean spore richness compared to that associated with the man-made forest systems.

More importantly, the larger-spored AMF (*Scutellospora* and *Gigaspora*) were lacking in soils sampled beneath the man-made tree plantations. The displacement of these larger-spored AMF species was further associated with a concomitant proliferation of small-spored (*Glomus*) species, indicating clearly that the community structure of soil mycorrhizal fungi was affected (Sene et al., 2012a, 2012b).

AMF specialists and generalists have been identified before (Oehl et al., 2010), and the data from Sene et al. (2012a, 2012b) also suggest that this group of fungi differs in niche breadth. Therefore, one can hypothesize that factors such as the environmental conditions created by the management options could be responsible to the distribution pattern of AMF rather than the plant species themselves.

In fact, while most experiments suggest that the tree presence induces an increase in soil microbial community (Ingleby et al., 1997; Diagne et al., 2001; Mwenda et al., 2011) details of the results may depend on how the plant species are managed and, if environmental conditions are not conductive, certain plants can favor the development of particular AMF taxa (Alguacil et al., 2010). Further stating such effects of vegetation changes on AMF communities are studies from Zhang et al. (2004), Guadarrama et al. (2008), and Stürmer and Siqueira (2011).

On the other hand, the differences in the community composition of the AMF may be due to soil properties, such as soil P and N contents, that greatly differed between the two land systems (Sene et al., 2012a).

According to the authors, the higher levels in some soil chemical components they observed in the reforested plots may have a negative impact on the AMF species diversity independently of host plants. This statement recalls earlier reports from Johnson et al. (1991), and from Johnson (1993), and recent works by Egerton-Warburton and Allen (2000), and by Alguacil et al. (2010). These authors have reported shifts in AMF communities along gradients of N deposition, soil organic C, or N and P additions. In addition, it has been shown by Oehl et al. (2010) that C and N contents were important factors partitioning niche space within the AMF genera.

## 4.4. Influence of the Man-Made Tree Species on the Mycorrhizal Inoculum Potential (MIP) of Soils

As AMF are believed to be obligatory symbionts, changes in a vegetation cover should be expected to influence the activity and viability of their propagules (Zhang et al., 2004), and this can lead to a significant increase or reduce of the quantity of viable mycelium for colonizing and infecting plant roots (van der Heijden et al., 2006). At first sight, data from Sene et al. (2012b) indicated that most plants found in their field study areas could support the AMF symbiosis (Table 3), which is why AMF propagules were ubiquitous in all soil samples. Nevertheless, the authors showed that the soil MIP decreased in soil samples from the man-made forest systems though there is increase of tree plant covers. This result was not in agreement with the traditional view that MIP was greater around the tree plants (Duponnois et al., 2001), but it conforms with a number of recent studies reporting increased MIP after deforestation (Zhang et al., 2004; Guadarrama et al., 2008), or in different land use areas including mature forest and sites converted to pasture (Stürmer and Siqueira, 2011). According to the authors, the most striking difference between the deforested and man-made forest systems was that the former had fewer trees (mainly remnant to the ancient woodland that once blanketed the area), and less different tree species than those in the latter (Sene et al., 2012a). In such a degraded land pattern, once the tree vegetation was removed, carbon and nutrient cycles can be disrupted and soils exposed to intensive radiation. Nevertheless, the authors have indicated that the deforested land supports dense herbaceous vegetation comprised of plant species highly mycotrophic dominated by *Cassia obtusifolia*, that have certainly evolved distinct mechanisms such as mycorrhizal symbiosis for nutrient cycling and uptake. AMF usually have a positive influence on their host plants when soil resources are limiting, allowing the most efficient plant-AMF combination to eventually dominate and cope with stressful conditions (van der Heijden et al., 1998). If this was the case, the herb vegetation compositions may have, in turn, maintained high numbers of AMF propagules in the soil. These annual herbaceous plants are known to be very efficient in AMF propagules multiplication in the soil surface (Duponnois et al., 2001; Zhang et al., 2004; Guadarrama et al., 2008), and might encourage build up of AMF propagule numbers. This is in accordance with their observation on field AM colonization showing clearly that, root systems of the dominant herbaceous species out of the tree plantations (*C. obtusifolia*, *D. ciliaris*, *Rhynchosia minima*, *Brachyaria racemosa* etc.) were highly colonized (Table 3). This observation was also confirmed by data obtained in *C. integrifolia* plantation where there is a poor survival of the tree plants, and this has lead to an open canopy areas, and more ground vegetation with concomitant inputs of AMF propagules. These data confirm earlier studies, where difference on spore concentrations between tree species was attributed to differences on the ground vegetation under these trees (Ingleby et al., 1997; Diagne et al., 2006). Research carried out in recent decades also corroborates such a statement and emphasizes the role of herbaceous species in sustaining soil mycorrhizal inoculums (Duponnois et al., 2001; Zhang et al., 2004; Diagne et al., 2006; Guadarrama et al., 2008).

**Table 3. Arbuscular mycorrhizal colonization of dominant plants from the deforested (ZHC) and man-made (RES) forest land (adapted from Sene et al., 2012a)**

| | Species | Deforested land | | Man-made forest land | |
|---|---|---|---|---|---|
| | | AM colonization | Confident limits $P < 0.05$ | AM colonization | Confident limits $P < 0.05$ |
| Herb plants | *Acalypha crenata* Hochst. ex A Ric. | 32.31A,def | ± 6.27 | 37.83A,bcd | ± 6.32 |
| | *Achyranthes aspera* L. | 31.70A,ef | ± 5.50 | 32.36A,cde | ± 5.67 |
| | *Brachyaria racemosa* L. | 47.24A,bc | ± 6.96 | 13.90B,gh | ± 3.79 |
| | *Cassia obtusifolia* L. | 79.32A,a | ± 11.71 | 61.63A,a | ± 9.84 |
| | *Dactyloctenium aegyptium* (L.) Wild. | 7.90A,g | ± 2.86 | 6.40A,i | ± 1.97 |
| | *Digitaria ciliaris* (Retz.) Cl. | 62.20A,ab | ± 10.72 | 46.65A,ab | ± 5.39 |
| | *Enteropogon prieurii* (Kunth) Cl. | 28.40A,f | ± 5.04 | 25.04A,ef | ± 4.50 |
| | *Mitracarpus villosus* Cham. and Schltdl. | 41.42A,cde | ± 5.30 | 32.92A,cde | ± 4.51 |
| | *Rhynchosia minima* (L.) DC. | 52.07A,bc | ± 5.60 | 41.18B,bc | ± 4.07 |
| Indigenous tree species | *Acacia nilotica* (L.) Wild. ex. Del. | 5.01B,g | ± 1.68 | 12.22A,ghi | ± 4.86 |
| | *A. senegal* (L.) Wild. | – | – | 14.01gh | ± 3.40 |
| | *A. seyal* Del. | 24.75A,f | ± 5.98 | 20.45A,fg | ± 5.74 |
| | *A. tortilis* (Forsk.) Hayne | – | – | 19.87fg | ± 5.09 |
| | *Celtis integrifolia* Lam. | – | – | 8.95hi | ± 3.06 |
| | *Grewia bicolor* Roth | – | – | 27.8def | ± 6.71 |
| | *Sclerocarya birrea* Hochst. | – | – | 16.38fgh | ± 5.45 |
| | *Sterculia setigera* Del. | – | – | 8.22hi | ± 3.61 |
| | *Zizyphus mauritiana* Lam. | – | – | 11.83ghi | ± 4.83 |
| Exotic tree species | *Azadirachta indica* A. Juss. | – | – | 50.18ab | ± 11.84 |
| | *Eucalyptus camaldulensis* Dehnh. | – | – | 6.05i | ± 1.78 |
| | *E. microtheca* F. Muell. | – | – | 10.55hi | ± 2.88 |
| | *Cassia siamea* Lam. | – | – | 13.50gh | ± 3.86 |
| | *Gliricidia sepium* (Jacq.) Kunth | – | – | 26.70ef | ± 4.74 |
| | *Hardwikia binata* Roxb | – | – | 10.42hi | ± 4.14 |
| | *Prosopis juliflora* (Sw.) DC. | – | – | 24.55ef | ± 4.63 |

For each species, different capital letters (line) indicate significant differences between areas (Tukey multiple means test, $P < 0.05$); for each area, different lowercase letters (column) indicate significant differences between species.

In contrast, the tree plants in the man-made forest systems grew close to each other, strongly limiting light penetration to the soil, reducing soil temperature and possibly limiting colonization and sporulation of obligate AMF (Sene et al. (2012a, 2012b). According to the authors, this could partially explain the poor colonization of the tree root systems and corollary, this may led to the decrease of soil MIP. Alternatively, high nutrient contents are also reported for soil samples from the man-made forest and this can generate "novel ecological niches", which may have large negative effects on root plant colonization (Sene et al., 2012a, 2012b). This observation recalls numbers of studies, which have shown that the concentration of nutrients (mainly P) in plant tissues affects root colonization by AMF (Smith and Read, 2008), because the benefit of AMF may become relatively smaller than the C drain they create. Under such circumstances, the feedback from AMF is previously reported to be negative (Johnson et al., 1997). These observations interestingly point to the need of further models and ideas for the man-made forest managers, with special emphasis on the soil microbial ecologies.

***Why is There Slower Mycorrhizal Propagule during the Rainy Season?***

For several tree plantations, Sene et al. (2012b) showed that the AMF propagules were significantly reduced in soil samples at the wet season compared to that of the dry season. However, spore density increased with no change observed in AMF spores richness. Additionally, further comparison on the relative abundance of each spore morphotype had shown little variation between the two seasons (Sene et al., 2012b). Although there is a strong evidence of seasonality in AMF communities (Dumbrell et al., 2011), seasonal niche of the different sources of AMF propagules (mycelium, spore), owing to environmental factors to which these propagules are most sensitive, may be the driving force in regulating the temporal dynamics of AMF inoculums. One can tentatively explain this seasonal variability of AMF propagules by evoking distinct processes according to the tree species in question.

For the first process, one can suspect release of toxins from the tree root exudation and/or leaf litters decomposition at the wet season. The chemical compounds of root exudation are not only different among species but can be modified according to the plant phenology (Cheng, 2008). The seasonal effect affects not only the plants but also the AMF community, altering the life cycle of different species (Oehl et al., 2009, 2010). These compounds, when accumulated into soil, could act as allelochemicals on AMF development (Stinson et al., 2006). This could also affect mycorrhiza through decreasing the underground herbaceous vegetation, as host infection sites and rhizospheres are not available. Such allelopathic interferences usually occur in the ecotone of exotic plant species (Kisa et al., 2007; Silva et al., 2007; Remigi et al., 2008; Sanon et al., 2009).

An equally consistent explanation is that during the rainy period, the factors influencing microbial activity in Sahelian areas were apparently promoted by the availability of water and decomposing cover residues in the plantations, leading to an increase in nutrient availability, as indicated by Sene et al. (2012a, 2013). As a consequence, this can reduce the membrane permeability of root cortex cells and root exudes and then can depress root colonization (Tawaraya et al., 1996). Accordingly, the nutrient availability together with the gap of light beneath the tree species (Sene et al., 20012b) can create parasitic associations between AMF and the ground vegetation because the costs of the symbiosis exceed the benefits (Johnson et al., 1997), and this corollary decreases soil MIP through decreasing root herbs availability (Sene et al., 2012b). In addition, beyond a certain threshold, a more abundant nutrient availability should possibly result in a weaker dependence of plants on AMF for optimum nutrition (Mendoza et al., 2011). Such statements should be expected in plantations of Sahelian acacia species, where increasing soil organic matter, N and P content might generate a "novel ecological niche", negatively effecting soil AMF propagules (Sene et al., 2012b). However, because both the soil chemical composition and the ground vegetation changed beneath these tree plantations, cause-and-effect relationships are difficult to attribute primarily to soil or ground vegetation characteristics, and data to date did not provided sufficient reasons to conclude on this.

## Conclusion and Further Directions

There is substantial evidence from the man-made forest systems in the semi-arid Sahel region that specific tree plantations can affect the natural LNR and AMF growth and diversities as well as the soil nutrients cycling. This chapter shows that the tree plantations of $N_2$ fixing species are more efficient in enriching the soil with LNR organisms. It clearly indicated that high numbers and genetic diversity of *Rhizobium* spp. were best fitted to these tree plantations, which likely suggest that enrichment of soil *Rhizobium* spp. population is host-specific. Nevertheless, there are few plantations of exotic tree legumes that harbored less number and genetic diversity of LNR compared to those of non $N_2$ fixing species. The chapter also highlighted that the man-made forest systems can greatly alter the AMF communities and thus keep soil MIP low. The scarcity of herbaceous layers beneath the tree plantations might account for the low number of AMF spores and the low soil MIP. These indicators are nonetheless insufficient to give all the information required by a man-made forest manager. Hence, more work will be necessary to elucidate how to maximize the biodiversity of soil LNR and AMF communities in man-made forest systems. To investigate this aspect, this chapter suggests that attention be given to the mixed-tree plantations, as recommended by Forrester et al. (2006), of $N_2$ fixing species with non $N_2$ fixing species. It will be interesting, therefore, to further investigate the processes and interactions that will affect the productivity in such mixed-species stands. An additional challenge with the man-made tree plantations is to elucidate how to maximize the above and belowground plant coexistence, with special emphasis on the herb legume hosts with a large spectrum for nodulation, and the herb plants having a high mycorrhizal dependency. This can lead to fully exploiting their potential uses by promoting the soil microbial communities.

## References

Akpo, L. E., Banoin, M., Grouzis, M. (2003). Effet de l'arbre sur la production et la qualité fourragère de la végétation herbacée : bilan pastoral en milieu sahélien. *Rev. Méd. Vét.*, 154, 619-628.

Alguacil, M. M., Lozano, Z., Campoy, M. J., Roldán, A. (2010). Phosphorus fertilization management modifies the biodiversity of AM fungi in a tropical savanna forage system. *Soil Biol. Bioch.*, 42, 1114-1122.

Andrade, D. S., Murphy, P. J., Giller, K. E. (2002). The diversity of *Phaseolus*-nodulating rhizobial populations is altered by liming of acid soils planted with *Phaseolus vulgaris* L. in Brazil. *Appl. Environ. Microbiol.*, 68, 4025-4034.

Bai, Z. G., Dent, D. L., Olsson, L., Schaepman, M. E. (2008). Global assessment of land degradation and improvement. 1. Identification by remote sensing. Report 2008/01, ISRIC–World Soil Information, Wageningen.

Bala, A., Giller, K. E. (2006). Relationships between rhizobial diversity and host legume nodulation and nitrogen fixation in tropical ecosystems. *Nutr. Cycl. Agroecosyst.*, 76, 319-330.

Bever, J. D., Westoven, K. M., Antonovics, J. (1997). Incorporation the soil community into plant population dynamics: the utility of the feedback approach. *J. Ecol.*, 85, 561-573.

Bilgo, A., Sangare, S. K, Thioulouse, J., Prin, Y., Hien, V., Galiana, A., Baudoin, E., Hafidi, M., Bâ, A.M., Duponnois, R. (2011). Response of native soil microbial functions to the controlled mycorrhization of an exotic tree legume, *Acacia holosericea* in a Sahelian ecosystem. *Mycorrhiza*, 22, 175-187.

Cardoso, I. M., Kuyper, T. W. (2006). Mycorrhizas and tropical soil fertility Agriculture. *Ecosyst. Environ.*, 116, 72-84.

Cheng, W. (2008). Rhizosphere priming effect: its functional relationships with microbial turnover, evapotranspiration, and C–N budgets. *Soil Biol. Biochem.*, 41, 1795-1801.

Curlevski, N. J. A., Xu, Z., Anderson, I. C., Cairney, J. W. G. (2010). Converting Australian tropical rainforest to native Araucariaceae plantations alters soil fungal communities. *Soil Biol. Biochem.*, 42, 14-20.

Diagne, O., Ingleby, K., Deans, J. D., Lindley, D. K., Diaité, I., Neyra, M. (2001). Mycorrhizal inoculum potential of soils from alley cropping plots in Sénégal. *Forest Ecol. Manag.*, 146, 35-43.

Diagne, O., Seck, A. S., Gassama-Dia, Y. K. (2006). Seasonal and spatial distribution of roots and arbuscular mycorrhizal fungi under live fences of *Acacia laeta* and *Acacia mellifera* in Sénégal. *Southern Afr. Forest J.*, 207(1), 41-46.

Diouf, D., Samba-Mbaye, R. T., Lesueur, D., Ba, A. T., Dreyfus, B., de Lajudie, P., Neyra, M. (2007). Genetic diversity of *Acacia seyal* rhizobial population indigenous to Senegalese soils in relation to salinity and pH of the sampling sites. *Microbial Ecol.*, 54, 553-566.

Diouf, D., Sougoufara, B., Neyra, M., Lesueur, D. (2002). Le reboisement au Sénégal : bilan des réalisations de 1993 à 1998. *Rev. For. Fr. LIV.*, 3, 227-238.

Dommergues, Y., Duhaux, E., Hoang, G.D. (1999). Les Arbres Fixateurs d'Azote : Caractéristiques fondamentales et rôle dans l'aménagement des écosystèmes méditerranéens et tropicaux. Y. Dommergues (ed). Édition espaces 34. Paris, 475p.

Donfack, P., Floret, C., Pontanier, R. (1995). Secondary succession in abandoned fields of dry tropical northern Cameroon. *J. Veg. Sci.*, 6, 499-508.

Dreyfus, B., Diem, H. G., Freire, J., Keya, S. O., Dommergues, Y. R. (1988). Nitrogen fixation in tropical agriculture and forestry. In Dasilva E.J, Dommergues YRNyns EJ, Ratledge C (Eds). Microbial technology in developping world. Oxford University Press, Oxford, pp. 1-7.

Dumbrell, A. J., Ashton, P. D., Aziz, N., Feng, G., Nelson, M., Dytham, C., Fitter, A. H., Helgason, T. (2011). Distinct seasonal assemblages of arbuscular mycorrhizal fungi revealed by massively parallel pyrosequencing. *New Phytol.*, 190(3), 794-804.

Duponnois, R., Colombet, A., Hien, V. Thioulouse, J. (2005). The mycorrhizal fungus *Glomus intraradices* and rock phosphate amendment influence plant growth and microbial activity in the rhizosphere of *Acacia holosericea*. *Soil Biol. Biochem.*, 37, 1460-1468.

Duponnois, R., Plenchette, C., Thioulouse, J., Cadet, P. (2001). The mycorrhizal soil infectivity and arbuscular mycorrhizal fungal spore communities in soils of different aged fallows in Senegal. *Appl. Soil Ecol.*, 17, 239-251.

Egerton-Warburton, L. M., Allen, E. B. (2000). Shifts in arbuscular mycorrhizal communities along an anthropogenic nitrogen deposition gradient. *Ecol. Appl.*, 10, 484-496.

Faye, A., Krasova-Wade, T., Thiao, M., Thioulouse, J., Neyra, M., Prin, Y., Galiana, A., et al. (2009). Controlled ectomycorrhization of an exotic legume tree species *Acacia holosericea* affects the structure of root nodule bacteria community and their symbiotic effectiveness on *Faiderbia albida*, a native Sahelian *Acacia*. *Soil Biol. Biochem.*, 41, 1245-1252.

Forrester, D. I., Bauhus, J., Cowie, A. L., Vanclay, J. K. (2006). Mixed-species plantations of *Eucalyptus* with nitrogen fixing trees: a review. *Forest Ecol. Manag.*, 233, 211-230.

Fortin, J. A., Plenchette, C., Piché, Y. (2008). Les Mycorhizes : la Nouvelle Révolution Verte. Édition multimondes, Québec Canada, 138 p.

Graham, P. H. (2008). Ecology of the root-nodule bacteria of legumes. In: Dilworth, M.J., James, E.K., Sprent, J.I., Newton, W.E. (Eds.), Leguminous Nitrogen-Fixing Symbioses. Springer, pp. 23-58.

Guadarrama, P., Castillo-Argüero, S., Ramos-Zapata, J. A., Camargo-Ricalde, S. L., Álvarez-Sánchez, J. (2008). Propagules of arbuscular mycorrhizal fungi in a secondary dry forest of Oaxaca, Mexico. *Int. J. Trop. Biol.*, 56(1), 269-277.

Hobbie, S. E., Ogdahl, M., Chorover, J., Chadwick, O. A., Oleksyn, J. et al. (2007). Tree Species Effects on Soil Organic Matter Dynamics: The Role of Soil Cation Composition. *Ecosyst.*, 10, 999-1018.

Hobbie, S. E., Reich, P. B., Oleksyn, J., Ogdahl, M., Zytkowiak, R. et al. (2006). Species effects on decomposition and forest floor dynamics in a common garden. *Ecol.*, 87(9), 2288-2297.

Ingleby, K., Diagne, O., Deans, J. D., Lindley, D. K., Neyra, M., Ducousso, M. (1997). Distribution of roots, arbuscular mycorrhizal colonisation and spores around fast-growing tree species in Senegal. *Forest Ecol. Manag.*, 90, 19-27.

Johnson, N. C. (1993). Can fertilization of soil select less mutulalistic mycorrhizae? *Ecol. Appl.*, 3, 749-757.

Johnson, N. C., Zak, D. R., Tilman, D., Pfleger, F. L. (1991). Dynamics of vesiculararbuscular mycorrhizae during old field succession. *Oecologia*, 86, 349-358.

Johnson, N.C., Graham, J.H. and Smith, F.A. (1997). Functioning and mycorrhizal associations along the mutualism-parasitism continuum. *New Phytol.*, 135, 575-586.

Jordan, N. R., Larson, D. L., Huerd, S. C. (2008). Soil modification by invasive plants: effects on native and invasive species of mixed-grass prairies. *Biol. Invasion.*, 10, 177-190.

Kahindi, J. H. P., Woomer, P., George, T., de Souza Moreira, F. M., Karanja, N. K., Giller, K. E. (1997). Agricultural intensification, soil biodiversity and ecosystem function in the tropics: the role of nitrogen-fixing bacteria. *Appl. Soil Ecol.*, 6, 55-76.

Kiers, ET, Rousseau, RA, West, SA, Denison, RF (2003) Host sanctions and the legume-*Rhizobium* mutualism. Nature (Lond.), 425, 78-81.

Kisa, A., Sanon, A., Thioulouse, J., Assigbetse, K., Sylla, S., Spichiger, R., Dieng, et al. (2007). Arbuscular mycorrhizal symbiosis can counterbalance the negative influence of the exotic tree species *Eucalyptus camaldulensis* on the structure and functioning of soil microbial communities in a sahelian soil. *FEMS Microbiol. Ecol.*, 62, 32-44.

Larsen, J. Bo (1995). Ecological stability of forests and sustainable silviculture. *Forest Ecol. Manag.*, 73, 85-96.

Lawson, K., Barnet, Y. M., McGilchrist, C. A. (1987). Environmental factors influencing numbers of *Rhizobium leguminosarum* bv. trifolii and its bacteriophages in two field soils. *Appl. Environ. Microbiol.*, 53, 1125-1131.

Lemma, B., Nilsson, I., Kleja, D. B., Olsson, M., Knicker, H. (2007). Decomposition and substrate quality of leaf litters and fine roots from three exotic plantations and a native forest in the southwestern highlands of Ethiopia. *Soil Biol. Biochem.*, 39, 2317-2328.

Lindström, K., Murwira, M., Willems, A., Altier, N. (2010). The biodiversity of beneficial microbe-host mutualism: the case of rhizobia. *Res. Microbiol.*, 161, 453-463.

Mascaro, J., Becklund, K. K., Hughes, R. F., Schnitzer, S. A. (2008). Limited native plant regeneration in novel, exotic-dominated forests on Hawai'i. *Forest Ecol. Manag.*, 256, 593-606.

Mendoza, R., Cabello, M., Anchorena, J., García, I., Marbán, L. (2011). Soil parameters and host plants associated with arbuscular mycorrhizae in the grazed Magellanic steppe of Tierra del Fuego. *Agr. Ecosyst. Environ.*, 140, 411-418.

Mungai, N. W., Motavalli, P. P., Kremer, R. J., Nelson, K. A. (2005). Spatial variation of soil enzyme activities and microbial functional diversity in temperate alley cropping systems. *Biol. Fertil. Soils*, 42, 129-136.

Mwenda, G. M., Kranja, N. K., Boga, H., Kahindi, J. H. P., Muigai, A., Odee, D. (2011). Abundance and diversity of legume-nodulating rhizobia in soils of Embu district, Kenya. *J. Trop. Subtrop. Agroecosystems*, 13, 1-10.

Ndoye, F, Kane, A., Mangaptché, E. L. N., Bakhoum, N., Sanon, A., Diouf, D., Sy, M. O., Baudoin, E., Noba, K., Prin, Y. (2012). Changes in land use system and environmental factors affect arbuscular mycorrhizal fungal density and diversity, and enzyme activities in rhizospheric soils of *Acacia senegal* (L.) Willd. *ISRN Ecol.* (in press) doi:10.5402/2012/563191.

Noel, K. D. (2009). Rhizobia. In: Encyclopedia of Microbiology, 3rd edition, ed. Schaechter M. Oxford: Elsevier, pp. 261-277.

Oehl, F., Laczko, E., Bogenrieder, A., Stahr, K., Bösch, R., van der Heijden, M., Sieverding, E. (2010). Soil type and land use intensity determine the composition of arbuscular mycorrhizal fungal communities. *Soil Biol. Biochem.*, 42, 724-738.

Oehl, F., Sieverding, E., Ineichen, K., Maïder, P., Wiemken, A., Boller, T. (2009). Distinct sporulation dynamics of arbuscular mycorrhizal fungal communities from different agroecosystems in long-term microcosms. *Agric. Ecosyst. Environ.*, 134, 257-268.

Peichl, M., Thevathasan, N. V., Gordon, A. M., Huss, J., Abohassan, R. A. (2006). Carbon sequestration potentials in temperate tree-based intercropping systems, southern Ontario, Canada. *Agroforest. Syst.*, 66, 243-257.

Remigi, P., Faye, A., Kane, A., Deruaz, M., Thioulouse, J., Cissoko, M., Prin, Y., Galiana, A., Dreyfus, B., Duponnois, R. (2008). The exotic legume tree species *Acacia holosericea* alters microbial soil functionalities and the structure of the arbuscular mycorrhizal community. *Appl. Environ. Microbiol.*, 74, 1485-1493.

Rillig, M. C., Mummey, D. L. (2006). Tansley review – mycorrhizas and soil structure. New Phytol., 171, 41-53.

Rivest, D., Cogliastro, A., Vanasse, A., Olivier, A. (2009). Production of soybean associated with different hybrid poplar clones in a tree-based intercropping system in southwestern Québec, Canada. *Agric. Ecosyst. Environ.*, 131, 51-60.

Sanon, A., Beguiristain, T., Cébron, A., Berthelin, J., Ndoye, I., Leyval, C., Sylla, S., Duponnois, R. (2009). Changes in soil diversity and global activities following invasions of the exotic invasive plant, *Amaranthus viridis* L., decrease the growth of native sahelian *Acacia* species. *FEMS Microbiol. Ecol.*, 70, 118-131.

Schnitzer, S. A., Klironomos, J. N., Hillerislambers, J., Kinkel, L. L., Reich, P. B., Xiao, K. et al. (2011). Soil microbes drive the classic plant diversity-productivity pattern. *Ecol.*, 92, 296-303.

Sene, G., Thiao, M., Samba-Mbaye, R., Ndoye, F., Kane, A., Diouf, D., Sylla, S. N. (2010). Response of three peanut cultivars toward inoculation with two *Bradyrhizobium* strains and an arbuscular mycorrhizal fungus in Senegal. *Afr. J. Microbiol. Res.*, 4(23), 2520-2527.

Sene, G., Samba-Mbaye, R., Thiao, M., Khasa, D. P., Kane, A., Mbaye, M. S., Manga, A., Sylla, S. N. (2012a). The abundance and diversity of legume-nodulating rhizobia and arbuscular mycorrhizal fungal communities in soil samples from deforested and man-made forest systems in a semiarid Sahel region in Senegal. *Europ. J. Soil Biol.*, 52, 30-40 http://dx.doi.org/10.1016/j.ejsobi.2012.05.005.

Sene, G., Thiao, M., Manga, A., Kane, A., Samba-Mbaye, R., Mbaye, M. S., Khasa, D. P., Sylla, S. N. (2012b). Arbuscular mycorrhizal soil infectivity and spores distribution across plantations of tropical, subtropical and exotic tree species: A case study from the forest reserve of Bandia, Senegal. *Afr. J. Ecol.*, 50, 218-232 doi: 10.1111/j.1365-2028.2011.01315.x.

Sene, G., Thiao, M., Samba-Mbaye, R., Khasa, D. P., Kane, A., Mbaye, M. S., Beaulieu, M.-È., Manga, A., Sylla, S. N. (2013). The abundance and diversity of legume-nodulating rhizobia in 28-year-old plantations of tropical, subtropical and exotic tree species: a case study from the forest reserve of Bandia, Senegal. *Microb. Ecol.*, 65(1), 128-144 doi: 10.1007/s00248-012-0094-y.

Silva, A. M. M., Silva, A. R., Pinheiro, A. M., Freitas, S. R. V. B., Silva, V. D. A., et al. (2007). Alkaloids from *Prosopis juliflora* leaves induce glial activation, cytotoxicity and stimulate NO production. *Toxicon*, 49, 601-614.

Smith, S. E., Read, D. J. (1997). Mycorrhizal symbiosis. Academic Press; 2nd ed. London, 605 p.

Smith, S. E., Read, D. J. (2008). Mycorrhizal Symbiosis, 3rd ed. Academic Press Inc., London, UK, 815 p.

Stinson, K. A., Campbell, S. A., Powell, J. R., Wolfe, B. E., Callaway, R. M., et al. (2006). Invasive plant suppresses the growth of native tree seedlings by disrupting belowground mutualisms. *PLoS Biol.*, 4, 727-731.

Stürmer, S. L. Siqueira, J. O. (2011). Species richness and spore abundance of arbuscular mycorrhizal fungi across distinct land uses in Western Brazilian Amazon. *Mycorrhiza*, 21(4), 255-267.

Tappan, G. G., Sall, M., Wood, E. C., Cushing, M. (2004). Ecoregions and land cover trends in Senegal. *J. Arid Environ.*, 59, 427-462.

Tawaraya, K., Watanabe, S., Yoshioda, E. (1996). Effect of onion (*Allium cepa*) root exudates on the hyphal growth of *Gigaspora margarita*. *Mycorrhiza*, 6(1), 57-59.

Tiffen, M., Mortimore, M. (2002). Questioning desertification in dryland sub-Saharan Africa. *Natural Resources Forum*, 26, 218-233.

van der Heijden, M. G. A., Bakker, R., Verwaal, J., Scheublin, T. R., Rutten, M., van Logtestijn, R., Staehelin, C. (2006). Symbiotic bacteria as a determinant of plant community structure and plant productivity in dune glassland. *FEMS Microbiol. Ecol.*, 56, 178-187.

van der Heijden, M. G. A., Bardgett, R. D., van Straalen, N. M. (2008). The unseen majority: soil microbes as drivers of plant diversity and productivity in terrestrial ecosystems. *Ecol. Lett.*, 11, 296-310.

van der Heijden, M. G. A., Klironomos, J. N., Ursic, M., Moutoglis, P., Streitwolf-Engel, R., et al. (1998). Mycorrhizal fungal diversity determines plant biodiversity, ecosystem variability and productivity. *Nature*, 396, 72-75.

Vincke, C., Diédhiou, I., Grouzis M. (2010). Long term dynamics and structure of woody vegetation in the Ferlo (Senegal). *J. Arid Environ.*, 74, 268-276.

Wolde-meskel, E., Terefework, Z., Frostegård, Å. sa, Lindström, K. (2005). Genetic diversity and phylogeny of rhizobia isolated from agroforestry legume species in southern Ethiopia. *Int. J. Syst. Evol. Microbiol.*, 55, 1439-1452.

Woomer, P., Singleton, P. W., Bohlool, B. B. (1988). Ecological indicators of native rhizobia in tropical soils. *Appl. Environ. Microbiol.*, 54, 1112-1116.

Xingjun, Y., Dan, Y., Zhijun, L., Keping, M.A. (2005). A new mechanism of invader success: exotic plant inhibits natural vegetation restoration by changing soil microbe community. *Chinese Sci. Bull.*, 50, 1105-1112.

Zhang, Y., Gui, L. D., Liu, R. J. (2004). Survey of arbuscular mycorrhizal fungi in deforested and natural forest land in the subtropical region of Dujiangyan, southwest China. *Plant Soil*, 261, 257-263.

In: Plantations
Editor: Hai Ren

ISBN: 978-1-62808-090-2


*Chapter 4*

# INDIGENOUS-OWNED PASTORAL LAND FORESTRY CARBON BIOSEQUESTRATION AND BIOENERGY OPTIONS IN ARID, SALT-AFFECTED WESTERN AUSTRALIAN REGIONS

***Mark P. McHenry*[1*] *and Julia Anwar McHenry*[2]**
[1]Faculty of Minerals and Energy, Murdoch University, Western Australia
[2]Faculty of Health Sciences, Curtin University Health Research Campus, Western Australia

## ABSTRACT

This research comprises a technical and economic feasibility study of geoengineering and revegetating 1,500 ha on an Indigenous-owned pastoral lease in an arid region (<250 mm mean annual rainfall) in the Mid West Region of Western Australia for forestry carbon biosequestration and bioenergy projects. A single tree species, *Eucalyptus camaldulensis*, an indigenous tree species of the region was used in two modeled scenarios. The carbon biosequestration and NPC (8% p.a. real discount rate) was modeled over 15 years, yielding an equivalent mitigation cost of AUD15.34 $tCO_2$-$e^{-1}$, assuming the total biosequestration (185,635 $tCO_2$-e) was forward sold in year 0 in the first scenario. In the second scenario, the stand was the primary feedstock for a biorefinery producing activated carbon, renewable electricity, and biodiesel produced from 20 t $day^{-1}$ (16% moisture) *E. camaldulensis*, with harvesting commencing in the sixth year after stand establishment. The final total biosequestration for both the unharvested and harvested stand scenarios were modeled as identical in the 15$^{th}$ year. The NPV of the biorefinery was modeled separately from stand biosequestration also discounted at 8% p.a. Depending on the mitigation assumptions, and the commercial costs/prices of inputs/outputs over the 15 years, the biorefinery NPV ranged from slightly below zero to AUD6.5 million, and mitigated a further 73,761 $tCO_2$-e, based on modeling assumptions.

* Corresponding Author address Murdoch University, 90 South St Murdoch, Western Australia, 6150. Email: mpmchenry@gmail.com.

**Keywords:** Carbon; forestry biosequestration; bioenergy; arid; salt; pastoral plantations

# INTRODUCTION

Livestock grazing land use activities on pastoral leaseholds occur on around 40% of the state of Western Australia (WA), representing approximately 98 million ha. Attempts to diversify pastoral leases since the 1800s into more intensive and/or productive uses other than grazing have been occurring in a minor sense over time, including agricultural and horticultural alternatives. However, most diversifications, apart from mining, have been commercially unsuccessful over time (Payne et al., 1998). The Land Administration Act 1997, various other environmental legislation, and the Pastoral Lands Board limit leasehold activities, and any diversification, conservation, or sale of any product of a non-pastoral nature currently requires a permit from the Board. Penalties range from fines of a few thousand dollars to lease termination. In addition to bureaucratic barriers, active land regeneration may require expensive geoengineering techniques including earthworks, stream flow modification, and replanting of native species in suitable ecological niches, which are all dependent on local hydrology, vegetation, soils, land surface characteristics, existing land uses, and local capacity (Payne et al., 1998; McHenry, 2010, 2011, 2012a, 2012b, 2012c, 2012d, 2013). Several regions are currently are overgrazed with considerable loss of perennial vegetation, biodiversity, and in places advanced degradation has resulted in loss of water and nutrient infiltration due to compacted soils (Curry et al., 2008). Therefore, there is a considerable challenge to achieving reasonable tree survival and growth rates in arid land rehabilitation, carbon (C) biosequestration, and bioenergy projects, quite separate from any commercially viable energy or conventional diversification alternative for pastoralists (Yamada et al., 1999; McHenry, 2009c, 2009a, 2009b, 2009d, 2012e). In arid lands of Australia, the availability of moisture and flows of water largely determine the growth and development of trees, which often uniquely utilize groundwater as a primary water source (Mensforth et al., 1994; Thorburn & Walker, 1994; Yamada et al., 1999; Shiono et al., 2004; Matsumoto et al., 2006; Tanouchi et al., 2006). As arid land rains are infrequent, irregular, and the majority of the volumes are lost by runoff and evaporation (Hirukawa et al., 2006), the soil and land characteristics are important determinants of improving soil water availability and water holding capacity for revegetation activities (Payne et al., 1998; Egashira et al., 2003). Locations with eroded soils and large siliceous layers (hardpans) inhibit tree growth as the hardpan barrier prevents root development for sourcing groundwater (Shiono et al., 2006), and a rapid recovery of the original vegetation may be impossible without mechanical intervention to break up the hardpans of around 1 m thickness (Yamada et al., 1999; Curry et al., 2008). In the main, the total biomass accumulation of native tree species in these region are low, with the exception of Eucalyptus species in niche ecologies, or geoengineered areas (Yamada et al., 1999). As tree growth in both natural and revegetated stands are related to the volume and specific path of water flows (Abe et al., 2003), small-scale geoengineering offers an additional means to recreate soil water holding capacities akin to, or even surpassing pre-European periods, enabling high growth and survival rates for rehabilitation projects using selected local species.

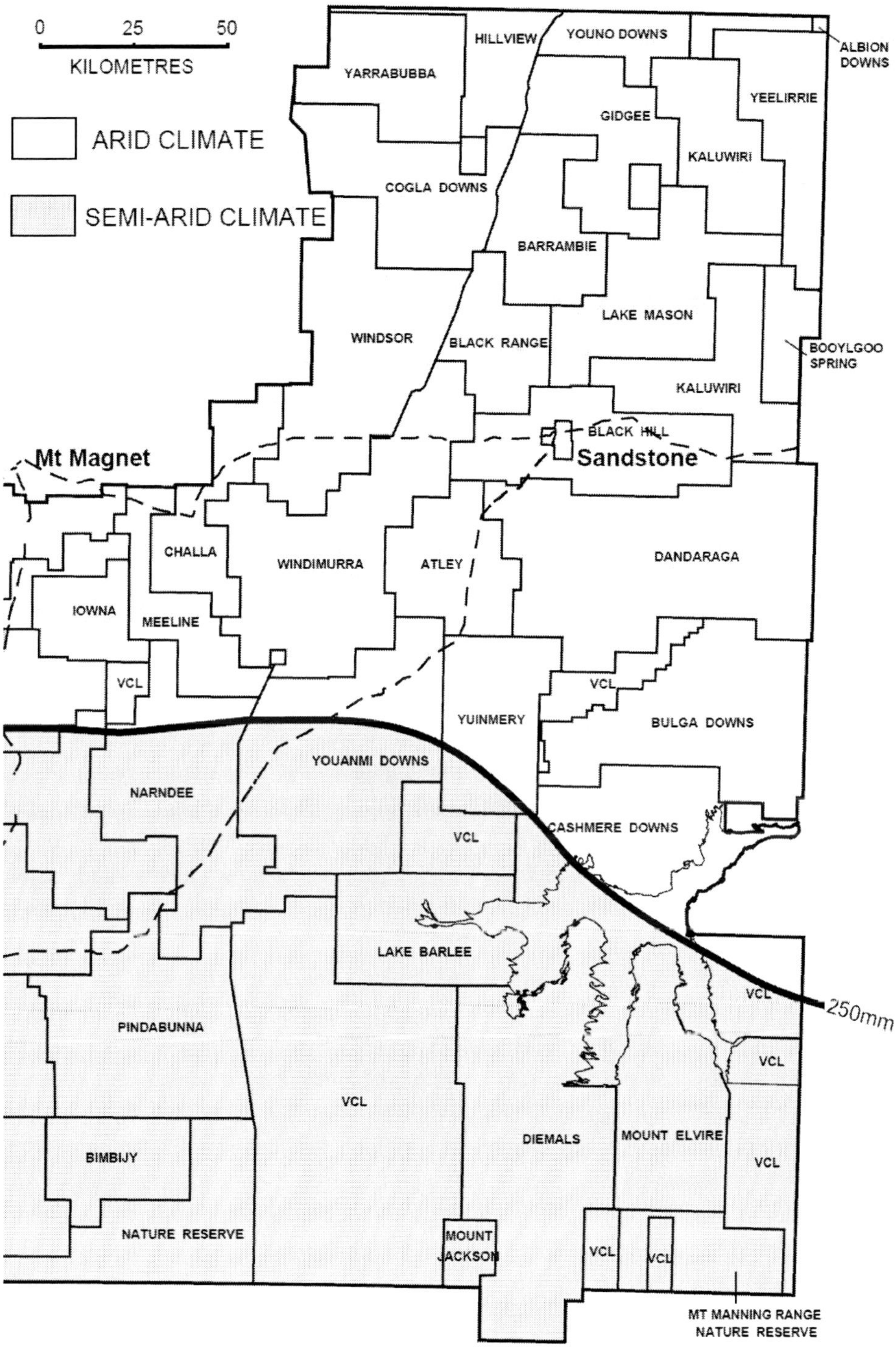


Figure 1. Location of the 250 mm (mean) isohyet dividing semi-arid and arid climatic zones. Source: (Payne et al., 1998).

## THE FOCUS REGION AND AVAILABLE BIOMASS GROWTH DATA

The focus pastoral region for this research is within the Murchison bioregion, and is climatically arid (Curry et al., 2008) (Figure 1). Geologically, the sub-region lies on Archean rock, and is part of the Yilgarn Craton (Payne et al., 1998). Linear to arcuate belts of greenstone and local gneissic rocks intruded by granitoid rock of Archaean age are overlaid with alluvial, colluvial, Aeolian, and lacustrine deposits of Cainozoic age (Payne et al., 1998; Yamada et al., 1999). The extensive surface landscape is largely comprised of flat colluvial-alluvial geologies that support various species, including dominant Acacia and Eucalyptus species (Payne et al., 1998). The region exhibits high annual rainfall variability, receiving between 60 to 500 mm, with higher rainfall years due to unpredictable summer storms. The area receives around 45% of the total annual mean rainfall of around 250-200 mm in the wet months and around 55% in the dry months (Payne et al., 1998; Yamada et al., 1999; Abe et al., 2003). The Cogla Downs pastoral station, Lat. (S) 27° 26', Lon. (E) 118° 56', at 500 m a.s.l., has 56 years of recorded rainfall observations. The lowest recorded rainfall was 69 mm in 1969, and the highest of 546 mm recorded in 1942 demonstrates the extreme variation from the median annual recorded rainfall of 191 mm (Payne et al., 1998). In terms of temperature, the mean monthly maximum and minimum temperatures of the hottest month (January) is around 37 and 22°C, respectively. The mean monthly maximum and minimum temperature of the coldest month (July) is around 18 and 6 °C, respectively (Yamada et al., 1999). Thus, evaporation dominates as an influence on flora and fauna systems of the region, often being in excess of 10 times the annual rainfall received (Payne et al., 1998). The technical and economic analyses in this research are based on longer-term research data derived from the Sturt Meadows pastoral lease, which were calibrated as a representative dataset using recent (2010) experimental tree plantings at Cogla Downs station to derive practical on-site planting information (Figure 2). Sturt Meadows is in the same Murchison bioregion and exhibits very similar geological, hydrological, and climatic conditions as Cogla Downs, and is located around 200 km SE from Cogla Downs, and 200 km N from Kalgoorlie (Figure 3). The total area of Cogla Downs pastoral station is 199,478 ha.

Since 1996, a research partnership between Japan Science and Technology Corporation (JST) and Core Research for Evolutional Science and Technology (CREST) have conducted comprehensive intensive revegetation trials, sampling, field experiments, and associated atmospheric and ecophysical monitoring on several sites on the Sturt Meadows pastoral lease (Curry et al., 2008). Arial photographic analysis of the Sturt Meadows station (a total area of 299,250 ha) in 1999 provided an estimate of 8.5 tDM ha$^{-1}$ for the aboveground biomass and a total 10.5 tDM ha$^{-1}$, assuming an average belowground root biomass component of around 24% (Yamada et al., 1999). Most of the species in the region exhibit slow growth, with the notable exception of the *E. camaldulensis* trees which can grow at around 30-35 kg DM yr$^{-1}$. This is around 9-10 times greater growth rates than other local species, which is attributed to both a species and research site difference (Yamada et al., 1999) (Table 1). Table 1 shows estimating average C biosequestration of tree species in arid regions is complicated by age uncertainty, often due to the episodic growth, germination, and plant establishment in years of above average rainfall (Yamada et al., 1999). Nonetheless, an average total potential natural biomass growth C biosequestration post-destocking for the Sturt Meadows lease was estimated at approximately 2.126 tC $ha^{-1}$ (Abe et al., 2003)[1]. Whilst a simple multiplication of the large lease area yields a final additional potential biosequestration of around 2.3 million $tCO_2$-e, destocking activities are ineligible as a mitigation activity in existing C market accounting methodologies, including the Kyoto Protocol. Therefore, active revegetation work was undertaken at the Sturt Meadows sites to determine several aspects of active revegetation of arid lands in WA.

A range of selected local tree species growth data were generated in revegetation trials on the Sturt Meadows pastoral lease, including total biomass and tree height for range of soil preparation methods and land topographies (Tables 2 and 3). The interaction of soil and nutrient flow and accumulation was found to have a major influence on biomass growth (Abe et al., 2003), with deeper soil layers created mechanically enabling additional water holding capacities and resulting in increased tree growth rates (Takahashi et al., 2003). In terms of topography, Abe et al. (2003) found that higher biomass densities were associated with very gentle slopes, and lower biomass levels were associated with steeper slopes, although this was not always the case. For the most part, the local biomass levels were maximized in gradients around 0.1%, and the local biomass minimum was in gradients around 1%.

Figure 2. Experimental tree planting preparations at Cogla Downs station. Note the poor condition of the existing degraded vegetation, and no perennial ground cover.

[1] See Abe et al. (2003) for an extremely detailed quantification methodology.

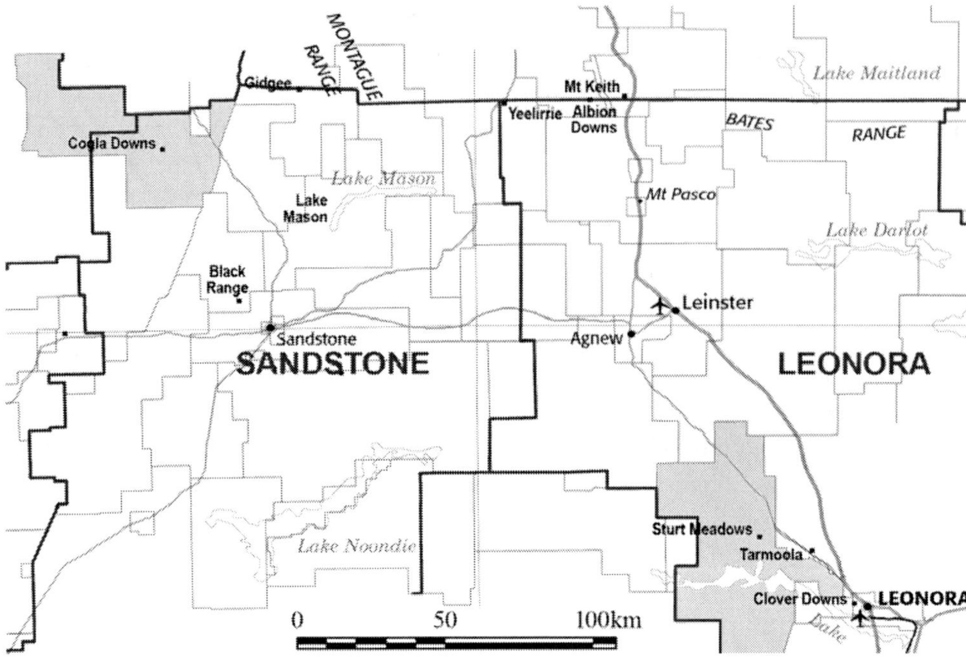


Figure 3. Map of Kalgoorlie, Sturt Meadows (bottom right) and Cogla Downs (top left). Source: Department of Land Administration, Western Australia.

**Table 1. Estimated aboveground DM, C content, and C productivity for a selected range of tree species and estimated ages. Source: (Yamada et al., 1999)**

| Species | Aboveground DM (kg) | Carbon (%) | Age (yrs) | Mean DM Productivity ($kg\ yr^{-1}$) | Mean C Productivity ($kg\ yr^{-1}$) |
|---|---|---|---|---|---|
| *Eucalyptus camaldulensis* | 682.4 | 47.7 | 19-22 | 31.0-35.9 | 14.8-17.1 |
| *Malaleuca sheathiana* | 286.35 | 49.9 | 43-83 | 3.5-6.7 | 1.8-3.3 |
| *Acacia aneura* (specimen a) | 4.4 | 48.6 | 23-26 | 0.17-0.19 | 0.08-0.09 |
| *Acacia aneura* (specimen b) | 97.2 | 47.0 | 42-55 | 1.8-2.3 | 0.8-1.1 |
| *Acacia aneura* (specimen c) | 280.8 | 51.7 | 61-123 | 2.3-4.6 | 1.2-2.4 |
| *Acacia ramulosa* | 64.9 | 43.3 | 22-34 | 1.9-3.0 | 0.82-1.3 |
| *Acacia tetragonophylla* | 27.4 | 46.0 | 24-31 | 0.8-1.14 | 0.4-0.5 |
| *Eremophila macmillaniana* | 0.72 | 47.8 | 10 | 0.07 | 0.03 |

## *EUCALYPTUS CAMALDULENSIS* GROWTH, C, AND WATER USE DATA

*E. camaldulensis* stands commonly grow in open woodlands where the top soil is thick and in small closed forest stands adjacent to creeks (Yamada et al., 2003; Suganuma et al., 2006). *E. camaldulensis* plantations in the region are able to biosequester around 16 kg C $tree^{-1}\ yr^{-1}$ or around 10 tC $ha^{-1}\ yr^{-1}$ at 625 trees $ha^{-1}$ spacing in the aboveground biomass (Yamada et al., 1999). However, this was based on a destructive analysis of a single tree, and does not include belowground C biosequestration. Further work by (Suganuma et al., 2006) on existing

Sturt Meadows *E. camaldulensis* stand destructive analyses found very large differences in total biomass in different *E. camaldulensis* sites, although very similar ratios of trunk/branch to leaf and root biomass (Table 4). The belowground proportions of trees in arid regions tend to be greater than wetter climates, although when arid trees are irrigated the root component is known to decrease as a consequence of greater water access. Furthermore, *E. camaldulensis* root components increase as trees age because of an increase in small roots, especially at the surface to exploit episodic rains (Barton & Montagu, 2006).

**Table 2. Approximate average heights of tree plots planted at ages 14 months (August 1999), 24 months (June 2000), and 36 months (June 2001), all against control plots with no mechanical soil preparation. Source: (Yamada et al., 2003)**

| Species | Height at 14 months (cm) | Height at 24 months (cm) | Height at 36 months (cm) |
|---|---|---|---|
| *Eucalyptus camaldulensis* | *160* | *280* | *450* |
| Control | 20 | 130 | 285 |
| *Eucalyptus lesouffei* | *20* | *120* | *210* |
| Control | 15 | 70 | 195 |
| *Eucalyptus torquata* | *120* | *150* | *195* |
| Control | 70 | 90 | 140 |
| *Eucalyptus salubris* | *30* | *110* | *190* |
| Control | 20 | 70 | 175 |
| *Eucalyptus griffithsii* | *30* | *100* | *190* |
| Control | 75 | 90 | 140 |
| *Acacia aneura* | *10* | *75* | *165* |
| Control | 10 | 80 | 150 |

**Table 3. Tree stems $ha^{-1}$ and biomass $ha^{-1}$, differentiated by height (Yamada et al., 1999)**

| Hydrology/ Site Topography | <4m Height Tree Stems (kg $ha^{-1}$) | >4m Height Tress Stems (kg $ha^{-1}$) | Total Stems ($ha^{-1}$) | <4m Height Tree Biomass (kg $ha^{-1}$) | >4m Height Tree Biomass (kg $ha^{-1}$) | Total Biomass (kg $ha^{-1}$) |
|---|---|---|---|---|---|---|
| Stream bank | 155 | 205 | 360 | 1,068 | 86,807 | 87,875 |
| Run on/natural depression | 1,719 | 672 | 2,391 | 1,233 | 86,230 | 87,463 |
| Run through/run off | 6,750 | 0 | 6,750 | 610 | 0 | 610 |
| Drainage line/run off | 3,488 | 263 | 3,751 | 8,766 | 38,725 | 47,491 |
| Road drain diversion | 4,013 | 175 | 4,188 | 7,914 | 21,750 | 29,664 |
| Broad drainage floor | 1,500 | 1,825 | 3,325 | 14,553 | 87,396 | 101,949 |

The biomass C factor determined by Saito et al. (2004) and Yamada et al. (1999) for each component of the tree was generally around 450 mgC $gDM^{-1}$ (or 45% by DM) for species *Casuarina obesa*, *E. camaldulensis*, *Eucalyptus torquate*, *Eucalyptus lesouffei*, *Acacia aneura*, and *Acacia tetragonophylla*. This research used a Yamada et al. (1999) *E. camaldulensis* C factor of 47.7% C $DM^{-1}$ to approximate total project C biosequestration potentials. Whilst over a 20 year period *E. camaldulensis* species are able to biosequester around 300 kg of C $tree^{-1}$, or around 10,000 kg of C $ha^{-1}$ $yr^{-1}$, this growth is highly dependent on local water availability, especially groundwater. Sap flow through trunks is related to the volumes of water obtained by the roots and transpired from leaves. Unlike many arid tree species in the region, *E. camaldulensis* trees avoid water stress by securing deep water sources, rather than a reducing sap activity in dry conditions (Tanouchi et al., 2006). Groundwater generally comprises in excess of 50% of the *E. camaldulensis* trees long-term water needs (Mensforth et al., 1994; Thorburn & Walker, 1994; Matsumoto et al., 2006). Whilst arid trees with some seasonal access to stream flows or periodic flooding generally vary their water use efficiency with the availability of variable stream flows, this does not occur with *E. camaldulensis* trees growing at distances greater than 15 m from the seasonal streams. These trees survive solely on groundwater, and surprisingly do not increase their water uptakes even during several weeks of inundation from flood events (Mensforth et al., 1994; Thorburn & Walker, 1994). The higher the C-to-water ratio, the higher the plants water requirements are, and Eucalyptus species generally require more water than Acacia species, which in turn require more than Casuarina species in the region to fix the same quantity of C in the biomass (Saito et al., 2004). Water use efficiency of *E. camaldulensis* was found to average 5 gDM $kgH_2O^{-1}$ over the year, and decreased during wet periods and higher soil water contents (Takahashi et al., 2003). Tanouchi et al. (2006) determined the C biosequestration-to-water use efficiency (gC $kg^{-1}H_2O$) of *E. camaldulensis* trees of 2.68 ± 1.14 and 1.32 ± 1.47 in the wet and dry season, respectively. This was approximately 15% and 90% more efficient in the wet and dry seasons than the *C. obesa* trees in the sample plots, respectively. However, very large trees in the region nevertheless require large root structures, which are often limited physically by minimal soil depths.

**Table 4. Destructive analyses of ten (10) *E. camaldulensis* plots, showing relative the DM proportion variability in small plots from the region. Source: (Suganuma et al., 2006)**

| *E. camaldulensis* plot no. and area | Trunk and branch DM (t $ha^{-1}$) | % | Leaf DM (t $ha^{-1}$) | % | Root DM (t $ha^{-1}$) | % | Total DM (t $ha^{-1}$) |
|---|---|---|---|---|---|---|---|
| Plot 1 (0.25 ha) | 69.339 | 53.99 | 5.572 | 4.34 | 53.516 | 41.67 | 128.427 |
| Plot 2 (0.40 ha) | 85.672 | 57.40 | 4.372 | 2.93 | 59.201 | 39.67 | 149.245 |
| Plot 3 (0.25 ha) | 57.120 | 55.86 | 3.658 | 3.58 | 41.469 | 40.56 | 102.247 |
| Plot 4 (0.25 ha) | 64.647 | 55.90 | 3.880 | 3.36 | 47.116 | 40.74 | 115.643 |
| Plot 5 (0.25 ha) | 72.691 | 55.22 | 4.496 | 3.42 | 54.453 | 41.37 | 131.640 |
| Plot 6 (0.25 ha) | 54.970 | 55.26 | 4.139 | 4.16 | 40.373 | 40.58 | 99.482 |
| Plot 7 (0.25 ha) | 47.208 | 57.47 | 2.657 | 3.23 | 32.275 | 39.29 | 82.140 |
| Plot 8 (0.25 ha) | 28.857 | 55.34 | 2.178 | 4.18 | 21.109 | 40.48 | 52.144 |
| Plot 9 (0.25 ha) | 22.925 | 54.48 | 1.675 | 3.98 | 17.479 | 41.54 | 42.079 |
| Plot 10 (0.25 ha) | 11.473 | 56.77 | 0.512 | 2.53 | 8.225 | 40.70 | 20.210 |
| % Averages | | 55.77 | | 3.57 | | 40.66 | |

## SITE GEOENGINEERING AND BLASTING RESEARCH RESULTS

An issue in the focus region is the inability for the *E. camaldulensis* roots to penetrate past the hardpan layers which exist fairly close to the surface of the soil. This prevents the survival and growth of smaller trees into large trees which biosequester more C per unit area (Yamada et al., 1999). The use of soil conditioners such as calcinated bauxite to increase water holding capacities by increasing the soil surface area, was trialed in the area with some success (Hamano et al., 2000b). Research by Hamano et al. (2000a) also explored soil preparation methods that included drilling and fracturing the soil layer to break the hardpan and to fill the hole with larger particles with high permeability and smaller intraparticles for water retention. Ground preparation research by Yamada et al. (2003) simply using explosives and the results described the growth and survival rates of trees planted for C biosequestration in plots with the tree holes (5 m diameter blast and a 5 m blast depth) created by dynamite against a control plot with no blasting. The researchers found that the additional $CO_2$-e emission created from the dynamite blast was only around one fortieth of the emissions expected to be absorbed by the trees by maturity. Hardpan blasting research was also undertaken by Shiono et al. (2006) with the equivalent of around 200 blast holes per ha, with one tree planted per blast hole. Each blast was created with approximately 10 kg of ammonium nitrate fuel oil explosive, and the resulting cracked hardpan blasted zone diameter was on average 2.4 m, and the average depth penetration was around 2.6 m (Shiono et al., 2006; Tanouchi et al., 2006). The survival ratios for the *E. camaldulensis* after blasting were approximately double that of the control plots, and the growth rates were also found to be improved (Shiono et al., 2006). Further geoengineering methods to increase water availability include surface water diversion to create water harvesting embankments which also capture eroded soil and nutrient runoff. Whilst series of embankments may reduce flows downstream, it may mitigate the occurrence of flood events commonly caused by degraded upstream land in the region while retaining local nutrients upstream (Abe et al., 2003). In terms of direct geoengineering of lands to restore water infiltration and root penetration, additional trials are needed to quantify the variety and impact of each mechanical manipulation for rehabilitation of arid regions (Yamada et al., 1999).

## C Biosequestration Extrapolations

The model used 200 surviving trees $ha^{-1}$ as the basis behind the biomass and associated C biosequestration potential. This was based on a tree planting density of around 300 $ha^{-1}$ at an estimated (medium-to-high) survival rate of 66%. The model assumed a more conservative growth rate than the Yamada et al. (1999) research, and 8 kg C $tree^{-1}$ $yr^{-1}$ over the 15 year interval was selected for the model[2]. This gave an average of 24 tC $ha^{-1}$, or 87.984 $tCO_2$-e $ha^{-1}$ of aboveground biosequestration at 200 surviving stems $ha^{-1}$. Assuming 40.66% belowground biomass based on research by Suganuma et al. (2006), and a consistent C content of both below and aboveground biomass of 47.7% based on the Yamada et al. (1999) results, the total biosequestration projection was 123.758 $tCO_2$-e $ha^{-1}$ over the 15 year interval. As the growth and associated biosequestration of the stand would not be a constant 8 kgC $yr^{-1}$, the author introduced a standardized tree growth model suitable for the tree species which sums to an annual average biosequestration of 8 kg C $tree^{-1}$ $yr^{-1}$ over the full 15 year interval. Figures 4 and 5 show the sum of the above and belowground dry biomass and C biosequestration

[2] This is half of the growth rate of the Yamada et al. (1999) research.

projection profile of the tree stand over the 15 year interval on a per ha basis. The total projected biosequestration for the 1,500 ha stand was 50,637 tC, or 185,635 $tCO_2$-e.

**Table 5. The model in table form of the projected scheduled annual discounted cashflow calculations for establishing, maintaining, and verifying C biosequestered in a 1,500 ha stand. (Total project NPC is in bold)**

| Establishment costs | | Rates/Prices/Area | |
|---|---|---|---|
| Soil preparation/ha | -$50.00 | Inflation rate | 3.00% |
| Seedling price | -$0.60 | Real discount rate | 8.00% |
| Seedling rate/ha | -$232.50 | Land lease/yr ($/ha/yr) | -$10.00 |
| Fence construction | $0.00 | Total ha | 1,500 |
| Planting cost/ha | -$210.00 | | |
| Planting subsidy | $0.00 | | |

| Annual values | |
|---|---|
| Annual fertiliser (MAP) cost/ha (**F**) - at $800/t applied | -$186 |
| Annual watering cost/ha (**W**) - 2 people, watering twice in the dry season | -$180 |
| Annual spraying costs/ha (**S**) - 1 person on foot with manual sprayer | -$60 |
| Verification cost/ha (**V**) - subcontractor | -$55 |
| General maintenance cost/ha (**G**) by owner (including fence repairs) | -$30 |
| Lease & insurance cost/ha (**L**) - medium-to-high average value | -$20 |

| Yr | Cashflow Data | Net $ Flow | Cumul. $ Flow | Discounted Net $ Flow | Discounted cumul. $ Flow |
|---|---|---|---|---|---|
| 0 | V, W, S | -$1,046,250 | -$1,046,250 | -$1,046,250 | -$1,046,250 |
| 1 | L, F, W, S, G | -$714,000 | -$1,760,250 | -$661,111 | -$1,707,361 |
| 2 | L, F, S, G | -$444,000 | -$2,204,250 | -$380,658 | -$2,088,020 |
| 3 | L, G | -$75,000 | -$2,279,250 | -$59,537 | -$2,147,557 |
| 4 | L, G | -$75,000 | -$2,354,250 | -$55,127 | -$2,202,684 |
| 5 | L, S, G, V | -$247,500 | -$2,601,750 | -$168,444 | -$2,371,129 |
| 6 | L, G | -$75,000 | -$2,676,750 | -$47,263 | -$2,418,391 |
| 7 | L, G | -$75,000 | -$2,751,750 | -$43,762 | -$2,462,153 |
| 8 | L, G | -$75,000 | -$2,826,750 | -$40,520 | -$2,502,673 |
| 9 | L, G | -$75,000 | -$2,901,750 | -$37,519 | -$2,540,192 |
| 10 | L, S, G, V | -$247,500 | -$3,149,250 | -$114,640 | -$2,654,832 |
| 11 | L, G | -$75,000 | -$3,224,250 | -$32,166 | -$2,686,998 |
| 12 | L, G | -$75,000 | -$3,299,250 | -$29,784 | -$2,716,782 |
| 13 | L, G | -$75,000 | -$3,374,250 | -$27,577 | -$2,744,359 |
| 14 | L, G | -$75,000 | -$3,449,250 | -$25,535 | -$2,769,894 |
| 15 | L, S, G, V | -$247,500 | -$3,696,750 | -$78,022 | **-$2,847,916** |

The 1,500 ha tree stand represents around 0.75% of the total area of the Cogla Downs lease. These growth rates are based on soil geoengineering using only deep ripping using a bulldozer to break up the hardpan to a depth of 1-1.5 m, and trees planted in holes dug to a depth of slightly below the surrounding ground level as shown in Figure 6. Post planting watering regimes were modeled to occur at planting, and twice each dry season for the first two years to achieve the estimated 66% survival rates. These survival rates will be heavily dependent on the season, and significantly less survival rates, even zero, may occur with insufficient watering regimes or grazing fencing. Figure 7 represents the same tree stand over the 15 year interval with harvesting of an average of 3.27 t $ha^{-1}$ of dry biomass (equivalent to around 1.56 tC $ha^{-1}$) from the stand annually, commencing in the $6^{th}$ year. The annual stand C biosequestration rate remains relatively high compared to the unharvested stand scenario due to reduced tree competition approximated in the model. The annual harvest rate was chosen to fulfill the requirements of the 20 t $day^{-1}$ (16% moisture content) biorefinery demand, and is roughly half of the total maximum potential harvest from the stand each year after the sixth year. Figure 8 shows the dry biomass growth projection of the tree stand over 15 years, with harvesting commencing in the sixth year.

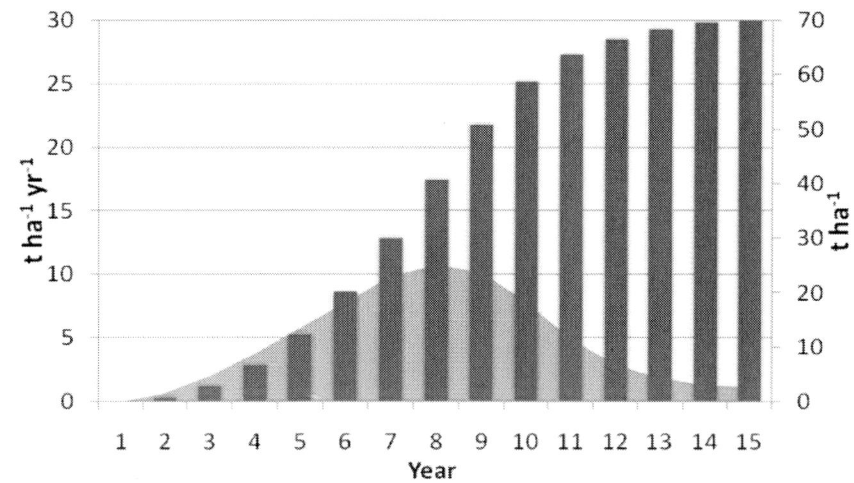


Figure 4. Annual modeled changes in stand dry biomass growth rate (light grey, t $ha^{-1}$ $yr^{-1}$), and total dry biomass (dark grey, t $ha^{-1}$).

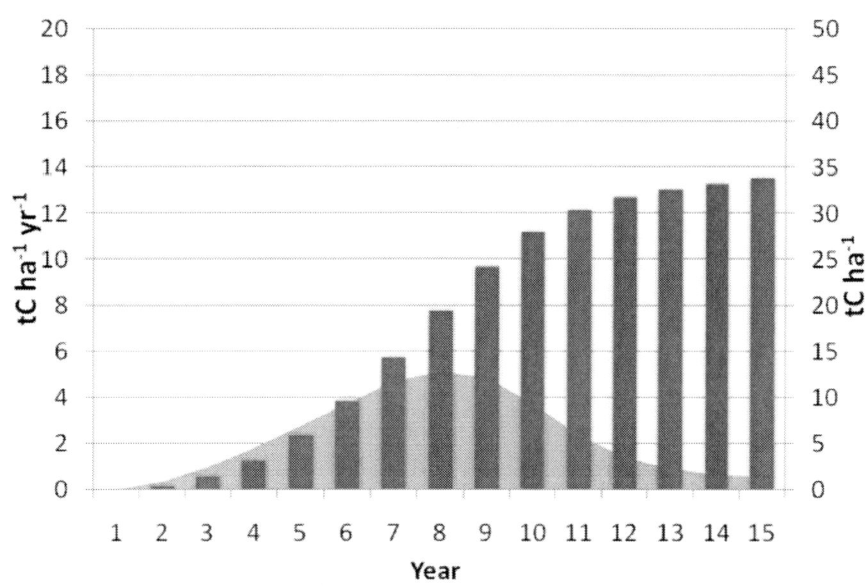


Figure 5. Annual modeled changes in stand C biosequestration rate (light grey, tC $ha^{-1}$ $yr^{-1}$), and total C biosequestration (dark grey, tC $ha^{-1}$).

Figure 6. Tree tubers planted slightly (~100 mm) below ground level on Cogla Downs.

The straightforward economic model consists of a spreadsheet detailing all scheduled costs and income on an annually averaged basis. All prices are based on actual costs in 2010 as determined by on site preliminary trials, works, and site characteristics of the Cogla Downs pastoral lease. The author was careful to avoid infringing the diversification, conservation, and clearing legislation imposed on WA leaseholder by designing the revegetation trial as a four ha fenced stock holding paddock near the existing yards and homestead. This enabled the primary establishment data to be collected, and after research completion, the vegetation which survives will be a suitable shade in the holding pen, maintaining the integrity of the grazing-only land use. The generous AUD10 land lease included in the model for each ha (totaling AUD15,000 p.a.) is a much higher return from land in the region than grazing, and can be viewed as the opportunity cost of not utilizing the land for conventional purposes.

Leasing of pastoral land also currently requires a permit from the Pastoral Lands Board. Where not available, prices were estimated using conservative costs based on the generally high cost of transport, capital, and acquiring both skilled and unskilled labor in the region. Table 5 shows the major activities occurring in each year in column two, and concisely describes the assumptions behind each costing.

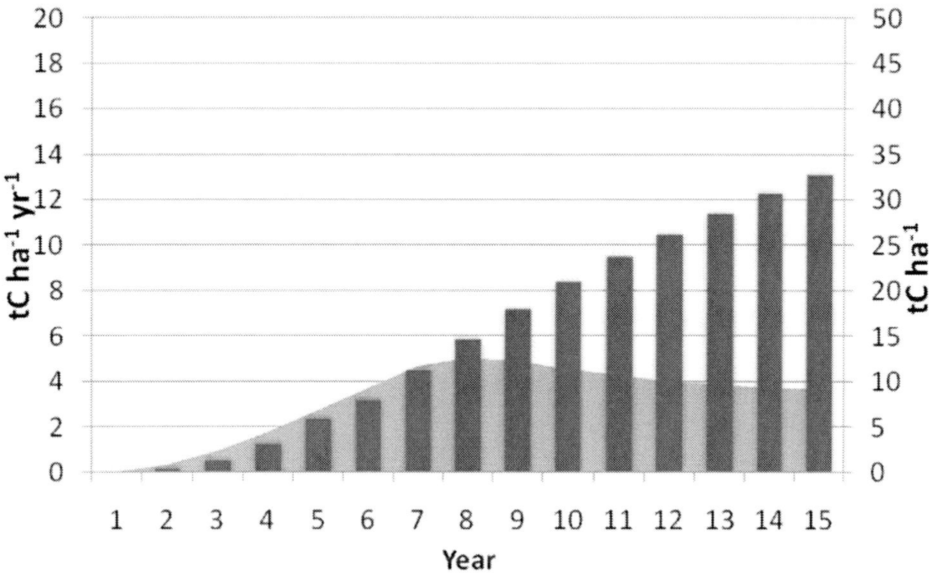


Figure 7. Annual modeled changes in the harvested stand C biosequestration rate (light grey, tC $ha^{-1}$ $yr^{-1}$), and total C biosequestration (dark grey, tC $ha^{-1}$).

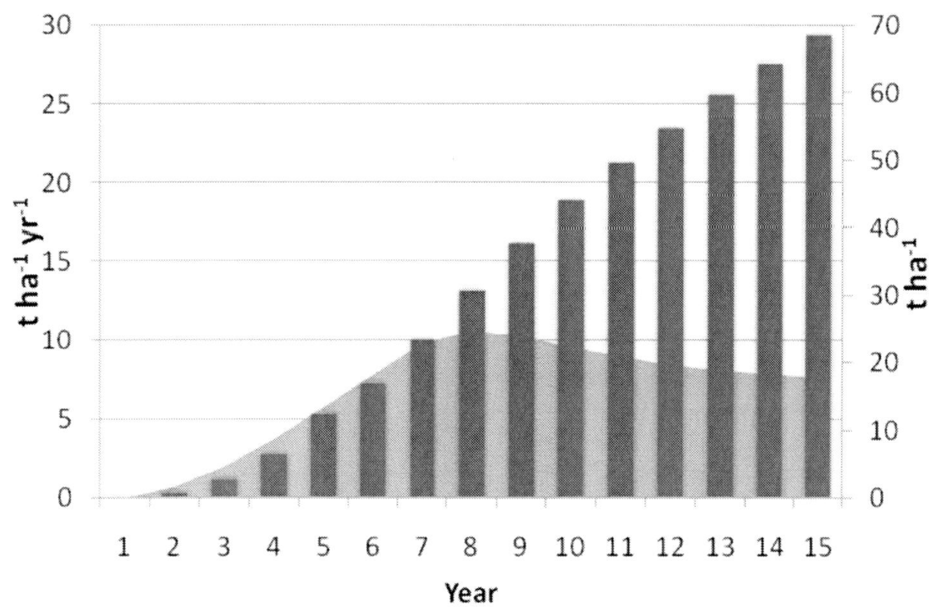


Figure 8. Annual modeled changes in the harvested stand dry biomass growth rate (light grey, tC $ha^{-1}$ $yr^{-1}$), and total dry biomass accumulation (dark grey, tC $ha^{-1}$).

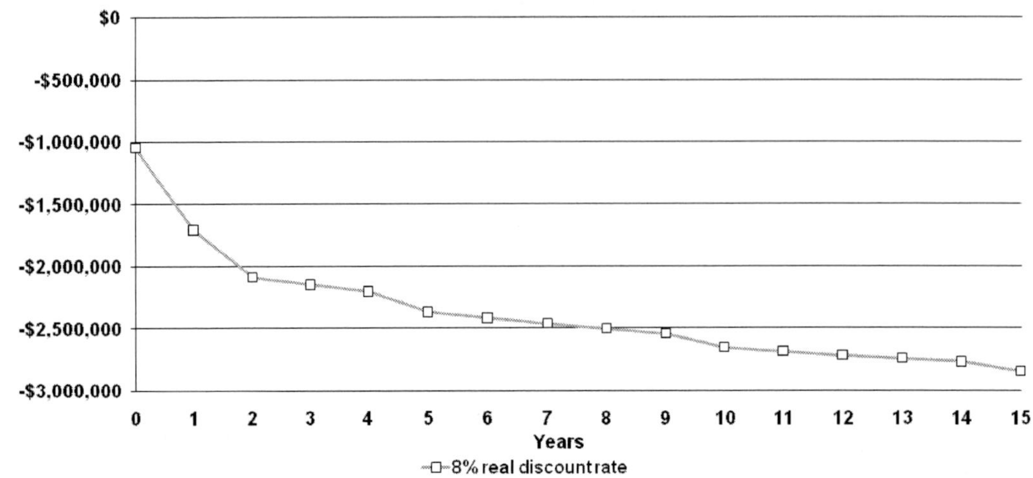


Figure 9. The cumulative costs of the 1,500 project, discounted at a real 8% p.a.

Using a large bulldozer to deep rip the tree lines at around 1-1.5 m depth on the land contour (to capture runoff and prevent erosion) required a speed of around three km $hr^{-1}$. Ripping the rows at average widths of seven meters between rows and avoiding existing vegetation where possible lead to the completion of one ha approximately every 15 minutes of continuous operation, costing around AUD50 $ha^{-1}$ at a AUD200 $hr^{-1}$ contract rate (inc. GST). Trees planted at approximately 4-5 m apart in the row, mostly in bare patches, yields approximately 300 stems $ha^{-1}$. The planting cost $ha^{-1}$ included using a stand-on "kanga digger" articulated drilling arm to break up the ripped clods of generally dry earth to create a soft soil hole of approximately 0.45 m diameter and 0.2 m deep in the deep ripped line. The hole preparation makes the use of a hand held tube tree planter feasible, conservatively taking two people around 3.5 hrs to prepare and plant one ha of *E. camaldulensis* tubers using the "kanga digger" and tree planter in combination. The assumed time included "watering-in" of trees using a standard ute-mounted fire-fighter tank and petrol-powered pump. The modeled planting regime was based on four teams of two workers, working 37.5 hrs in a week, taking around 36 weeks to complete the entire 1,500 ha revegetation site, avoiding working in the hottest months (December to February). All establishment emissions were excluded from the project mitigation potential calculations, as they are relatively minor. The planting costs occurred in year zero in the model, and were based on a simple AUD210 payment $ha^{-1}$, reflecting the above calculations paid at the equivalent of a AUD30 $hr^{-1}$ wage. The fertilizer calculations were based on two applications (in year one and two) of 775 g MAP per tree, scattered slightly away from the stem in the hole. The fertilizer cost was assumed to be AUD800 $t^{-1}$ delivered, based on the scale of the requirements of around 700 t total MAP for the applications over the two years delivered to the remote site. Figure 9 shows the cumulative discounted costs of the project over the 15 year interval.

The market potential of the project was calculated by dividing the projected total NPC (AUD2,847,916) by the total projected mitigation potential (185,635 $tCO_2$-e) for the unharvested 1,500 ha stand to determine a C credit price that the project will require to break even. Assuming all C biosequestered was forward sold in year zero, the minimum breakeven price of AUD15.34 $tCO_2\text{-}e^{-1}$ is required by the project proponent. Any additional C price, biosequestration yield, or cost increase/decrease is to the advantage/detriment of the project proponent and/or investors. This relatively low C price required, combined with the relatively conservative biomass growth yields when compared to the peer-reviewed literature is a promising finding. Furthermore, many of the "costs" involved in the tree propagation, stand establishment, and land management are themselves opportunities for small-scale business operators, which require only moderate capital outlays, skill and labor requirements appropriate for many remote arid areas of Australia, which may not require full-time on-site involvement.

## BIOREFINERY MODELLING RESULTS AND DISCUSSION

In contrast to the relatively clear-cut task of establishing and maintaining a tree revegetation project, developing and operating a bioenergy facility requires highly skilled individuals, maintaining system logistics and input/output flows full-time, and entails high CAPEX, and medium OPEX. The modeled system was designed to displace current fossil fuel-based mineral diesel fuel, and fossil fuel electricity demand of a theoretical facility in the region with 100% renewable biodiesel and electricity. Additional investment value was derived from Renewable Energy Certificates (RECs), representing 1 MWh of certified renewable electricity production, and activated C sales. The technology proposed for the model was based on technical specifications provided to the author by the technology supplier[3]. The technology, sometimes known as a biorefinery, can utilize various biomass inputs other than vegetative forestry or agricultural wastes, including municipal solid waste, plastics, tyres, paper, animal products (etc.), in addition to the active removal of contaminants (usually heavy metals) from the direct processing of soils. Technologies such as these would enable processing of waste materials co-fired alongside the primary input of harvested, chipped, and dried *E. camaldulensis* biomass, derived from sustainable harvesting of the 1,500 ha revegetation project. If alternative waste/biomass sources are used as inputs, this will displace the equivalent wood requirement, as each alternative waste source will exhibit a different LHV.

The specific biorefinery selected for this research was based on a two-step process of anaerobic gasification to produce syngas, which is converted to a commercial grade biodiesel by Fischer-Tropsch synthesis. The offgas from the Fischer-Tropsch process is used for direct electricity generation. See Figure 10 for a basic flow diagram of the system. The model assumed 20 t day$^{-1}$ wood-only input at 16% moisture content, representing 16.8 t day$^{-1}$ DM biomass input. On an annual basis at a capacity factor of 0.8 (or 292 days yr$^{-1}$) a total average of 4,905.6 t of DM biomass, or 3.27 tDM ha$^{-1}$ year$^{-1}$ from the 1,500 ha project was harvested for bioenergy. In the model, the 20 t day$^{-1}$ wood chips, yield around 3,470 L day$^{-1}$ of biodiesel, a net export of 350 kW of electricity, 0.4 t day$^{-1}$ solid residues consisting of activated C produced at 1,100 $^{o}$C with steam, alongside some inactivated C and minor ash components. The thermal waste exhaust energy of 64,601 MJ day$^{-1}$ was used to dry the wood

[3] Since this research was undertaken, the technology provider no longer offers this commercial unit, and has asked to remain anonymous. Whilst this technically renders this research hypothetical rather than a pre-commercial feasibility study, this experience is indicative of many relatively immature renewable energy systems at present. Nonetheless, many new scalable bioenergy conversion technologies are currently under development, and have the potential to reduce capital costs and increase conversion efficiencies significantly.

chips to a 16% moisture content from the initial freshly harvested *E. camaldulensis* moisture content from around 38% found in the region (Yamada et al., 1999). The continuous operation of the biorefinery would provide a relatively smooth operating schedule. As the 6$^{th}$ year of the biosequestration stand was the first year of the biorefinery project, both of the 15 year projects could theoretically be successively undertaken in a 20 year period.

The quoted full system CAPEX included design, fabrication, and supervised test runs and commissioning. Additional costs included site works, installation labor, utility connection, and transport to site. The system installation lead times for these technologies were around one year after order placement, including the one month of on-site installation. All prices, factors, costs, and outputs for the system are noted in Table 6. Table 6 also includes all discounted cashflows, estimated on an annual basis, and also introduces current market prices for labor, electricity, diesel, and activated C, and also emission factors for the displaced fossil fuel electricity and diesel consumption for the region. Furthermore, the intrinsic C content remaining in the activated C was included in the spreadsheet model to enable a full market-based direct mitigation potential specific to the operation of the biorefinery. The model included the assumption of the remaining C sequestered in the activated C was used as a biochar soil amendment, and incorporated an estimate of a 70% stable C percentage in the activated C after undergoing decomposition in the soil. The biorefinery model included an identical real discount rate of 8% p.a. over the 15 year interval as the revegetation project model.

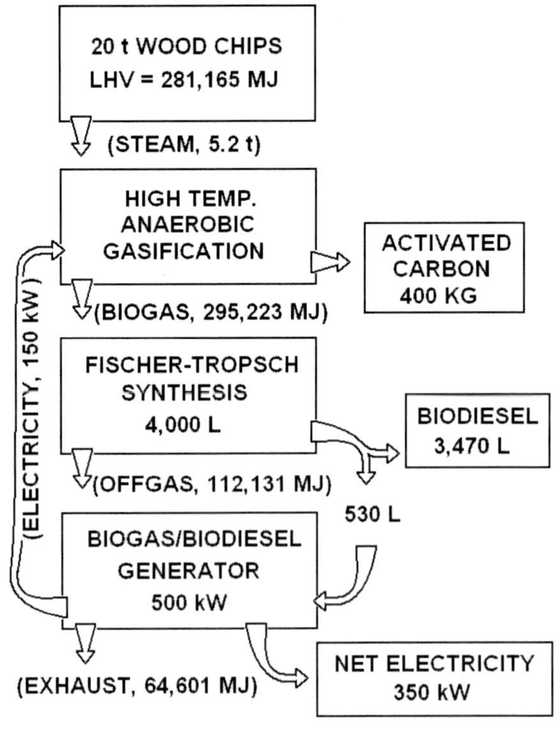


Figure 10. A simple biorefinery system sizing and daily average output flow chart.

**Table 6. The biorefinery discounted cashflow and mitigation potential projections for a 15 year operation, and decommissioning (NPV is in bold)**

| Initial costs & rates | | Rates/Prices/Factors | | | Annual costs & production | |
|---|---|---|---|---|---|---|
| Inflation rate | 3.00% | Carbon price ($/tCO2-e) | 0.00 | | Electricity price for import ($/kWh, inc GST) | $0.2083 |
| Real discount rate | 8.00% | Wood price ($/T, inc. GST) | -$30.00 | | Electricity displaced (kWh) (80% CF) | 2,452,800 |
| Transp.+works (est.) | -$700,000.00 | Wood cost per year | -$175,200.00 | | Displaced electricity value ($/yr) | $510,918.24 |
| Full plant+commissioni | -$9,600,000.00 | SWIS EF. (kg CO2-e / kWh) | 0.82 | (Scope 2) | REC value (at an average of $40/MWh) | $98,112.00 |
| Capital subsidy | $0.00 | Diesel EF. (kg CO2-e / L) | 2.57232 | (Scope 1) | Activated biochar production (t/yr) | 116.800 |
| | | Annual OPEX (est.) | -$300,000.00 | | Activated biochar value (at $3.5/kg) ($/yr) | $408,800.00 |
| | | Annual labour (est.) | -$350,400.00 | | Stable tCO2-e in biochar @70% soil stability | 299.732 |
| | | | | | Biodiesel production (L/yr) | 1,013,240 |
| **Total** | **-$10,300,000** | | | | Displaced biodiesel value (at $1/L, excl. tax) | $1,013,240.00 |

| Yr | Costs | Income | Net $ Flow | Cumul. $ Flow | D.Net $ Flow | D.Cumul. $ Flow | tCO2 Baseline | tCO2 New | Net. tCO2 Mit. | Cumul. tCO2 Mit. |
|---|---|---|---|---|---|---|---|---|---|---|
| 0 | -$10,300,000 | $0 | -$10,300,000 | -$10,300,000 | -$10,300,000 | -$10,300,000 | 4,617.674 | -299.732 | 0.000 | |
| 1 | -$825,600 | $2,031,070 | $1,205,470 | -$9,094,530 | $1,116,176 | -$9,183,824 | 4,617.674 | -299.732 | 4,917.406 | 4,917.406 |
| 2 | -$825,600 | $2,031,070 | $1,205,470 | -$7,889,060 | $1,033,496 | -$8,150,327 | 4,617.674 | -299.732 | 4,917.406 | 9,834.811 |
| 3 | -$825,600 | $2,031,070 | $1,205,470 | -$6,683,589 | $956,941 | -$7,193,386 | 4,617.674 | -299.732 | 4,917.406 | 14,752.217 |
| 4 | -$825,600 | $2,031,070 | $1,205,470 | -$5,478,119 | $886,057 | -$6,307,330 | 4,617.674 | -299.732 | 4,917.406 | 19,669.623 |
| 5 | -$825,600 | $2,031,070 | $1,205,470 | -$4,272,649 | $820,423 | -$5,486,907 | 4,617.674 | -299.732 | 4,917.406 | 24,587.028 |
| 6 | -$825,600 | $2,031,070 | $1,205,470 | -$3,067,179 | $759,651 | -$4,727,256 | 4,617.674 | -299.732 | 4,917.406 | 29,504.434 |
| 7 | -$825,600 | $2,031,070 | $1,205,470 | -$1,861,708 | $703,380 | -$4,023,876 | 4,617.674 | -299.732 | 4,917.406 | 34,421.840 |
| 8 | -$825,600 | $2,031,070 | $1,205,470 | -$656,238 | $651,278 | -$3,372,598 | 4,617.674 | -299.732 | 4,917.406 | 39,339.245 |
| 9 | -$825,600 | $2,031,070 | $1,205,470 | $549,232 | $603,035 | -$2,769,563 | 4,617.674 | -299.732 | 4,917.406 | 44,256.651 |
| 10 | -$825,600 | $2,031,070 | $1,205,470 | $1,754,702 | $558,366 | -$2,211,197 | 4,617.674 | -299.732 | 4,917.406 | 49,174.057 |
| 11 | -$825,600 | $2,031,070 | $1,205,470 | $2,960,173 | $517,006 | -$1,694,191 | 4,617.674 | -299.732 | 4,917.406 | 54,091.462 |
| 12 | -$825,600 | $2,031,070 | $1,205,470 | $4,165,643 | $478,709 | -$1,215,482 | 4,617.674 | -299.732 | 4,917.406 | 59,008.868 |
| 13 | -$825,600 | $2,031,070 | $1,205,470 | $5,371,113 | $443,249 | -$772,233 | 4,617.674 | -299.732 | 4,917.406 | 63,926.274 |
| 14 | -$825,600 | $2,031,070 | $1,205,470 | $6,576,583 | $410,416 | -$361,818 | 4,617.674 | -299.732 | 4,917.406 | 68,843.679 |
| 15 | -$825,600 | $2,031,070 | $1,205,470 | $7,782,054 | $380,014 | **$18,197** | 4,617.674 | -299.732 | 4,917.406 | 73,761.085 |
| **Remaining System Value** | | | **$0** | | | | | | | |

With the above assumptions, the biorefinery investment returned approximately 8% to the project proponent, or a NPV of close to zero (Figure 11). Furthermore, the reliance on four separate income streams (displacing diesel and electricity requirements, and selling RECs and activated C) exposes investors to several market price reduction risks. Thus, the risk tolerance of potential investors must be relatively high when attempting to integrate several bioenergy production streams into one system, and the unique circumstances of particular regions will likely determine the risk profile and general feasibility of the system for specific applications. For comparative interest, a AUD2 million (19.4%) CAPEX subsidy was introduced into the model (theoretically, this could occur from a direct investment from forward selling of the C credits in the revegetation project, or by government contribution), in addition to the estimate of a further AUD2 million in remaining value of the biorefinery post decommissioning. The resulting discounted NPV of the investment yields AUD6,648,680. With these assumptions the NPV of the new scenario becomes zero by year seven (Figure 12.) In a simple manner, these NPV comparisons demonstrate the medium-to-high investment sensitivity to the CAPEX, OPEX, timelines, and cashflows of biorefinery facilities generally. In the model, the most sensitive factor which primarily determined the cost-effectiveness of the investment is the price of diesel. As the model excluded fuel tax (AUD0.38143 $L^{-1}$), a potentially high cost-component if required to pay, potential investors would require prior knowledge of their specific circumstances relating to claiming fuel tax rebates.

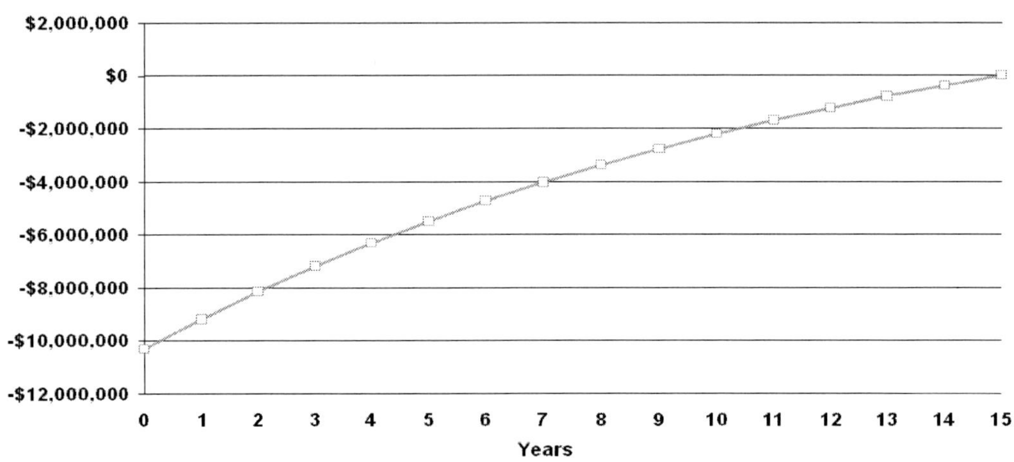


Figure 11. The discounted cumulative cashflow of the biorefinery over the 15 years.

Research by Matsumoto and Kojima (2006) indicated that wide-scale salt accumulation problems in regions with the average rainfall regime of around 217 mm $yr^{-1}$ may occur in around 70 years, inhibiting vegetative growth of tree species such as *E. camaldulensis*. However, with additional average rainfall (561 mm $yr^{-1}$), the projected interval increases to around 90 years, although with less rainfall (54 mm $yr^{-1}$) this period decreases to around 50 years. Nonetheless, *E. camaldulensis* is known for its salt-tolerance (Nasim et al., 2008), with the maximum saltwater tolerance of around 20,000 mg $L^{-1}$ (Peter & Glen, 1993), as cited by Matsumoto and Kojima (2006). Therefore, access to freshwater streams are not required for *E. camaldulensis* survival, even in areas with moderate groundwater salinity (Thorburn & Walker, 1994). However, the hydrology/salinity changes from extensive *E. camaldulensis* revegetation will require significant research, and future options to reduce of NaCl and other minerals from harvested pastoral areas will also require investment to reduce these considerable uncertainties associated with these enormous parcels of land.

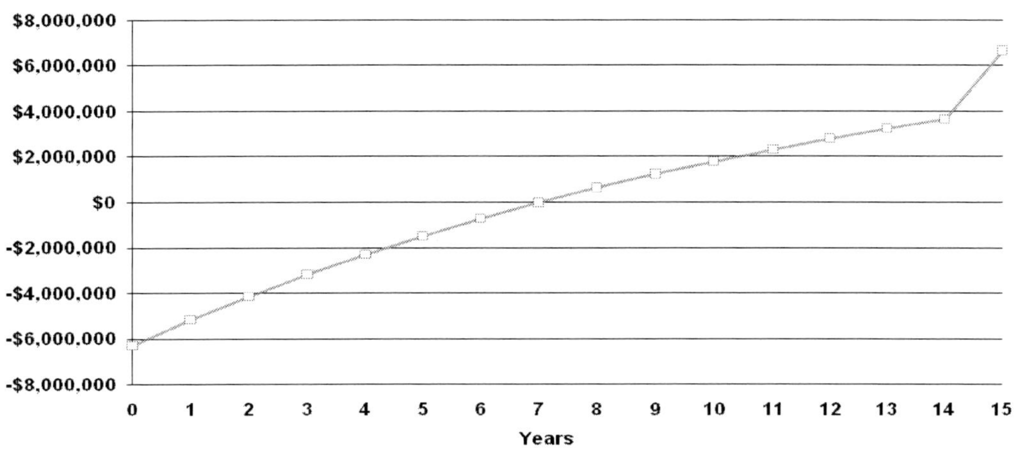


Figure 12. Biorefinery discounted cashflow over 15 years, with new assumptions of AUD2 million capital subsidy, and AUD2 million in system salvage value at decommissioning.

As the areas associated with arid pastoral production are large, it is wholly impractical to undertake individual tree measurements, and alternative methods to determine stand C biosequestration and biomass growth are required (Shiono et al., 2004). However, quantitative data regarding rangeland vegetation growth and C biosequestration is rare (Curry et al., 2008). Available Landsat imagery provides a medium spatial resolution, although estimation accuracy is relatively low (Suganuma et al., 2006). Canopy projected cover only, or canopy projected cover and tree height regression functions were both found to be a good predictor of stand biomass for the *E. camaldulensis* forests in the region (Suganuma et al., 2004; Curry et al., 2008). The canopy projected cover is able to be determined accurately even with aerial photographs (Suganuma et al., 2004). Shiono et al. (2004) found a high positive correlation between canopy projected cover and stand biomass growth with a co-efficient of determination ranging from 0.68 and 0.82, and field measurements of plant growth and C fractions can be used to calibrate the Landsat imagery (Curry et al., 2008). These monitoring activities will likely be improved by using data from the Western Australian Rangeland Monitoring System (WARMS). WARMS is comprised of around 1,600 fixed monitoring sites, and the database maintains around 5,400 records and photos (Watson et al., 2001). These records, while often incomplete, may enable an estimation of original vegetative quality as a potential baseline reference for biosequestration activities. The WARMS data may also improve the regional understanding of medium-term salinity influences on vegetation. Collation of these disparate sources of information and on-ground experiments will reduce the uncertainties faced by investors and governments when assessing the prospects of extensive revegetation of arid lands for effective C biosequestration and bioenergy projects and for achieving mitigation targets.

## Conclusion

Assuming grazing lands in WA are able to biosequester 8 tC $ha^{-1}$ $yr^{-1}$ for 15 years, a mitigation potential of 121,275,000 $tCO_2$-e can be achieved if only 1% of the 98 million ha of grazed pastoral lands were revegetated. The average annual WA emissions over the next 15 years are projected to be approximately 80,000,000 $tCO_2$-e, and a 1% land use change from grazing to revegetation in pastoral WA could roughly mitigate 10% of the total WA state emissions over the 15 years. At an approximate value of AUD15 $tCO_2$-$e^{-1}$, this potential represents around AUD1.8 billion in gross revenue flowing into remote and pastoral regions over the 15 years, a major new source of income for regional WA. This opportunity, especially for understocked Indigenous owned pastoral leases like Cogla Downs, is an entirely new means of increasing productivity, enabling many small-scale synergistic entrepreneurial opportunities and associated meaningful jobs outside of social welfare system. Many of these land management activities are synergistic with existing Indigenous cultural tourism development initiatives in remote Australia. Finally, this research indicates that further research and development is required to reduce technical and economic uncertainties by focusing on the impacts of major changes to physical flows of C, energy, water, and elemental compositions in pastoral ecosystems of WA. However, as all WA pastoral leases expire on the 30 June 2015, the fate of new lease activities and lease holders themselves remains uncertain, and a decrease in administrative barriers to pastoral diversification may facilitate greater local innovation.

## Acknowledgements

Thanks to John Law for passing on the majority of the published reference material. Also thanks to Dave, Danny, Kylie, and the kids for helping us clean up the site, drive in pickets and posts, and straining the fence and watering trees while we constructed and planted. Thanks to Brendin Flanigan, Wayne McDonald, and Robert McEachern at the Mid West Development Commission, and Alan Thompson and Ron Bradfield from Yulella for your assistance in supporting the revegetation trial.

## References

Abe, Y., Taniguchi, M., Suganuma, H., Saito, M., Kojima, T., Egashira, Y., Yamamoto, Y., & Yamada, K. (2003). Comparative analysis between biomass and topographic features in arid land, Western Australia. *Journal of Chemical Engineering of Japan,* 36(4), 376-382.

Barton, C. V., & Montagu, K. D. (2006). Effect of spacing and water availability on root:shoot ratio in *Eucalyptus camaldulensis*. *Forest Ecology and Management,* 221(1-3), 52-62.

Curry, P., Zdunic, K., Wallace, J., & Law, J. (2008). *Landsat monitoring of woodland regeneration in degraded mulga rangelands: implications for arid landscapes managed for carbon sequestration.* Paper presented at the 14th ARSPC, Darwin.

Egashira, Y., Tomii, D., Ueyama, K., Takahashi, N., Kojima, T., Abe, T., Saijo, T., & Yamada, K. (2003). Development of integrated simulator of water transport and plant growth as an evaluation tool for arid land afforestation of $CO_2$ fixation. *Journal of Chemical Engineering of Japan,* 36(4), 383-390.

Hamano, H., Egashira, Y., & Kojima, T. (2000a). Numerical prediction of water movement in Western Australian soil for large scale afforestation. *Journal of Arid Land Studies,* 10S, 25-28.

Hamano, H., Kato, S., Shimizu, T., Kojima, T., & Yamada, K. (2000b). A study on possibility of bauxite utilization to improve soil properties for afforestation of arid land. *Journal or Arid Land Studies,* 10S, 81-84.

Hirukawa, T., Asaka, N., Hamano, H., Yamada, K., & Kojima, T. (2006). A modelling methodology of large scale water balance and salt accumulation for afforestation in arid land. *Journal of Arid Land Studies,* 15(4), 247-250.

Matsumoto, M., Kato, S., Abe, T., & Kojima, T. (2006). Estimation of water availability condition for afforestation in desert of Western Australia using carbon stable isotope ratio analysis *Journal of Arid Land Studies,* 15(4), 243-246.

Matsumoto, M., & Kojima, T. (2006). Simulations of salt accumulation at soil surface under different annual precipitation amounts in arid Leonora area, Western Australia. *Journal of Arid Land Studies,* 16(1), 53-60.

McHenry, M. P. (2009a). Agricultural bio-char production, renewable energy generation and farm carbon sequestration in WA: Certainty, uncertainty & risk. *Agriculture, Ecosystems and Environment,* 129, 1–7.

McHenry, M. P. (2009b). Policy options when giving negative externalities market value: Clean energy policymaking and restructuring the Western Australian energy sector. *Energy Policy,* 37, 1423-1431.

McHenry, M. P. (2009c). Remote area power supply system technologies in Western Australia: New developments in 30 years of slow progress. *Renewable Energy,* 34, 1348-1353.

McHenry, M. P. (2009d). Why are remote Western Australians installing renewable energy technologies in stand-alone power supply systems? *Renewable Energy,* 34, 1252-1256.

McHenry, M. P. (2010). Carbon-based stock feed additives: a research methodology that explores ecologically delivered C biosequestration, alongside live-weights, feed-use efficiency, soil nutrient retention, and perennial fodder plantations. *Journal of the Science of Food and Agriculture,* 90, 183-187.

McHenry, M. P. (2011). Australian carbon biosequestration and clean energy policy market co-evolution: Mechanisms, mitigation, and convergence. *Australian Forestry Journal,* 75(2) 901-913.

McHenry, M. P. (2012a). Australian carbon biosequestration and bioenergy policy co-evolution: mechanisms, mitigation, and convergence. *Australian Forestry Journal* 75(2), 901-913.

McHenry, M. P. (2012b). Harnessing landholder's knowledge for efficient environmental monitoring, accounting, and management for new environmental markets: lessons from plantation forestry carbon sequestration in Western Australia. In: TBA (Ed.), *Plantations: Seasonal variation, growth and ecological impacts.* Hauppauge, USA Nova Science Publishers.

McHenry, M. P. (2012c). Hybrid microalgal biofuel, desalination, and solution mining systems: increased industrial waste energy, carbon, and water use efficiencies. *Mitigation and Adaptation Strategies for Global Change,* DOI: 10.1007/s11027-012-9361-y.

McHenry, M. P. (2012d). Practicalities of establishing forestry carbon sequestration projects in the agricultural sector: a technical and economic analysis with implications. In: B. J. Ryan & D. E. Anderson (Eds.), *Carbon sequestration: technology, measurement technologies and environmental effects*. Hauppauge, New York, USA: Nova Science Publishers. 978-1-62081-081-7.

McHenry, M. P. (2012e). Small-scale (≤6 kWe) stand-alone and grid-connected photovoltaic, wind, hydroelectric, biodiesel, and wood gasification system's simulated technical, economic, and mitigation analyses for rural regions in Western Australia. *Renewable Energy,* 38, 195-205.

McHenry, M. P. (2013). How farming and forestry converge: enhancing the interface between agricultural production, and tree biomass/bioenergy systems to improve farm-scale productivity in Western Australia. In: M. C. Allard (Ed.), *Bioenergy systems, biological sources and environmental impact*. Hauppauge, New York, USA: Nova Science Publishers. 978-1-62417-330-1.

Mensforth, L. J., Thorburn, P. J., Tyerman, S. D., & Walker, G. R. (1994). Sources of water used by riparian *Eucalyptus camaldulensis* overlying highly saline groundwater. *Oecologia,* 100, 21-28.

Nasim, M., Qureshi, R. H., Aziz, T., Saqib, M., Nawaz, S., Talib Sahi, S., & Pervaiz, S. (2008). Growth and composition of salt-stressed *Eucalyptus camaldulensis* and *Eucalyptus tereticornis*. *Pakistan Journal of Botany,* 40(2), 799-805.

Payne, A. L., Van Vreeswyk, A. M. E., Pringle, H. J. R., Leighton, K. A., & Hennig, P. (1998). *An inventory and condition survey of the Sandstone-Yalgoo-Paynes Find area, Western Australia. Technical Bulletin No. 90*. Perth, Western Australia: Department of Agriculture, Western Australia.

Peter, J. T., & Glen, R. W. (1993). The source of water transpired by *Eucalyptus camaldulensis*: soil, groundwater, or stream? In: *Stable isotope and plant carbon-water relations* (pp. 511-527). San Diego: Academic Press Inc.

Rengasamy, P. (2006). Wold salinization with emphasis on Australia. *Journal of Experimental Botany,* 57(5), 1017-1023.

Saito, N., Kojima, T., Kato, S., Hamano, H., Yamada, K., & Takahashi, F. (2004). Water/carbon ratio measurement for selection of trees to fix carbon by arid land afforestation (pp. 6): Seikei University and Shinshu University.

Shiono, K., Kanri, Y., Abe, Y., Tanouchi, H., Kojima, T., & Yamada, K. (2004). Stand growth estimation of representative plant communities in arid land of Western Australia. *Journal of Arid Land Studies,* 14S, 207-210.

Shiono, K., Suganuma, H., Abe, Y., Tanouchi, H., Utsugi, H., Saito, M., Takahashi, N., Kojima, T., & Yamada, K. (2006). Biomass growth estimation of an afforestation site and natural forests in an arid land of Western Australia. *Journal of Arid Land Studies,* 15(4), 251-254.

Suganuma, H., Abe, Y., Taniguchi, M., Saito, M., & Yamada, K. (2004). Fundamental research on detection of stand biomass change in an arid rangeland. *Journal of Arid Land Studies,* 14S, 85-88.

Suganuma, H., Abe, Y., Taniguchi, M., Tanouchi, H., Utsugi, H., Kojima, T., & Yamada, K. (2006). Stand biomass estimation method by canopy coverage for application to remote sensing in an arid area of Western Australia. *Forest Ecology and Management,* 222, 75-87.

Takahashi, N., Tahara, K., Utsugi, H., Kojima, T., Egashira, Y., Abe, Y., Saito, M., & Yamada, K. (2003). Water use efficiency of *Eucalyptus camaldulensis* growing in arid regions in Western Australia. *Journal of Chemical Engineering of Japan,* 36(4), 391-400.

Tanouchi, H., Utsugi, H., Takahashi, N., Hamano, H., Kawarasaki, S., Kojima, T., & Yamada, K. (2006). Water use efficiency of trees in arid lands: plasticity to water conditions. *Journal of Arid Land Studies,* 15(4), 267-270.

Thorburn, P. J., & Walker, G. R. (1994). Variations in stream water uptake by *Eucalyptus camaldulensis* with differing access to stream water. *Oecologia,* 100, 293-301.

Watson, I., Blood, D., Novelly, P., Thomas, P., & van Vreeswyk, S. (2001). *Rangeland monitoring, resource inventory, condition assessment and lease inspection activities in Western Australia conducted by the Department of Agriculture. Report prepared for the Rangeland Theme of the National Land and Water Resources Audit. Version 1.0*. Perth, Western Australia: Department of Agriculture, Western Australia.

Yamada, K., Kojima, T., Abe, Y., Saito, M., Egashira, Y., Takahashi, N., Tahara, K., & Law, J. (2003). Restructuring and afforestation of hardpan to sequester carbon. *Journal of Chemical Engineering of Japan,* 36(3), 328-332.

Yamada, K., Kojima, T., Abe, Y., Williams, A., & Law, J. (1999). Carbon sequestration in an arid environment near Leonora, Western Australia. *Journal of Arid Land Studies,* 9(2), 143-151.

In: Plantations
Editor: Hai Ren
ISBN: 978-1-62808-090-2


*Chapter 5*

# HARNESSING LANDHOLDER'S KNOWLEDGE FOR ENVIRONMENTAL MONITORING AND MANAGEMENT FOR NEW ENVIRONMENTAL MARKETS: LESSONS FROM PLANTATION FORESTRY CARBON SEQUESTRATION IN WESTERN AUSTRALIA

*Mark P. McHenry*[*]
Faculty of Minerals and Energy, Murdoch University, Western Australia

## ABSTRACT

Leveraging limited environmental monitoring and management funding using landholder knowledge and capacity can reduce total costs of implementing plantation forestry environmental market mechanisms by both reducing duplication and enhancing on-ground activities to assimilate environmental and production system decisionmaking. This chapter explores the integration of high accuracy on-site vegetative and ecological methods and metrics (on-site verification, allometrics, expansion factors, carbon factors, root-to-shoot ratios, etc.) into landscape and macro-scale plantation carbon and land monitoring models to assist the development of innovative ecosystem service markets derived from forestry plantation and carbon sequestration developments. Several limitations of current methods and policies are outlined, requiring primarily information tools and communication pathways to redress, opening the possibility for formally capturing lifetimes of local landholder knowledge of forestry land use and environmental system changes. This work suggests parallel advancements of on-site landholder and remote land monitoring and management has the potential to achieve a multitude of efficiencies. These include: cost-effective skilled environmental management jobs in regional and remote areas; streamlined administration and research expenditures for on-site environmental monitoring and supporting costs (transport, accommodation, etc.); locally appropriate conservation activities sustained over time, and; direct communication between landholders, researchers, and policymakers.

[*] Corresponding Author address Murdoch University, 90 South St Murdoch, Western Australia, 6150. Email: mpmchenry@gmail.com.

**Keywords**: Land; biomass; allometry; modelling; policy; plantation forestry

## INTRODUCTION

Emerging climate change, carbon (C), biodiversity, and water markets bring to attention the multitude of functions that ecological systems perform (Fromm, 2000; Griffiths et al., 2000; Nunes & van den Bergh, 2001; Australian Greenhouse Office, 2006). While afforestration, reforestation, or revegetation seem to offer a simple prescriptive solution to achieve environmental and climate change objectives from a policymakers perspective, those on the land must manage a range of complex biological processes that compete for limited resources (Van Vreeswyk et al., 2004). By aiming to simultaneously improve land productivity, biodiversity, water quality (etc.), and mitigate greenhouse gas (GHG) emissions, progress can be made where transdisciplinary government and landholder objectives overlap (Antheaume, 2004).

On-site physical ecological accounting by inhabitants at the landscape level provide calibration and verification of satellite monitoring and model assessment methods, enhance quantitative analyses of vegetative physical characteristics, and offer first-hand multigenerational knowledge of the land (Landsberg & Crowley, 2004; Balilla et al., 2012). Landholder on-site accounting presents a relatively simple method to monitor changes in complex ecological functions that bolster effective environmental market development. For example, biomass, C, surface or groundwater, vapour, heat, or endemic species (etc.) fluxes may be used as proxy indicators that markets can deem to represent functional ecosystem changes at the landscape level. However, such deeming mechanisms may be imprecise, exhibit many approximations, assumptions, and conceal uncertainty in complex biological responses to change, and high uncertainties in real values generally result in lower market values for deeming activities and products (McHenry, 2009b, 2010, 2011). Nonetheless, the two crucial points in relation to deeming are essentially economic in nature: the net value of deeming market participation to private landholders, and the administrative efficiency of 'soft' approaches that capture landholder capacity to protect public goods at the landscape scale. Essentially, deeming may be an imperfect mechanism, but it need not be inaccurate or an expensive market mechanism to tip the balance towards better land use policy outcomes.

In terms of new C markets, to effectively engage landholders to invest in changing the productive balance of their land requires knowledge of the sequestration potential of specific species and the likely costs and benefits to their existing production systems (McHenry, 2009a). New suites of tree species will influence the lands groundwater tables, stream flows, weed and fauna management regimes, fire control measures, and the landowner's economic situation (Clifton et al., 2007). Planning for new complex activities requires a very high resolution of often site-specific biological, geological, sociological, and economic information. While larger regional or landscape scale site models can broadly inform land use change (LUC) policymakers and researchers regionally, in practice there is often little reciprocal communication with those who actively manage and live on the land. Such inhabitants will likely be able to enhance the appropriateness of input data, assumptions and refine resulting policy outputs and conclusions.

To sustainably operate at the landscape scale, landholders require very high resolution and extremely reliable information with a focus on seasonal fluxes and environmental extremes, rather than annually averaged data. Of particular interest to landholders are the maintenance and thresholds of critical systems that sustain productivity, and often use specific indicator species characteristics to monitor biological responses to seasonal and annual landscape change. Thus seasonally differentiated functional ecosystem change datasets are a fundamentally appropriate foundation for researchers and policymakers to develop and maintain (Landsberg & Crowley, 2004; Stoneham, 2009). While potentially more costly initially, information systems and models with multiple utilities will have a greater reciprocal value to land managers, government planners, and industry security generally in periods of increased climatic variability than much of the annually averaged data currently generated and used. Market mechanisms can be harnessed to attract new investment to either directly or indirectly assist the ability of governments to initiate the collaborative development of monitoring and verification activities. However, investors will require certainty a priori (Stoneham, 2009).

New markets will likely create new land ownership rights to maintain investor certainty. Thus, landholders will also require long-term certainty regarding possible claims of rights to the land from governments, ecological product holders, and how a diversified ownership structure may influence their livelihood. In 2003, the Western Australian Parliament passed the Carbon Rights Act 2003, which gave landowners a statutory basis to register, and trade the C rights earned since 1 January 1990 as a separate interest in the land (The Parliament of Western Australia, 2003). At present, what transpires on the land after C trading is dependent on the legal obligations imposed under transfer of such rights, which are in-turn dependent on each particular trading mechanism design. New markets will likely require similar new property rights, and will thus require a similar legislative basis, although the monitoring and verification regimes will likely require much additional but complimentary ecological input data. While quality datasets enable the development of stable and sustained markets with a high level of transparency, such datasets require an additional level of resolution, transparency, and efficiency in combining regional models and on-site verification than is available today. However, analogous developments have enabled the development of C markets, which with augmentation, may be a suitable foundation for various ecological markets.

## THE INTERFACE BETWEEN C MODELS AND ON-SITE VERIFICATION

The development of the Australian National Carbon Accounting System (NCAS) was required to decrease uncertainty, improve transparency, and improve verification of national land use, land-use change, and forestry (LULUCF) assets and fluxes under the United Nations Framework Convention on Climate Change (UNFCCC) and Kyoto Protocol commitments (Australian Greenhouse Office, 1999b, 1999a; Turner et al., 1999; Polglase et al., 2000; Snowdon et al., 2000; Richards, 2001; Australian Greenhouse Office, 2002; Furby, 2002; Furby & Woodgate, 2002; Paul et al., 2002; Richards et al., 2002; Booth et al., 2003; Australian Greenhouse Office, 2005b, 2005a). The NCAS methodology was developed to be more rigorous than the minimum requirements of the Intergovernmental Panel on Climate Change (IPCC) guidelines for forest C monitoring. The IPCC guidelines were based on forests of trees with the potential to reach a minimum of 2 metres in height at maturity, a tree crown cover (or equivalent) of more than 20% and a minimum area of 0.2 ha (Australian Greenhouse Office, 2002, 2005b). To quantify the total amount of C in woody tissue both the volume and density of tissue must be known (Ilic et al., 2000). (See Table 1 for an example of vegetative class partitioning and density data).

**Table 1. Partitioning of biomass and wood density values by major vegetation group class**

| Vegetation | Stem | Branches | Bark | Leaves | Coarse Root | Fine Root | Wood Density (DMkg m$^{-3}$) |
|---|---|---|---|---|---|---|---|
| Eucalyptus Tall Open Forests | 0.67 | 0.09 | 0.10 | 0.02 | 0.03 | 0.08 | 550 |
| Eucalyptus Open Forest | 0.65 | 0.07 | 0.07 | 0.01 | 0.05 | 0.15 | 625 |
| Eucalyptus Low Open Forest | 0.45 | 0.12 | 0.10 | 0.02 | 0.05 | 0.25 | 550 |
| Eucalyptus Woodland | 0.36 | 0.15 | 0.10 | 0.02 | 0.06 | 0.31 | 890 |
| Acacia Forest and Woodland | 0.36 | 0.15 | 0.10 | 0.02 | 0.06 | 0.31 | 940 |
| Callitris Forest and Woodland | 0.36 | 0.15 | 0.10 | 0.02 | 0.06 | 0.31 | 650 |
| Casuarina Forest and Woodland | 0.36 | 0.15 | 0.10 | 0.02 | 0.06 | 0.31 | 860 |
| Melaleuca Forest and Woodland | 0.36 | 0.15 | 0.10 | 0.02 | 0.06 | 0.31 | 660 |
| Other Forest and Woodland | 0.36 | 0.15 | 0.10 | 0.02 | 0.06 | 0.31 | 800 |
| Tropical Eucalyptus Woodland/Grassland | 0.36 | 0.16 | 0.10 | 0.02 | 0.06 | 0.30 | 830 |
| Eucalyptus Open Woodland | 0.36 | 0.15 | 0.10 | 0.02 | 0.06 | 0.31 | 890 |
| Acacia Open Woodland | 0.22 | 0.165 | 0.10 | 0.025 | 0.07 | 0.42 | 940 |
| Mallee Woodland and Shrubland | 0.22 | 0.165 | 0.10 | 0.025 | 0.07 | 0.42 | 1060 |
| Low Closed Forest and Close Shrubland | 0.22 | 0.165 | 0.10 | 0.025 | 0.07 | 0.42 | 1000 |
| Acacia Shrubland | 0.22 | 0.165 | 0.00 | 0.025 | 0.07 | 0.42 | 940 |
| Other Shrublands | 0.22 | 0.165 | 0.10 | 0.025 | 0.07 | 0.42 | 940 |
| Heath | 0.00 | 0.30 | 0.18 | 0.03 | 0.07 | 0.42 | 900 |
| Chenopod Shrub, Samphire Shrub & Forbland | 0.00 | 0.30 | 0.18 | 0.03 | 0.07 | 0.42 | 900 |
| Unclassified Native Vegetation | 0.39 | 0.14 | 0.09 | 0.02 | 0.06 | 0.30 | 780 |

Source: (Australian Greenhouse Office, 2002)

The NCAS is based on resource inventories, site specific studies, multi-temporal remote sensing, and sequential modelling land cover change, LUC, soil type, forest type, management, climate, biomass growth, and litter inputs to scenario projections of the C balance of LUC projects, or at the national level (Booth et al., 2003; Australian Greenhouse Office, 2005b). The NCAS includes C quantification verification, which is based on the Full Carbon Accounting Model (FullCAM), that provides a 25 m grid resolution of C fluxes over monthly time steps (Richards, 2001; Australian Greenhouse Office, 2002; Booth et al., 2003; Australian Greenhouse Office, 2005b, 2006). FullCAM is capable of C accounting for afforestation, reforestation, revegetation, and deforestation, as well as agroforestry systems, and is comprised of a combination of Excel-based accounting tools that model biological processes. Many of the model parameters are user-defined to reflect land management decisions and rates of C transfer between pools, or the atmosphere (Richards, 2001). The various modelled biological processes include land cover change, land use and management, climate input, crop growth and plant parameters, biomass stock and growth incremements, tree parameters, forest parameters, soil C, and the model framework (Australian Greenhouse Office, 2002, 2005b). The five sub-models incorporated into FullCAM are 3-PG (forest physiological growth model), CAMFor (forest C accounting model), CAMAg (cropping and grazing C accounting model), GENDEC (microbial decomposition model), and the Roth-C (soil C model) (Richards, 2001; Australian Greenhouse Office, 2002). The CAMFor component assists C accounting and projections by calculating the C flows associated with forest and tree stand changes, and also any wood product harvesting activities (Brack et al., 2002; Brack & Richards, 2002). CAMFor converts annual changes in stem wood using parametrics that approximate annual changes in tree components alongside C content. The parametrics of importance are stem wood density and the variation of other tree components relative to the stem, which vary with species, age, size, and on-site conditions (Booth et al., 2003). Growth in the bark, branch, leaf, twig and root pools are modelled as an age dependent increase relative to stem growth (Brack & Richards, 2002). FullCAM incorporates these and other biological processes in relation to biogeographic information in the FullCAM database (Booth et al., 2003).

To generate emissions and removals estimates for reforestation projects, a component of the NCAS, the Carbon Accounting Toolbox (NCAT) was collaboratively developed by the Australian Greenhouse Office, CSIRO, and the Australian National University. NCAT uses the same data and models as the NCAS to generate net GHG removals and emission estimates without requiring input data relating to stem volume and wood density. Thus, forest entities using NCAT only require forest data such as tree species, years of establishment, thinning measures, rotation intervals and any fertiliser applications as input data. While the Australian Government does not yet prescribe the use of the NCAT, they do recommend its use for forestry C emission and removal modelling to utilise the continually updated climate, management activity, remote sensing, soil data, and forest type databases (Department of Climate Change, 2008). Crucially, much of this updating incorporates on-site assessment, as it is important that local data sets are not solely reliant on satellite imagery as to limit interpretive errors. The on-site data greatly increases image interpretation and validation of remote sensing results (Furby & Woodgate, 2002; Clifton et al., 2007).

## On-site Verification to Improve Model Effectiveness

Some on-site measurement methods are expensive, but accurate, while others are inexpensive and less accurate. Remote sensing is one relatively inaccurate method, but has the advantage of being inexpensive (Richards et al., 2002). Physiological models based on remote sensing require relatively expensive field calibration from a network of quality audited permanent plots (Australian Greenhouse Office, 1999a; Turner et al., 1999; Richards, 2001; Richards et al., 2002). Researching economically efficient means to calibrate and verify models through longitudinally maintained site-specific plot data sets is an important area for policymakers to facilitate. The temporal distinction between calibration and verification practices and data enable a longer term comprehension of changes in ecosystems from historically continuous data. One serious gap in environmental inventories for verification and calibration purposes relates to the dearth of private land information held in the public sphere. This restricts the reliability of the increasingly important land use policy (Richards, 2001).

Development of systems such as the NCAS are undertaken recognising the imperfect nature of the input data and the output. The system focuses on functional aggregated accuracy and objectivity rather than attempt to pursue an ultimate level of precision of biomass growth (Australian Greenhouse Office, 2002). The net primary productivity (NPP) of forests is the rate at which solar energy is converted to biomass, and its quantification is essentially the objective of such models. Destructive sampling is one method of determining NPP, and while it is suitable for individual plots of grass and shrub vegetation, it is rarely used for forest ecosystems (Snowdon et al., 2002; Kesteven et al., 2004). There are two basic methods for destructively estimating aboveground biomass of individual trees: the complete harvest method, and; the partial harvest methods (Snowdon et al., 2002). However, determining the biomass and C in standing, unharvested trees is accomplished by using calculations that model the tree components such as stems, crowns, roots and tops for different sizes, ages and tree spacing (Australian Greenhouse Office, 1999a; Ritzon & Sochacki, 2002; Bi et al., 2004). Using easily measured tree characteristics to estimate tree biomass is called allometry, and thousands of allometric equations for individual species in specific areas have been developed using slightly different methods (Australian Greenhouse Office, 1999a; Eamus et al., 2000; Richards et al., 2002; Specht & West, 2003; Bi et al., 2004).

## Allometrics and Expansion Factors: Utility and Limitations for New Markets

Allometric equations are an important technique to estimate tree and forest biomass and sufficient information is available to estimate growth rates for forests (Australian Greenhouse Office, 1999a; Turner et al., 1999; Richards et al., 2002). Relatively few allometric equations are required to adequately describe the aboveground biomass of some vegetation types, especially where one or two species dominate site basal area (the cross sectional area of trees in a forest in $m^2$ $ha^{-1}$) (Australian Greenhouse Office, 1999a; Richards et al., 2002). On-site sampling of basal area tree height, and volume are currently used to refine NCAS estimates of mature forest and regrowth of various ages (Australian Greenhouse Office, 1999a, 2006). Basal area is often recommended as a standard measure for allometric equations as it was often the only forestry variable recorded (Richards et al., 2002). Aboveground biomass can be estimated from stand basal areas using ratios established for major Australian forest types (Australian Greenhouse Office, 2006). However, basal area only allometric equations are less useful than those that incorporate diameter at breast height (DBH) measurement (Richards et al., 2002). Stem diameter can include DBH, or DBHOB (diameter at breast height over bark), which are generally 130 cm from the ground unless stated otherwise (Brokaw & Thompson, 2000; Eamus et al., 2000; Keith et al., 2000; Snowdon et al., 2002; Zianis & Mancuccini, 2004). If possible all diameter measurements should be undertaken for both over and under the bark. This can either be undertaken by removing the bark, or using a bark depth gauge to estimate the under bark diameter. One use of measuring stem diameters both over and under the bark is to estimate dry weight of dead trees and also to determine biomass information about the bark itself. It is possible to get a good indication (generally within 2% difference) of stem bark thickness by measuring only two opposite points on the stem. Other common stem measurements are diameters at 10 cm ($d_{10}$) and 30 cm ($d_{30}$), which are used for small trees or shrubs prone to forking. When there is more than one stem on the sample tree, calculating the diameter of a circle of area equal to the sum of the cross-section areas of all stems at the same height is known as the diameter equivalent ($d_e$) (Snowdon et al., 2002). Average biomass estimates across a wide range of sites can be adequately estimated when using two variables, such as height and stem diameter for specific species and their sizes (Snowdon et al., 2000). Tree height estimation on standing trees commonly use trigonometric or hypsometers at right angles from any lean with two measurements from the opposite side of the tree (Snowdon et al., 2002).

A study by Ritzon and Sochacki (2002) developed a model for predicting the biomass and C for *Pinus pinaster* (Maritime pine) in the southwest of WA. This included destructive sampling of trees aged between 1 and 47 years to determine dry matter biomass and C content to develop allometric equations. Similar to comparable research, the study found that the aboveground biomass of the tree contained very close to 50% C and the belowground tree mass C content was around 48%, and that using more than one measurement provided the most reliable allometric estimations (Ritzon & Sochacki, 2002). They developed equations that predicted biomass and C for each component of the Maritime pines adequate for various management practices, such as pruned for sawlogs, unpruned single paddock trees, or grown in a windbreak.

**Table 2. Linear regressions of the form Y = a + *b*D tables to estimate physical attributes of 10 four year old *Eucalyptus astringens* (Brown Mallet) trees using DBHOB measurements**

| Y Value | D Value[1] | a | b | $R^2$ | P Value |
|---|---|---|---|---|---|
| Stem (kg) | DBHOB (m) | -2.5999 | 206.16 | 0.84 | <0.0001 |
| Height (m) | DBHOB (m) | 1.6300 | 70.924 | 0.87 | <0.0001 |
| Branches and fruit (kg) | DBHOB (m) | -0.6967 | 65.47 | 0.69 | <0.0018 |
| Leaves (kg) | DBHOB (m) | -0.4723 | 82.01 | 0.77 | <0.0005 |
| Total biomass (kg) | DBHOB (m) | -3.770 | 353.65 | 0.83 | <0.0001 |

[1] DBHOB is diameter at breast height over bark.

Source: (Eamus et al., 2000)

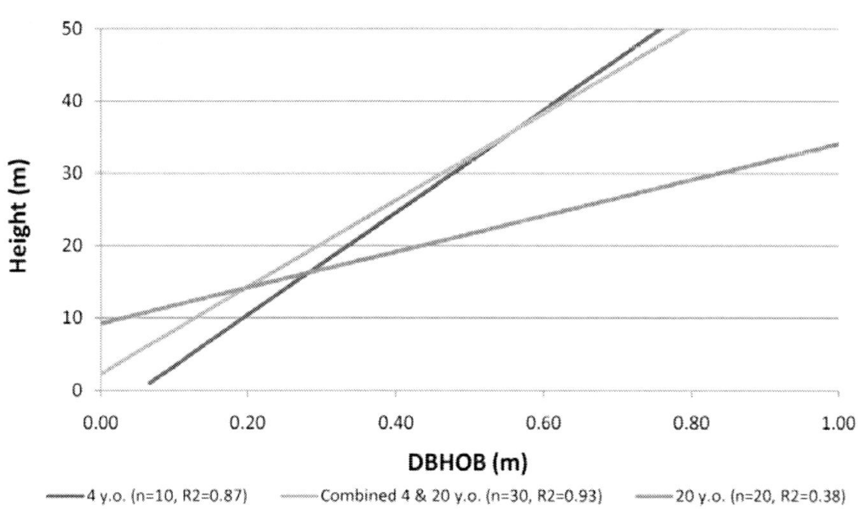


Figure 1. Linear regression functions derived from 30 *Eucalyptus astringens* (Brown Mallet) trees of four and twenty years of age to approximate height from DBHOB.

**Table 3. Selected available regression analyses of the form $ln(M) = a * ln(DBH) + b$ of tree species grown in WA**

| Name | Common Name | DBH Details[1] | DBH Range | n | a | b | $R^2$ | Author(s) |
|---|---|---|---|---|---|---|---|---|
| Eucalyptus marginata | Jarrah | AG (kg) DBH (cm) | - | 10 | 2.84 | -3.680 | 0.994 | Hingston et al. (1981) |
| Corymbia calophylla | Marri | AG (kg) DBH (cm) | - | 10 | 2.74 | -3.370 | 0.982 | Hingston et al. (1981) |
| Banksia grandis | Bull Banksia | AG (kg) DBH (cm) | - | 16 | 2.50 | -2.260 | 0.963 | Hingston et al. (1981) |
| Eucalyptus diversicolor | Karri | DBH (cm) at 30cm height | - | - | 2.107 | 4.501 | 0.99 | Grove (1998) |
| Eucalyptus diversicolor | Karri | DBH (cm) at 130cm height | - | - | 2.128 | 5.146 | 0.99 | Grove (1998) |
| Bossiaea laidlawiana | Waterbush subsp. | DBH (cm) at 30cm height | - | - | 2.70 | 4.998 | 0.99 | Grove (1998) |
| Bossiaea laidlawiana | Waterbush subsp. | DBH (cm) at 130cm height | - | - | 2.814 | 4.268 | 0.99 | Grove (1998) |
| Trymalium spathulatum | Karri Hazel | DBH (cm) at 30cm height | - | - | 2.722 | 4.284 | 0.99 | Grove (1998) |
| Trymalium spathulatum | Karri Hazel | DBH (cm) at 30cm height | - | - | 2.795 | 3.849 | 0.99 | Grove (1998) |
| Chlorilaena quercifolia | Karri Oak | DBH (cm) at 3cm height | - | - | 2.665 | 4.187 | 0.96 | Grove (1998) |
| Lasiopetalum floribundum | Free Flowering L. | DBH (cm) at 3cm height | - | - | 2.663 | 4.267 | 0.98 | Grove (1998) |
| Pimelea clavata | Rope Banjine | DBH (cm) at 3cm height | - | - | 3.146 | 3.027 | 0.98 | Grove (1998) |
| Acacia pulchella | Prickly Moses | DBH (cm) at 3cm height | - | - | 2.992 | 4.445 | 0.97 | Grove (1998) |
| Hibbertia cuneiformis | Cutleaf Hibbertia | DBH (cm) at 3cm height | - | - | 2.681 | 3.591 | 0.93 | Grove (1998) |
| Corymbia maculata site A | Spotted Gum | AG (kg) DBHOB (cm) | 2–24.5 | 10 | 2.47 | -2.51 | 0.95 | Ward and Pickersgill (1985) |
| Corymbia maculata site B | Spotted Gum | A&BG (kg) DBHOB (cm) | 2.0–11.5 | 11 | 1.87 | -1.1 | 0.98 | Ward and Pickersgill (1985) |
| Eucalyptus resinifera site A | Red Mahogany | AG (kg) DBHOB (cm) | - | 8 | 2.44 | -2.54 | 0.97 | Ward and Pickersgill (1985) |
| Eucalyptus resinifera site B | Red Mahogany | DBH (cm) | 2.0-11.5 | 10 | 1.74 | -1.12 | 0.98 | Ward and Pickersgill (1985) |
| Corymbia calophylla site A | Marri | DBH (cm) | - | - | 2.04 | -1.54 | 0.99 | Ward and Pickersgill (1985) |
| Corymbia calophylla site B | Marri | DBH (cm) | - | - | 1.64 | -0.92 | 0.89 | Ward and Pickersgill (1985) |

[1] AG is aboveground, A&BG is above and belowground, DBH is diameter at breast height, DBHOB is diameter at breast height over bark.

Source: (Eamus et al., 2000)

The fundamental importance of biomass estimation for C and other ecological markets necessitates a comprehensive suite of predictive equations for landscapes and species (Australian Greenhouse Office, 1999a). It is possible to find simple linear allometric equations of the form Y = a + *b*D, where *a* and *b* are the scaling coefficients, Y the total aboveground tree dry biomass or tree height, and D is the DBH (See Table 2 and Figure 1). However, the most commonly used form for biomass studies is $M = aD^b$ (Zianis & Mancuccini, 2004). The variability of M is largely related to variability of D measurement, while values of *a* and *b* vary with species, stand age, site, climate, and stand density. As the relationship between M and D corresponds well to a power function, the raw data is usually transformed logarithmically (Zianis & Mancuccini, 2004). (See Table 3 for a list of logarithmically transformed linear allometric equations). Coefficients *a* and *b* are usually determined by least-square regression of log-transformed data for DBH, and M measured from destructive sampling (Eamus et al., 2000; Zianis & Mancuccini, 2004).

Unfortunately, there are fundamental issues with much of the available allometric data and collection methods. Uncorrected estimates of biomass derived from least squares regression on logarithmically transformed data should not be used to estimate biomass as they seriously underestimate biomass without correction factors (Eamus et al., 2000). If available, correction factors allows landholders to accurately quantify tree biomass from allometric equations by measuring the relevant variables in the equations, such as the DBH (Eamus et al., 2000; Bi et al., 2004). However, without correction factors biomass is likely to be underestimated due to the DBH being relatively large and being skewed by the antilogarithmic transformation[1] (Eamus et al., 2000). The primary untransformed data is of much greater utility to landholders and new markets, although very little original data exists, and/or is available. Furthermore, allometric equations technically only apply to the forest stand where they were derived, and equations will likely be unsuitable outside of the tree species, tree size class, and the region of which they were obtained (Snowdon et al., 2000; Richards et al., 2002). Despite these fundamental limitations, the primary source of error with available biomass estimation equations derives from regression analyses. Statistical comparisons and the combination of equations is constrained by a lack of standard mathematical form, differences in independent variables, lack of statistical transformation information, and deficiency in details of the original data (Australian Greenhouse Office, 1999a).

For larger areas of trees there are a range of what are called 'expansion factors', which are used with sample results to more accurately determine biomass and sequestration of larger numbers of trees in accordance with NCAS protocols (Snowdon et al., 2002). Expansion data assist accurate useful for modelling when the location and volumes are known, and must be of sufficient detail, which may impinge on commercial sensitivities (Australian Greenhouse Office, 1999a). Expansion factors vary according to species, canopy cover, region (etc.) (Australian Greenhouse Office, 1999a; Furby, 2002). There are also differences between commercial forest and unmanaged forests in terms of allometric regression expansion factors (Australian Greenhouse Office, 1999b). These differences in land use should be reflected by differences in allometric and expansion factor data. However, the main source of error in upscaling biomass estimates using expansion factors relates to the number and adequacy of sampled representative plots and the level of natural heterogeneity of particular forest species

[1] Research by Eamus et al. (2000) includes a useful overview of allometric correction factors.

(Australian Greenhouse Office, 1999a). Conversion of measurements to total forest biomass requires wood density measurements and expansion factors that account for the stem as a proportion of the total tree including stem, bark, branches, twigs, leaves and root data (Australian Greenhouse Office, 1999a; Keith et al., 2000). Non-destructive sampling methods at a site commonly entails removing 0.12 cm increment core from 110 cm aboveground to ascertain whole tree density. The assessment of wood density also requires recording of the location, altitude, age, and species of the tree, the height of the tree and sample points, diameter over bark at the sampling point, compass direction of the sample, median and seasonal distribution of rainfall, soil type, and a measure of site quality (Ilic et al., 2000). Such details are rarely available, especially in primary form, free from statistical transformation.

## Carbon Content and Complexities in Belowground Biomass Estimation

Biomass estimates are converted to C by applying C factors, and in contrast to allometrics and expansion factors, turn out to be remarkably consistent between species and regions (Australian Greenhouse Office, 2006). (Table 4 shows general C content and turnover rates). Australian research by Gifford (2000a) found that overall C contents of all tree tissues and species was 50%, while leaves exhibit slightly higher C content of 52.8%, and leaf litter C content was 54.3%. Gifford recommended for when a single %C value is required to represent all aboveground components of all species, a value of 50±2% be used. When leaves were analysed separately a value of 53±2% be adopted, and for Australian native species, a value of 50±2% for woody components is appropriate. Deep wood near the centre of the trunk exhibited around 2% higher C contents than sapwood due to lower mineral contents (Gifford, 2000a). Gifford (2000b) also obtained an averaged value for the C content of coarse wood root of 49±1%, based on an analysis of 23 species. The range of these 23 species was 46.7 to 51.2%. When a continental value for forest roots including fine roots, 48±2% is suggested, and when a single figure is required that represents the C content of all woody components above and belowground (including branches and coarse roots), 49±2% is suggested. Significantly, there is often less variability between tree species than between tissues (Gifford, 2000b). While carbon factors themselves add little uncertainty to C estimates, the errors of biomass carried through assessments remain the greatest source of uncertainty.

Biomass estimation through allometric relationships can be complicated by seasonal and annual variability in NPP, unique underground vegetative structures, harvesting, fire, or unusual environmental conditions (Turner et al., 1999). In mallee vegetation (low growing, multi-stemmed Eucalypt species) individual lignotubers may be very large and coppicing or fire can stimulate lignotuber growth as the tissues grow and the lignotuber merges with the stem base (Kalin Arroyo et al., 1995; Snowdon et al., 2002). In southern Australia, the soil water holding capacity over the period in summer and autumn (December to May), often determines the growth and survival of trees (Hingston et al., 1998). In water limited regions the relationship between age and biomass may not be strong due to the opportunistic growth rate of vegetation in the more arid systems (Turner et al., 1999; Van Vreeswyk et al., 2004). As root biomass growth patterns generally reflect water availability, sampling for minimum root biomass should correspond to the driest period of the year and vice versa (Snowdon et al., 2002). Species and regional variability will undoubtedly require on-site sampling to refine estimations for many forest and ecological market requirements. This is especially so in relation to belowground biomass, often estimated using 'root-to-shoot' (RS) ratios (Australian Greenhouse Office, 1999a; Turner et al., 1999; Richards et al., 2002).

**Table 4. Carbon contents and annual turnover rates of tree components**

| Tree Component | Dry Matter Carbon Content | Component Turnover Rate |
|---|---|---|
| Leaves and Twigs | 0.52 | 0.0470 |
| Branches | 0.47 | 0.0056 |
| Bark | 0.49 | 0.0083 |
| Stem | 0.50 | - |
| Coarse Roots | 0.50 | 0.0560 |
| Fine Roots | 0.48 | 0.1042 |

Source: (Eamus et al., 2000; Australian Greenhouse Office, 2002)

RS ratios are used to estimate belowground biomass from aboveground data (Australian Greenhouse Office, 2006). The availability and resolution of appropriate RS ratio allometric equations and expansion factors will determine both the costs and the accuracy of total biomass measurement and estimation. Unfortunately, there is variable availability of allometrics and expansion factor details for RS ratios and for larger trees over 50 cm DBH (Australian Greenhouse Office, 1999a; Richards et al., 2002). Australian native species are renowned for their high variability of RS ratios within and between forest types, which is attributable to species, age, soil and climate (etc.) conditions (Snowdon et al., 2000; Richards et al., 2002). Temporal variations in fine root biomass for *Eucalyptus marginata* (Jarrah) forests in Western Australia can vary up to four-fold between summer and winter months, which is around 20% of the total root biomass (McKenzie et al., 2000; Snowdon et al., 2002). This complexity is also compounded by the relationship between above and belowground biomass from fires, coppicing, and seasonal variation (Snowdon et al., 2002).

## Management Complexities and Opportunities for Integrating Local Knowledge

There is also a paucity of data regarding Australian tree harvest characteristics in privately managed forests and environmental plantings. As fuelwood is often sourced from state forests, obtaining data regarding these activities can only occur at the retail and landholder level (Australian Greenhouse Office, 1999a). These woody C flows into various product classes and physical residence times needs be quantified, as the total net Australian pool of C sequestered as wood in the national housing stock alone is estimated to be around 10 MtC (Australian Greenhouse Office, 1999a; Jaakko Poyry Consulting, 2000). As policymakers seek to refine the accuracy and precision of national C accounting, there is likely to be a growing need for reliable data of this nature, and its collection will require the involvement of both researchers and local landholders (Australian Greenhouse Office, 1999a). Regional information of this nature can also be used to calculate the energy available for use in the expanding renewable energy markets in relation to woodheater demand, for cofiring in conventional generators, or the growing market for small bioenergy units cogenerating useful electricity and heat in regional areas (McHenry, 2009a).

In relation to environmental markets attributed to vegetative management, identifying the timing and implementation of any land use activity will likely require very detailed biological modelling, especially when concerning fire management. This quantification will require both the concurrent agents of change, and the separated impact of change to be determined (Richards, 2001). However, modelling data in inappropriately high resolutions may presume a sense of false precision for market participants, and a balance between market value, transparency, and practicality should evolve (Australian Greenhouse Office, 1999a).The most likely form that individual landholders will be able to take part in future environmental markets will be through 'pooling'. In a similar manner to supply market quantities of niche products, or more recently, C pooling involves combining a number of small registered areas in order to generate sufficient marketable product volume to obtain a positive return by sharing costs of technical measurement, certification and marketing of the credits (The CRC for Greenhouse Accounting & Tony Beck Consulting Services Pty Ltd, 2003; McHenry, 2012a).

Documented evidence required to establish eligibility of C sequestration projects generally include restrictions on land uses where the project resides, a regime of maintenance for ongoing compliance, and a risk management strategy (Independent Pricing and Regulatory Tribunal, 2008). Aerial photographs or satellite imagery with on-site data are required as a bare minimum to enable landowners to register land use and land management changes. Accredited forestry sequestration activities and verification are often allowed to be undertaken roughly every five years, with a corresponding credit for each period. C assessments are calculated using models consistent with a corresponding standard, and sometimes include an appropriate uncertainty analysis. Uncertainty analyses are often used to reduce the final amount to ensure a conservative estimate (The New South Wales Government, 2005). Vegetation that involves rotational harvesting at some point requires a more sophisticated C modelling and accounting system than simple conservation plantings. Eligible certificate creation for rotational harvesting usually requires C stocks never to fall below the threshold of certificates created (Independent Pricing and Regulatory Tribunal, 2008).

## Environmental and Land-Use Change Policy Ambiguity

The final forest C offset rules that will apply in Australia's climate change mitigation policies are yet to be determined. The Commonwealth Government has indicated its intention to review the development of longer term arrangements (The New South Wales Government, 2008). The previously proposed CPRS White Paper stated that offsetting will be likely allowed from sectors outside C policies (Cosier et al., 2009), and this is likely to be the approach followed by successor policies. As the agricultural sector will likely be excluded from the C policy liabilities, landholders will be eligible to create C offsets for captured sectors. Landholders, certain leaseholders and C property rights holders can apply to become accredited forest entities, which will be published by any scheme regulator. To be eligible for the first Kyoto Protocol (KP) commitment period permit entitlements, forests were to be registered before 1 January 2013 (Cosier et al., 2009).

There is likely to be continued developments occurring in the creation on new forest and ecological management markets, of which have some precedence in the KP requirements for LULUCF inclusion into national inventories. The COP 13 in Copenhagen secured further interest in market mechanisms to improve forest management under the Reducing Emissions from Deforestation and Forest Degradation in Developing Countries (UN-REDD Programme). Forest management is defined as "a system of practices for stewardship and use of forest land aimed at fulfilling relevant ecological (including biological diversity), economic and social functions of the forest in a sustainable manner" (United Nations Framework Convention on Climate Change, 2002) p 58. The KP's Article 3.4 includes the option of accounting for forest management, revegetation, cropland management, and grazing land management to be included in national targets. Of particular interest in Australia is ecological market developments relating to grazed woodlands and rangelands, which are accounted under the Grazing Lands component of Article 3.4 (Richards, 2001). As Australia currently chooses not to account for article 3.4 emissions, there is a possibility that some of these management activities will become involved in future markets under international frameworks. Thus, it is only logical to compile and develop monitoring and quantitative methods of proxy ecological indicators of such expanding management and conservation policies in addition to simply biomass and C NPP. It is interesting to contrast how far land use policies have come in Australia over a relatively short space of time. This may also be a source of concern to landholders seeking to undertake long-term commitments, as Australian policymakers, even in the same jurisdiction, have tended to provide little consistency over time (McHenry, 2012a).

## Lessons Learned From History: Land-use Market Risks

The WA Government's Conditional Purchase (CP) crown land release scheme was first introduced in early WA settlement, but was reinvigorated in 1961 and continued until 1982. Each property in the CP scheme was purchased for only around 35 cents per ha (in 2010 Australian dollars), which included the government provision of land surveys, road access, and telephone services. This attracted many unfamiliar with land management to implement the minimum conditions of the CP scheme, which were to clear two-thirds of the property and fence the boundary within a given period. This resulted in high rates of farm failure, and a resultant over-clearing from successful neighbours to recover costs of overextending themselves by acquiring the failed properties. Despite the known soil, biodiversity, productivity losses and associated salinity issues at the time, the amount of clearing facilitated by the WA Government between 1980 and 1990 in four significant bioregions totalled between 80,000 and 160,000 ha each year (Australian Greenhouse Office, 2000). (Table 5) The combined impact of government land development schemes, tax incentives and easy finance lead to a scarcity of uncleared arable land. Thus, native vegetation clearing policies were replaced by staunch clearing control legislation. The WA Government policy now favours conservation and a greater landholder involvement (Australian Greenhouse Office, 2000; Van Vreeswyk et al., 2004).

**Table 5. Clearing in four significant Biogeographic regions (IBRA) between 1980 and 1990**

| IBRA region | Estimated clearing (ha $yr^{-1}$) |
|---|---|
| Wheatbelt | 10,000 – 20,000 |
| South-eastern | 50,000 – 100,000 |
| Coastal | 10,000 – 20,000 |
| South-western | 10,000 – 20,000 |
| Total | 80,000 – 160,000 |

Source: (Australian Greenhouse Office, 2000).

## Environmental Market Development Guidance from Carbon Market Land-Use Experience

An indication of the order of value of future environmental markets can be approximated by expected values for C sequestration. The Cooperative Research Centre for Greenhouse Accounting (CRCGA) undertook a case study of the profitability of 'C farming' in Western Australia. The CRCGA used a 60 year discounted cashflow analysis that included the commercial wood products produced, the additional potential value of C, and the administration and other costs associated with C forestry. The study combined the local government regional data for average productivity per ha with a forestry project analysis to obtain an 'operating return per effective ha' for landholders engaging in a C sequestration activities. The results for the high rainfall areas (900 mm of annual rainfall) were encouraging with Australian dollars (AUD) 10 per tonne C credits supplementing the net present value (NPV) timber income by 16% for a *Eucalyptus globulus* (blue gum) plantation, 62% for a *Eucalyptus saligna* (Sydney blue gum) plantation, and AUD560 $ha^{-1}$ for unidentified native forest species revegetation (The CRC for Greenhouse Accounting & Tony Beck Consulting Services Pty Ltd, 2003). However, the range of monitoring, verification and certification cost estimates for C farming from some selected studies show the great variability in potential expenditures. Specht and West (2003) included information on the likely costs associated with a sampling program of the type required for C trading using professional measurements teams. To achieve the required level of sequestration confidence on the 350 ha of estates would require assessment of around 150 tree plots and 80 soil sites. This would cost around AUD5,000 or AUD240 $ha^{-1}$, which included the team's accommodation costs (Specht & West, 2003). However, the CRCGA chose in their modelling that an appropriate cost would be in the range of AUD12.50 $ha^{-1}$ per analysis. Whatever C accounting costs become, they will need to be assessed every five years, or at the start and end of each commitment period (Booth et al., 2003; The CRC for Greenhouse Accounting & Tony Beck Consulting Services Pty Ltd., 2003). The CRCGA study stated that while the C price is AUD5 to AUD10 $tCO_2$-e, it is not profitable to establish C sinks in their own right anywhere in WA. At AUD15 $tCO_2$-e the CRCGA found it would be profitable to establish trees for conventional timber products and C, even across the highly productive agricultural lands in Great Southern and into the

Southwest regions, only when the cost of land is not taken into account. The study concluded that C farming/forestry activities could not compete commercially with current landholder activities, but may be complementary to such activities or offset potential losses from establishment costs of windbreaks and revegetation (The CRC for Greenhouse Accounting & Tony Beck Consulting Services Pty Ltd., 2003). Effectively achieving this complementarily will require low C accounting costs, an intimate knowledge of the land, and its productive potential (McHenry, 2013).

## BIOLOGICAL COMPLEXITY AND ENVIRONMENTAL MARKETS USING A FORESTRY EXAMPLE

At the landscape scale, soil types, soil texture, local topography, water availability and the condition and exposure of individual trees influence many aspects of vegetative NPP (Kort & Turnok, 1999). Open-spaced trees are subjected to greater mechanical wind stress and respond by increasing the thickness of stems, branches, and particularly the root system (Ritzon & Sochacki, 2002). The proportion of total tree biomass in roots is generally between 30 - 50% of aboveground biomass (Kort & Turnok, 1999). However, it is known that planting trees in higher densities, such as in a windbreak or plantations, reduce the percentage of tree biomass in root systems. The amount of aboveground biomass also increases as the trees age relative to the biomass in roots. This is more pronounced with open spaced trees than close-spaced trees (Ritzon & Sochacki, 2002). For reasons of practicality, estimating the biomass of smaller shrubby plants is best related to shelterbelt volume; as the space occupied by the total leaf area available for photosynthesis is independent of the number of trees in the windbreak because the lateral growth is limited by their neighbours (Kort & Turnok, 1999).

Additional intricacies discovered by site analyses are the variability of soil organic C (SOC) below plantations and adjacent land. Short-term declines in SOC following LUC from pasture to plantation in wetter regions requires landholders and modellers to account for these losses in particular bioregions. The SOC levels would be expected to recover and possibly exceed the pre-plantation level (Kirschbaum, 2000; Specht & West, 2003). Such assumptions require on-site verification to refine models that aggregate data for market mechanisms and formal accounts, such as Australia's National Greenhouse Gas Inventory (NGGI) (Brack & Richards, 2002; Booth et al., 2003).

There are significant biological complexities in establishing large vegetative stocks for new markets which are poorly understood by markets and current biometric models. Although external to any C market transaction, negative environmental consequences may occur in relation to water availability in sensitive catchments due to large reforestation activities consuming sizeable quantities of water (Department of Climate Change, 2008; Cosier et al., 2009; McHenry, 2012b). This will require estimation, planning and management by local and state governments if such activities are projected to become widespread within their planning horizons. The development of transparent mechanisms to quantify the performance of activities will be crucial.

All biological conservation-based projects are subject to additional market risk, but are also impacted by physical risks associated with fire, drought, or pests, which must be reasonably accounted for (The CRC for Greenhouse Accounting & Tony Beck Consulting

Services Pty Ltd., 2003). Compounding these issues are the policy change concerns for landholders, including permanence issues arising from participation in conservation activities, most of which are likely to be around 100 years or even permanently on titles or leases (Booth et al., 2003; The CRC for Greenhouse Accounting & Tony Beck Consulting Services Pty Ltd., 2003; Independent Pricing and Regulatory Tribunal, 2008). The risks of limiting future management options is crucial for landholders, which may result in future compensation payments to carbon rights holders if contractual obligations are annulled by landholders (if this is possible) in addition to associated transaction costs (Booth et al., 2003; The CRC for Greenhouse Accounting & Tony Beck Consulting Services Pty Ltd., 2003).

## CONCLUSION

The integration of high accuracy on-site vegetative and ecological data into landscape and macro-scale land monitoring models will enable new market mechanisms that internalise environmental externalities from land management activities by providing both accurate and precise on-site accounting and verification. This approach has the potential to achieve several efficiencies including new skilled conservation jobs in remote areas, reduced capital and infrastructure duplication for on-site environmental monitoring, streamlined government administration and research ancillary expenses, locally appropriate conservation activities that are sustained over time, and direct communication between landholders, academic institutions, and land management policymakers.

Striking an operative balance between the accuracy of market product value, the genuine value of on-ground activities, and the associated market administrative and monitoring costs will necessitate on-site activities on both private and government lands. Policymakers should be aware that landholders, due to their relationship with the land, generally hold the majority of high resolution data regarding the efficacy and nuances of particular land use and management alternatives (Stoneham, 2009). Experience has shown that land management activities undertaken by governments or external entities in regional areas often under-perform, and initiatives are rarely funded sufficiently to be sustained effectively (Van Vreeswyk et al., 2004; Stoneham, 2009). The option of subcontracting landholders who live and work in such regions may be a more effective solution to increasing travel and accommodation costs of centralised contractors. An open relationship between policymakers and landholders will therefore assist the development of appropriate management practices that provide private landholder security, public good protection and enhancement, alongside efficiency in terms of administering limited environmental funding. This will more effectively harness the considerable existing knowledge and capacity in rural, regional, and remote Australia and assist the generation of detailed knowledge of ecosystem functional change.

Pressures from industry expansion, urban encroachments, and conservation objectives on existing regional food production systems and culturally sensitive lands will likely intensify over-time. Improving the interface between landholders and policymakers through rigorous monitoring mechanisms provides a much needed additional communication channel between landholders and those not on the land, but with an external interest in it. With sufficient knowledge and resources, private on-site landholders are often best placed to quantify the productive potential of their land, and develop appropriate and cost-effective management activities. An internalisation of non-monetary land systems can generate a more effective and sustained land management and environmental outcome that do not require acquisition or expensive maintenance regimes by governments. Therefore, it will be prudent for policymakers and researchers to work more transparently with private long-term landholders to maximise benefits and minimise costs from new market opportunities, maintain ecological services, and retain food productive capacity.

## REFERENCES

Antheaume, N. (2004). Valuing external costs - from theory to practice: implications for full cost environmental accounting. *European Accounting Review,* 13(3), 443-464.

Australian Greenhouse Office. (1999a). *Implementation plan for the 1990 baseline*. Canberra: Australian Greenhouse Office.

Australian Greenhouse Office. (1999b). *Setting the frame. Part 1: system framework, Part 2: project specification 1990 baseline estimation, land use change and forestry sector.* Canberra: Australian Greenhouse Office.

Australian Greenhouse Office. (2000). *Land clearing: a social history.* Canberra: Australian Greenhouse Office.

Australian Greenhouse Office. (2002). *Greenhouse gas emissions from land use change in Australia: an integrated application of the National Carbon Accounting System.* Canberra: Australian Greenhouse Office.

Australian Greenhouse Office. (2005a). *Development plan 2004-8.* Canberra: Australian Greenhouse Office.

Australian Greenhouse Office. (2005b). *Greenhouse gas emissions from land use change in Australia: results of the National Carbon Accounting System 1988-2003.* Canberra: Australian Greenhouse Office.

Australian Greenhouse Office. (2006). *Greenhouse action in regional Australia: Planning forest sink projects - a guide to forest sink planning, management and carbon accounting.* Canberra: Australian Greenhouse Office.

Balilla, V. S., Anwar McHenry, J., McHenry, M. P., Parkinson, R. M., & Banal, D. T. (2012). Aeta Magbukún of Mariveles: traditional Indigenous forest resource use practices and the sustainable economic development challenge in remote Philippine regions. *Journal of Sustainable Forestry,* 31(7), 687-709

Bi, H., Turner, J., & Lambert, M. J. (2004). Additive biomass equations for native eucalypt forest trees of temperate Australia. *Trees,* 18(4), 467-479.

Booth, T. H., Kirschbaum, M., Richards, G., Zhang, X. Q., Domrongphol, P., Chittachumnonk, P., Hieu, N. K., Snowdon, P., Paul, K., Polglase, P., Raison, J., Teiveyanathan, T., Waterworth, R., Elliott, A., & Jovanovic, T. (2003, 24 February 2003). *Carbon Accounting in Forests.* Paper presented at the Facilitating International Carbon Accounting in Forests, Canberra.

Brack, C., Coops, N., Furby, S., James, R., Kesteven, J., Landsberg, J., Raison, J., Richards, G., & Turner, B. (2002). *Report No. 27 - Biomass estimation: approaches for assessment of stocks and stock change.* Canberra: Australian Greenhouse Office.

Brack, C. L., & Richards, G. P. (2002). Carbon accounting model for forests in Australia. *Environmental Pollution,* 116, 187-194.

Brokaw, N., & Thompson, J. (2000). The H for DBH. *Forest Ecology and Management,* 129, 89-91.

Clifton, C., Cossens, B., & McAuley, C. (2007). *Project REM1 - A framework for assessing the environmental water requirements of groundwater dependent ecosystems report 1 assessment toolbox* Kent Town, South Australia: Resource & Environmental Management Pty Ltd.

Cosier, P., Flannery, T., Harding, R., Karoly, D., Lindenmayer, D., Possingham, H., Purves, R., Saunders, D., Thorn, B., Williiams, J., & Young, M. (2009). *Optimising carbon in the Australian landscape - How to guide the terrestrial carbon market to deliver multiple economic and environmental benefits*: Wentworth Group of Concerned Scientists.

Department of Climate Change. (2008). *Carbon Pollution Reduction Scheme White Paper: Australia's low pollution future (Volume 1).* Canberra: Commonwealth of Australia.

Eamus, D., McGuinnes, K., & Burrows, W. (2000). *Report No. 5a - Review of allometric relationships for estimating woody biomass for Queensland, the Northern Territory and Western Australia.* Canberra: Australian Greenhouse Office.

Fromm, O. (2000). Ecological structure and functions of biodiversity as elements of its total economic value. *Environmental and Resource Economics,* 16, 303-328.

Furby, S. (2002). *Land cover change: specification for remote sensing analysis.* Canberra: Australian Greenhouse Office.

Furby, S., & Woodgate, P. (2002). *Remote sensing analysis of land cover change - pilot testing of techniques.* Canberra: Australian Greenhouse Office.

Gifford, R. G. (2000a). *Carbon content of woody roots - revised analysis and a comparison with woody shoot components.* Canberra: Australian Greenhouse Office.

Gifford, R. M. (2000b). *Carbon contents of above-ground tissues of forest and woodland trees.* Canberra: Australian Greenhouse Office.

Griffiths, B. S., Ritz, K., Bardgett, R. D., Cook, R., Christensen, S., Ekelund, F., Sorensen, S. J., Baath, E., Bloem, J., de Ruiter, P. C., Dolfing, J., & Nicolardot, B. (2000). Ecosystem response of pasture soil communities to fumigation-induced microbial diversity reductions: an examination of the biodiversity - ecosystem function relationship. *Oikos,* 90, 279-294.

Hingston FJ, Galbraith JH, & Dimmock GM. (1998). Application of the process-based model BIOMASS to Eucalyptus globulus subsp. globulus plantations on ex-farmland in south western Australia - I. Water use by trees and assessing risk of losses due to drought. *Forest Ecology and Management,* 106, 141-156.

Ilic, J., Boland, D., McDonald, M., Downes, G., & Blakemore, P. (2000). *Wood density phase 1 - state of knowledge.* Canberra: Australian Greenhouse Office.

Independent Pricing and Regulatory Tribunal. (2008). Compliance and operation of the NSW Greenhouse Gas Reduction Scheme 2007. Retrieved 19 October, 2009, from http://www.greenhousegas.nsw.gov.au/documents/SchRep07.pdf

Jaakko Poyry Consulting. (2000). *Analysis of wood product accounting options for the National Carbon Accounting System.* Canberra: Australian Greenhouse Office.

Kalin Arroyo, M. T., Zedler, P. H., & Fox, M. D. (1995). *Ecology and biogeography of mediterranean ecosystems in Chile, California, and Australia.* New York, USA: Springer-Verlag.

Keith, H., Barrett, D., & Keenan, R. (2000). *Report No. 5b - Review of allometric relationships for estimating woody biomass for New South Wales, the Australian Capital Territory, Victoria, Tasmania and South Australia*. Canberra: Australian Greenhouse Office.

Kesteven, J., Landsberg, J., & UrS Australia. (2004). *Developing a national forest productivity model*. Canberra: Australian Greenhouse Office.

Kirschbaum, M. U. F. (2000). What contribution can tree plantations make towards meeting Australia's commitments under the Kyoto Protocol? *Environmental Science & Policy*, 3, 83-90.

Kort, J., & Turnok, R. (1999). Carbon reservoir and biomass in Canadian prairie shelterbelts. *Agroforestry Systems*, 44, 175-186.

Landsberg, J., & Crowley, G. (2004). Monitoring rangeland biodiversity: plants as indicators. *Austral Ecology*, 29, 59-77.

McHenry, M. P. (2009a). Agricultural bio-char production, renewable energy generation and farm carbon sequestration in WA: Certainty, uncertainty & risk. *Agriculture, Ecosystems and Environment*, 129, 1–7.

McHenry, M. P. (2009b). Synergies between conventional soil organic carbon, farm productivity, soil sequestration and soil carbon market risk in Australia. In: E. T. Nardali (Ed.), *In No-Till Farming: Effects on Soil, Pros and Cons, and Potential*. Hauppauge, New York, USA: Nova Science. 978-1-60741-402-5.

McHenry, M. P. (2010). Carbon-based stock feed additives: a research methodology that explores ecologically delivered C biosequestration, alongside live-weights, feed-use efficiency, soil nutrient retention, and perennial fodder plantations. *Journal of the Science of Food and Agriculture*, 90, 183-187.

McHenry, M. P. (2011). Soil organic carbon, biochar, and applicable research results for increasing farm productivity under Australian agricultural conditions. *Communications in Soil Science and Plant Analysis*, 42, 1187-1199.

McHenry, M. P. (2012a). Australian carbon biosequestration and bioenergy policy co-evolution: mechanisms, mitigation, and convergence. *Australian Forestry Journal* 75(2), 901-913.

McHenry, M. P. (2012b). Practicalities of establishing forestry carbon sequestration projects in the agricultural sector: a technical and economic analysis with implications. In: B. J. Ryan & D. E. Anderson (Eds.), *Carbon sequestration: technology, measurement technologies and environmental effects*. Hauppauge, New York, USA: Nova Science Publishers. 978-1-62081-081-7.

McHenry, M. P. (2013). How farming and forestry converge: enhancing the interface between agricultural production, and tree biomass/bioenergy systems to improve farm-scale productivity in Western Australia. In: M. C. Allard (Ed.), *Bioenergy systems, biological sources and environmental impact*. Hauppauge, New York, USA: Nova Science Publishers. 978-1-62417-330-1.

McKenzie, N., Ryan, P., Fogarty, P., & Wood, J. (2000). *Sampling, measurement and analytical protocols for carbon estimation in soil, litter and woody biomass*. Canberra: Australian Greenhouse Office.

Nunes, P. A. L. D., & van den Bergh, J. C. J. M. (2001). Economic valuation of biodiversity: sense or nonsense? *Ecological Economics*, 39, 203-222.

Paul, K., Polglase, P., Coops, N., O'Connell, T., Grove, T., Mendham, D., Carlyle, C., May, B., Smethhurst, P., & Baillie, C. (2002). *Modelling change in soil carbon following afforestation or reforestation: preliminary simulations using GRC3 and sensitivity analyses*. Canberra: Australian Greenhouse Office.

Polglase, P. J., Paul, K. I., Khanna, P. K., Nyakuengama, J. G., O'Connell, A. M., Grove, T. S., & Battaglia, M. (2000). *Change in soil carbon following afforestation or reforestation: review of experimental evidence and development of a conceptual framework*. Canberra: Australian Greenhouse Office.

Richards, G., Brack, C., Coops, N., Furby, S., James, R., Kesteven, J., Landsberg, J., Raison, J., & Turner, B. (2002). *Biomass estimation: approaches for assessment of stocks and stock change*. Canberra: Australian Greenhouse Office.

Richards, G. P. (2001). *The FullCAM carbon accounting model: development, calibration and implementation for the National Carbon Accounting System*. Canberra: Australian Greenhouse Office.

Ritzon, P., & Sochacki, S. (2002). Measurement and prediction of biomass and carbon content of Pinus pinaster trees in farm forestry plantations, south-western Australia. *Forest Ecology and Management,* 175, 103-117.

Snowdon, P., Eamus, D., Gibbons, P., Khanna, P., Keith, H., Raison, J., & Kirschbaum, M. (2000). *Synthesis of allometrics, review of root biomass and design of future woody biomass sampling*. Canberra: Australian Greenhouse Office.

Snowdon, P., Raison, J., Keith, H., Ritzon, P., Grierson, P., Adams, M., Montagu, K., Bi, H. Q., Burrows, W., & Eamus, D. (2002). *Report No. 31 - Protocol for sampling tree and stand biomass*. Canberra: Australian Greenhouse Office.

Specht, A., & West, P. W. (2003). Estimation of biomass and sequestered carbon on farm forest plantations in northern New South Wales, Australia. *Biomass and Bioenergy,* 25(4), 363-379.

Stoneham, G. (2009). *Creating markets for environmental goods and services: a mechanism design approach*. Canberra: Land and Water Australia.

The CRC for Greenhouse Accounting & Tony Beck Consulting Services Pty Ltd. (2003). *Opportunities for the Western Australian land management sector arising from greenhouse gas abatement*. Perth, Western Australia: Western Australian State Government.

The New South Wales Government. (2005). GGAS guide to applying to become an accredited abatement certificate provider - Greenhouse Gas Benchmark Rule (Carbon Sequestration) No. 5 of 2003 [Electronic Version], 2009 from http://www.greenhouse.nsw.gov.au/documents/ GtA-CS.pdf.

The New South Wales Government. (2008). *Transitional arrangements for the NSW Greenhouse Gas Reduction Scheme - Consultation paper* Sydney, NSW: NSW Department of Water and Energy.

The Parliament of Western Australia. (2003). *Carbon Rights Act 2003*. Perth, Western Australia: Parliamentary Counsel's Office of Western Australia.

Turner, B., Wells, K., Bauhus, J., Carey, G., Brack, C., & Kanowski, P. (1999). *Woody biomass methods for estimating change*. Canberra: Australian Greenhouse Office.

United Nations Framework Convention on Climate Change. (2002). Conference of the Parties on its seventh session (COP 7). *United Nations Framework Convention on Climate Change* Retrieved 9 January, 2010, from http://unfccc.int/ressource/docs/cop7/13a01.pdf.

Van Vreeswyk, A. M. E., Payne, A. L., Leighton, K., & Hennig, P. (2004). *Technical Bulletin No 92 - An inventory and condition survey of the Pilbara region, Western Australia* Perth, Western Australia: Department of Agriculture.

Zianis, D., & Mancuccini, M. (2004). On simplifying allometric analyses of forest biomass. *Forest Ecology and Management,* 187, 311-332.

In: Plantations
Editor: Hai Ren

ISBN: 978-1-62808-090-2


*Chapter 6*

# SEDIMENTARY ORGANIC CARBON DYNAMICS IN A NATIVE AND AN EXOTIC MANGROVE PLANTATION BASED ON DUAL CARBON ISOTOPIC ANALYSES

***Qianmei Zhang[1], Jinping Zhang[1], Lianlian Yuan[1], Chengde Shen[2] and Hai Ren[1*]***

[1]Department of Ecology, South China Botanical Garden, Chinese Academy of Sciences, Guangzhou, China

[2]Key Laboratory of Isotope Geochronology and Geochemistry, Guangzhou Institute of Geochemistry, Chinese Academy of Sciences, Guangzhou, China

## ABSTRACT

Sedimentary organic carbon (SOC) was quantified in two mangrove plantations and a nearby barren site based on measurements of C and C isotopes in sediment cores. Relative to the barren site, the native *Kandelia candel* plantation and the exotic *Sonneratia apetala* plantation greatly affected values for SOC content, $\delta^{13}C$, and $^{14}C$ age from 0 to about 60 cm but not at greater depths. SOC densities to a 60-cm depth were 155 and 174 Mg ha$^{-1}$ in the native and exotic plantation, respectively. Change of SOC $\delta^{13}C$ with depth was similar in both plantations, and the new contributions to the total SOC in the upper 16 cm were also similar (45–47%). At 16–60 cm, however, the new C contributions were only 9% in the native plantation but as high as 31% in the exotic plantation, i.e., the exotic *S. apetala* stored more mangrove-derived new C in its anaerobic sub-surface than the native *K. candel*. From the perspective of SOC storage, *S. apetala* rather than *K. candel* is recommended for restoring the degraded coastal area in Southern China.

**Keywords:** Mangroves, sedimentary organic carbon, carbon isotope

* Corresponding author. E-mail address: renhai@scib.ac.cn. Tel: +8620-37252916. Fax: +8620-37252916.

## INTRODUCTION

In coastal ecosystems, mangroves provide significant ecological services, such as storm and cyclone buffering, timber and fisheries production, and prevention of coastal erosion (Barbier et al. 2008; Shaifullah et al. 2008). Despite these values, mangroves are seriously threatened worldwide (Alongi 2002; Yanagisawa et al. 2009). About 35% of mangrove forests have been degraded or lost over the past several decades (Barbier et al. 2008). Most mangroves along the shorelines of China are threatened or degraded as a result of pollution, aquaculture, over-cropping for timber, and coastal urbanization (Benfiled et al. 2005; Ren et al. 2008). Recently, the establishment of mangrove plantations for sustainable coastal development has attracted widespread attention (Matsui et al. 2008). Attractive characteristics of mangroves forests/plantations include high productivity and a low ratio of decomposition to net primary production (Gonneea et al. 2004; Jennerjahn and Ittekkot 2002). Consequently, mangroves sequester significant amounts of carbon (C) in biomass and sediment (Bouillon et al. 2008; Fujimoto 2004; Twilley et al. 1992), and are regarded as potentially efficient pools or sinks for decreasing the concentration of atmospheric $CO_2$ and mitigating global warming.

Over the past two decades, many studies have increased our knowledge about the dynamics of C in mangrove forests. We still lack, however, a complete understanding of the dynamics of sedimentary organic carbon (SOC) in the mangrove forests because SOC dynamics are controlled by a large number of physical, chemical, and biological factors at a variety of timescales (Aller and Blair 2004; Kristensen et al. 2008).

Two powerful tools for studying ecological processes like SOC dynamics in mangrove forests are stable and radioactive C isotopic analyses (Bouillon and Bottcher 2006; Staddon 2004; West et al. 2006). For instance, the alteration in SOC $\delta^{13}C$ values caused by C decay provides clues for separating stages of SOC decomposition and thereby helps clarify SOC dynamics (Chen et al. 2002). Radiocarbon measurement of SOC reflects the mixture of new and old organic materials. When used with a simple two-source mixing model, $^{14}C$ measurements are helpful for detecting and quantifying the contributions of new and old C components to the total SOC (Rethemeyer et al. 2004). Unfortunately, few studies have simultaneously reported SOC content and its stable and radioactive C isotopic signatures from mangrove sediments (Kristensen et al. 2008).

The east coast of the Leizhou Peninsula, Southern China, presents a unique opportunity to compare SOC dynamics in two mangrove plantations. The native mangrove *Kandelia candel* and the exotic mangrove *Sonneratia apetala* were planted on initially barren flats to restore the degraded coastal area. *K. candel* was planted in one plantation in 1993, and *S. apetala* was planted in the other plantation in 2001. Relative to *K. candel*, *S. apetala* is highly adaptable and grows rapidly (Ren et al. 2008). It was introduced into China in 1985 and has been planted on over 1000 ha in China since 1991 (Ren et al. 2008). Because the introduced *S. apetala* grows faster than the native *K. candel*, we hypothesized that *S. apetala* plantation would store more SOC than the native *K. candel* plantation but that the vertical patterns (depth patterns) of dual C isotopes in both plantations would be similar. To test this hypothesis, we (1) quantified SOC accumulation in both plantations and in a reference barren flat located between the plantations; (2) separated SOC decomposition stages in both plantations based on vertical variations of $\delta^{13}C$ values; and (3) estimated the contributions of new C to the total SOC in both plantations.

# Materials and Methods

## Study sites

The study area is located along the east coast of Leizhou Bay and is part of the Zhanjiang Mangrove National Nature Reserve, which is the largest mangrove reserve in China. One sediment core was collected from each of three sites: the *K. candel* plantation, the *S. apetala* plantation, and a reference barren flat located between the plantations (Figure 1). The reference barren flat was considered to be representative of the initial pre-plantation sediments, assuming negligible tidal flushing of organic matter. For all sites, the average annual precipitation is 1711 mm, and precipitation usually falls as rain from April to October. The average annual temperature is 22.9 °C, and the average temperature in the coldest month (January) is 15.5 °C. Average annual relative humidity is 80%. The tidal regime is diurnal, and the average tidal amplitude is 2.52 m. The soil is mangrove acid soil with three typical layers (0–10 cm: black, with organic matter, 10–40 cm: grey, and >40 cm: yellow or mixed colour). The pH value for top soils is 6.9. The typical vegetation is the *Kandelia candel* - *Avicennia marina* (Forsk.) community (He et al. 2007).

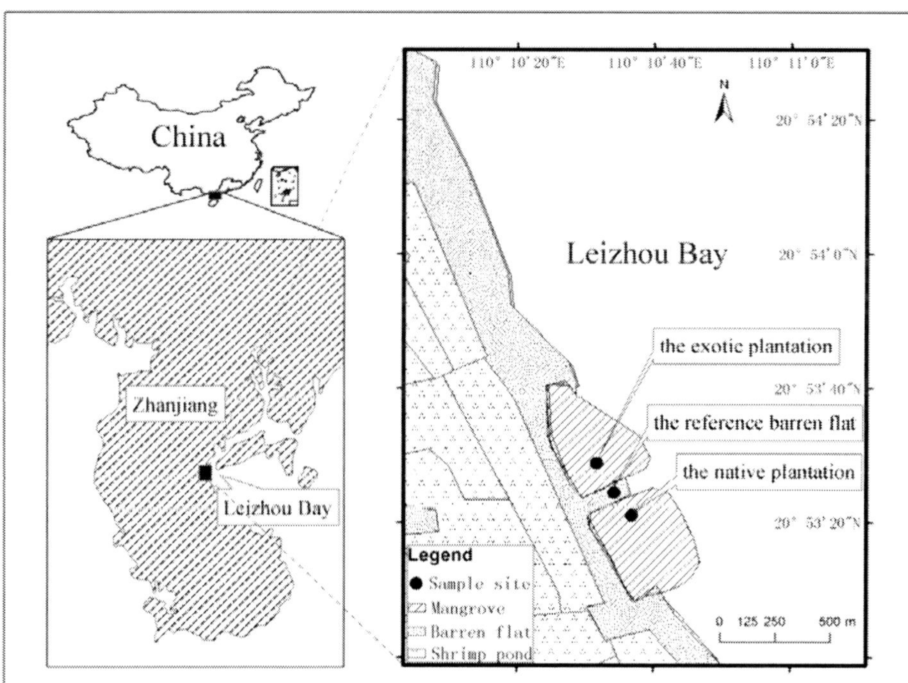


Figure 1. Locations of soil cores in the native *Kandelia candel* plantation, the exotic *Sonneratia apetala* plantation, and the reference barren flat in Zhanjiang, Southern China.

## Sediment Sampling

During ebb tide in June 2006, one sediment core (11 cm in diameter and 100 cm in length) was excavated in the barren flat, the native plantation, and the exotic plantation. The three cores were divided into segments based on depth and the sediment characteristics. In order to get detailed vertical distribution data, the sampling interval was 2 cm at 0–20 cm depth, 5 cm at 20–50 cm depth, 10 cm at 50–80 cm depth, and 20 cm at 80–100 cm depth. The collected sediment segments were transported to the Carbon Isotope Laboratory of the Guangzhou Institute of Geochemistry, where they were frozen at -26 ° C for later analyses.

## C Isotope Analyses

C isotope analyses of all samples were conducted in 2006. The frozen sediment segments were thawed to room temperature. After visible roots and fragmentary stones were removed, a subsample (20-35 g) of each segment was freeze-dried under vacuum for 48 h, ground using a mortar and pestle, and then passed through a 1-mm screen to remove rootlets and coarse sand. The samples were treated with 2 *M* HCl to remove carbonate and then rinsed repeatedly with distilled water until a neutral pH was reached. Water was removed by oven-drying at 90 °C for 24 h (Stuiver and Polach 1977).

Samples were then loaded into sealed, evacuated quartz tubes and combusted with CuO at 860 °C for 2 h. The $CO_2$ thus generated was cryogenically purified with dry-ice and liquid nitrogen. SOC content was calculated from the quantity of $CO_2$ generated from the subsamples. $\delta^{13}C$ values were determined in one portion of the generated $CO_2$ with a Finnigan MAT-251 (Finnigan Mat, Germany) mass spectrometer with a precision of 0.2% at the State Key Laboratory of Loess and Quaternary Geology, Chinese Academy of Sciences. Results were reported in per mil deviations from the International Pee Dee belemnite standard as: $\delta^{13}C = [(^{13}C/^{12}C)_{sample} / (^{13}C/^{12}C)_{standard} - 1] \times 1000$. The other portion of the generated $CO_2$ was catalytically reduced to graphite Accelerator Mass Spectrometry targets using the method of Vogel et al. (1987). Radioactive C isotope ratios were determined in the generated graphite using AMS at the Institute of Heavy Ion Physics of Peking University. $^{14}C$ data were all normalized for isotopic fractionation of $^{13}C$ with $\delta^{13}C$ assuming to be -25‰. Results were reported as the conventional radiocarbon ages (yr BP) (Stuiver and Polach 1977).

## Thc two-Source Isotope Mixing Model

First, based on the definition of the conventional radiocarbon age, $F^{14}C$ can be calculated by the radiocarbon age as:

$$F^{14}C = \exp(-(\frac{t}{8033} + \frac{(y-1950)}{8267})) \quad (1)$$

where

$$F^{14}C = \frac{\left(\frac{^{14}C}{^{12}C}\right)_{sample}}{\left(\frac{^{14}C}{^{12}C}\right)_{s\tan dard}} \tag{2}$$

and $t$ is the conventional radiocarbon age and $y$ is the measuring year, which was 2006 in this study.

The two-source isotope mixing model (Rethemeyer et al. 2004; Trumbore 2000) was used to calculate the contributions of new C to the total SOC in each of the sediment segments based on $F^{14}C$ values. The $F^{14}C$ value can be expressed as:

$$F^{14}C_{sample} = F^{14}C_{new} \times f + F^{14}C_{old} \times (1-f) \tag{3}$$

where $F^{14}C_{sample}$ is the $F^{14}C$ of the sample; $F^{14}C_{new}$ is the $F^{14}C$ of the new C component; $f$ is the contribution of new C to the total SOC; and $F^{14}C_{old}$ is the $F^{14}C$ of the initial C substrate (Levin and Kromer 2004).

Assuming the new C component is fresh litter derived from the mangrove plantation, we took the average $F^{14}C$ value of atmospheric $^{14}C$ concentration at the period of mangrove growth as $F^{14}C_{new}$, and we took the $F^{14}C$ value at the reference barren flat as $F^{14}C_{old}$. Based on the recorded annual atmospheric $^{14}C$ concentration (see Levin and Kromer 2004; Levin et al. 2008), we calculated the average $F^{14}C$ values of atmospheric $^{14}C$ concentration at the growth period, which was 1.093 for the native plantation and 1.077 for the exotic plantation. The value of $f$ was obtained once the above three $F^{14}C$ values (sample, new, and old C) were put into Eq. (3).

## RESULTS

Neither the native nor the exotic plantation affected SOC content below 60 cm, i.e., SOC content below 60 cm was similar in the plantations and in the barren flat (Figure 2). At 0-60 cm depth, SOC contents in the barren flat, the native plantation, and the exotic plantation ranged from 0.18 to 0.83%, 0.46 to 1.82%, and 0.66 to 1.63%, respectively. SOC contents at 0-16 cm depth fluctuated greatly in the native plantation but were relatively stable in the exotic plantation. SOC contents gradually decreased between 16 to 60 cm depth in both plantations.

Neither the native nor the exotic plantation affected SOC $\delta^{13}C$ values below 70 cm, i.e., SOC $\delta^{13}C$ values below 70 cm were similar in the plantations and in the barren flat (Figure 3). At 0-70 cm depth, however, SOC $\delta^{13}C$ values were smaller in the plantations than in the barren flat. At 0-70 cm depth, SOC $\delta^{13}C$ values in the barren flat changed only slightly with depth, with an average value of -21.98‰. In contrast, SOC $\delta^{13}C$ values in both plantations were relatively stable from 0 to 18 cm but increased from 18 to 70 cm. At 0–70 cm, SOC $\delta^{13}C$ values ranged from -27.37 to -23.38‰ in the native plantation and from -26.66 to -23.18‰ in the exotic plantation.

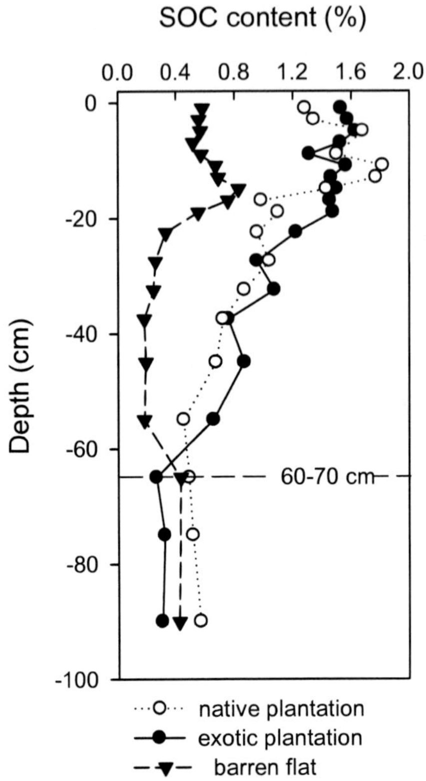


Figure 2. Changes in SOC content as a function of depth in the native *Kandelia candel* plantation, the exotic *Sonneratia apetala* plantation, and the reference barren flat, Southern China. Data at 6–8 cm in the native plantation and at 70–80 cm at the barren flat are missing due to experimental problems.

**Table 1. The contributions of new C to the total SOC in the native *Kandelia candel* and exotic *Sonneratia apetala* plantations, Southern China**

| Depth (cm) | New C contribution (%) | |
|---|---|---|
| | Native *K. candel* | Exotic *S. apetala* |
| 0-2 | 41 | 55 |
| 2-4 | 49 | 54 |
| 4-6 | 53 | 52 |
| 6-8 | 46 | 52 |
| 8-10 | 38 | 46 |
| 10-12 | 49 | 39 |
| 12-14 | 46 | 41 |
| 14-16 | 36 | 37 |
| 16-18 | 11 | 44 |
| 18-20 | 20 | 43 |
| 20-25 | 4 | 29 |
| 25-30 | 11 | 13 |
| 30-35 | 8 | 33 |
| 35-40 | 2 | 36 |
| 40-50 | 11 | 27 |
| 50-60 | 3 | 20 |

Relative to the reference barren flat, neither plantation affected SOC $^{14}C$ ages below about 60 cm (Figure 4). At 0–60 cm, however, SOC $^{14}C$ ages in the barren flat, the native plantation, and the exotic plantation ranged from 1397 to 2608 yr BP, 255 to 2247 yr BP, and 391 to 1657 yr BP, respectively. In both plantations, SOC $^{14}C$ ages fluctuated without obvious trend at 0-16 cm but increased with depth from 16 to 60 cm.

Table 1 summarized the contributions of new C to the total SOC as affected by sediment depth and plantation. The new C contributions in the plantations were relatively similar in the upper 16 cm, with 45 ± 5% in the native *K. candel* and 47 ± 6% in the exotic *S. apetala*. At 16–60 cm, however, the new C contribution was much greater in the exotic *S. apetala* plantation (31 ± 8%) than in the native *K. candel* plantation (9 ± 4%).

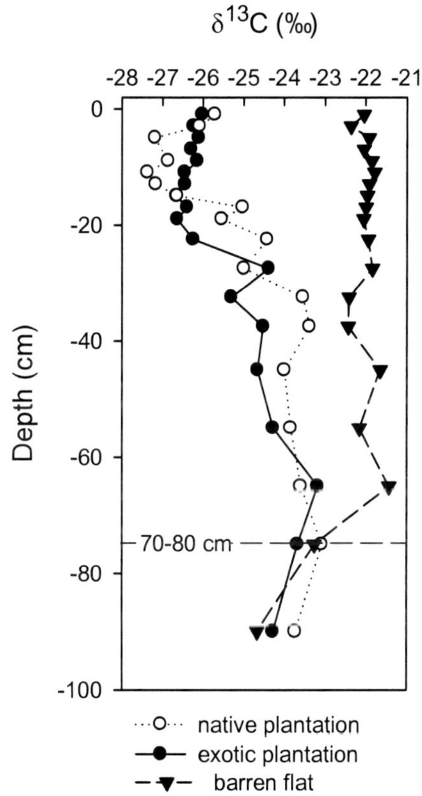


Figure 3. Changes in SOC $\delta^{13}C$ as a function of depth in the native *Kandelia candel* plantation, the exotic *Sonneratia apetala* plantation, and the reference barren flat, Southern China. The value of SOC $\delta^{13}C$ at 6–8 cm in the native plantation is missing because of experimental problems.

# Discussion

## SOC Accumulation in the Mangrove Plantations

SOC contents in most soil layers between 0–60 cm were dramatically higher in both plantations than in the reference barren flat (Figure 2), indicating substantial input and accumulation of C in the mangrove plantations. SOC density can be calculated by multiplying

SOC content by the bulk density and thickness of the sediment segment. The bulk densities of three cores differed only slightly at all soil layers. Based on the average bulk density of all soil cores, which was $2.65 \times 10^3$ kg $m^{-3}$ (unpublished data), SOC densities at 0–60 cm in the reference barren flat, the native plantation, and the exotic plantation were 57, 155, and 174 Mg $ha^{-1}$, respectively. After the initial SOC density (the density in the barren flat) is subtracted, the SOC densities increased by 98 Mg $ha^{-1}$ in the native plantation and by 117 Mg $ha^{-1}$ in the exotic plantation, representing increases of 173% and 206%, respectively, of the initial SOC density of the pre-plantation sediments. The results therefore demonstrate that the mangrove plantations can accumulate significant quantities of SOC in their sediments. As reported by others, the production of substantial quantities of leaf litter and root litter combined with the slow decomposition of litters and organic matter in sediments all contribute to high SOC accumulation in plantations (Fujimoto 2004; Kristensen et al. 2008; Zan et al. 2001). Moreover, the current study found that the exotic *S. apetala* could store slightly more C than the native *K. candel* in the plantation sediments.

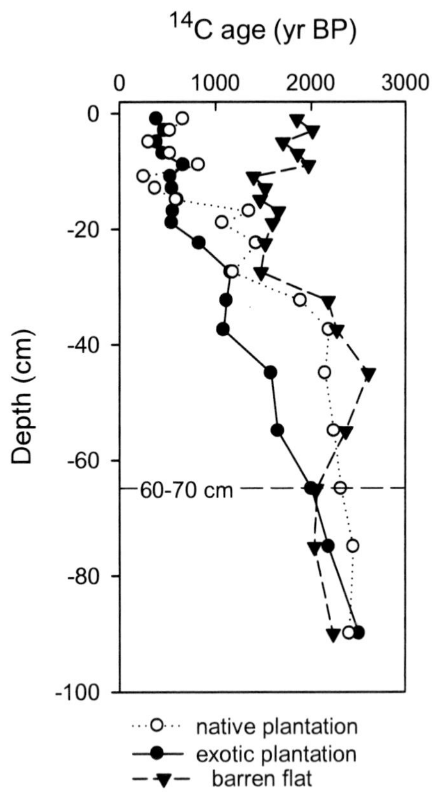


Figure 4. Changes in $^{14}$C age as a function of depth in the native *Kandelia candel* plantation, the exotic *Sonneratia apetala* plantation, and the reference barren flat, Southern China.

SOC densities (Mg $ha^{-1}$) in the mangrove plantations in this study were lower than those in the natural mangrove forests within the same climate, and those in the Asia-Pacific region. SOC densities to a depth of 50 cm were 432 Mg $ha^{-1}$ in an *Aegiceras corniculatum* pure natural forest and 244 Mg $ha^{-1}$ in a *Bruguiera gymnorrhiza* + *K. candel* mixed natural forest in the Zhanjiang Mangrove National Nature Reserve (Zhang et al. in press). This difference probably reflects age, i.e., the two natural forests were more than 80 years old when sampled (Zhang et al. 2009) whereas the two plantations in the current study were less than 16 years old when sampled. In the sediments of various types of mangroves from different sites of the Asia-Pacific region, the average SOC density was 887 Mg $ha^{-1}$ with a range of 220 to 4200 Mg $ha^{-1}$; many of these sediments were sampled as deep as 100 cm and some even as deep as 200 cm (Fujimoto 2004). Because our estimates of SOC densities in sediments are based on data only from the top 60 cm, the SOC densities in this study were underestimated relative to the estimates based on greater depths.

## SOC Decomposition in the Mangrove Plantations as Indicated by $\delta^{13}C$ Values

The vertical patterns of SOC $\delta^{13}C$ (the changes in SOC $\delta^{13}C$ with depth) were similar in the two plantations but differed greatly between the plantations and the reference barren flat (Figure 3). At 0–70 cm, SOC $\delta^{13}C$ values were generally enriched with depth probably as a result of C isotope fractionation during SOC decomposition in both plantations, but the values were relatively stable with depth in the barren flat. SOC $\delta^{13}C$ values in both plantations were more negative than those in the barren flat at each soil layer possibly because the input of $\delta^{13}C$-depleted mangrove-derived organic matter lowered the $\delta^{13}C$ value of the total SOC in the plantations.

In most terrestrial soil ecosystems, SOC $\delta^{13}C$ usually shows enrichment with depth, i.e., the values become less negative as soil depth increases (Balesdent et al. 1993; Boutton 1996; Ehleringer et al. 2000; Wynn et al. 2006; Wynn 2007). The pattern is different in peat soils, however, in that SOC $\delta^{13}C$ values tend not to change with depth (Agren et al. 1996; Chen et al. 2002; Collins et al. 1997). The pattern was also different in the mangrove sediments in the current study in that SOC $\delta^{13}C$ values were relatively stable in the top 18 cm and then became gradually enriched as the depth increased from 18 to 70 cm (Figure 3). High productivity and rapid organic matter deposition in mangrove forests might account for the relatively stable pattern at the surface. The increase in SOC $\delta^{13}C$ values between 18–70 cm, which probably reflected decomposition, could be divided into two stages: (1) from 18 to about 35 cm, SOC $\delta^{13}C$ values increased sharply probably as a result of C isotope fractionation resulting from the decomposition of the labile SOC components; (2) from about 35 to 70 cm, SOC $\delta^{13}C$ values increased slowly probably because of the slow but steady decomposition of the moderately resistant SOC components.

The enrichment of $\delta^{13}C$ with depth has been related to increases in SOC decay and SOC age with depth (Balesdent et al. 1990; Stout et al. 1981). Our results are consistent with this statement in that the increases in SOC $\delta^{13}C$ value with sediment depth (between 18–70 cm) were associated with decreases in SOC content and increases in SOC $^{14}C$ age (see Figure 2–4). Balesdent et al. (1993) reported that the enrichment in SOC $\delta^{13}C$ with depth was similar in eutrophic soils with rapid C decay and in oligotrophic soils with slow C decay. Our study suggests that the increase in SOC $\delta^{13}C$ with depth in mangrove plantations is independent of mangrove species (see Figure 3).

### Contribution of New C to the Total SOC in the Mangrove Plantations as Indicated by $^{14}C$ Age

The contributions of new C to the total SOC ranged from 2 to 53% in the native *K. candel* plantation and from 13 to 55% in the exotic *S. apetala* plantation (Table 1). From 0 to 16 cm, the contributions of new C differed slightly between the plantations (45 ± 5% in the *K. candel* plantation and 47 ± 6% in the *S. apetala* plantation) but from 16 to 60 cm, the contributions differed markedly between two plantations (9 ± 5% in the *K. candel* plantation and 31 ± 8% in the *S. apetala* plantation). In other words, much higher contributions of new C were transported and stored at depths between 16 and 60 cm in the *S. apetala* plantation than in the *K. candel* plantation. Thus, from the perspective of C sequestration, *S. apetala* rather than *K. candel* should be recommended as a key reforestation species in Southern China. Our results are consistent with those of Ren et al. (2009), who pointed out that because *S. apetala* grows rapidly and readily adapts to poor habitats, *S. apetala* plantations could sequester significant amounts of C in their sediments.

Our results and inferences have some limitations. First, the calculated new C contributions in the mangrove plantations were approximate because we assumed that the SOC was a mixture of only new organic material (C in the mangrove-derived fresh litter) and old organic material (C in the pre-plantation sediment). This simplification, i.e., the partitioning of SOC into only two components, ignores the input of C from exogenous, tide-eroded oceanic organic matter. Next we will take into account control plots to reduce the over simplification. A second limitation is that, because of the high cost of dual C isotope measurements, we sampled only one soil core in each plantation. Sampling with replication will be necessary to draw more convincing conclusions.

## Conclusion

The results of this paper, which were based on C and C isotopic measurements of the sediment cores in a native plantation, an exotic plantation, and in an adjacent barren flat in Southern China, indicated that establishment of the mangrove plantations resulted in the accumulation of a large amount of SOC. The exotic *S. apetala* plantation stored slightly more SOC than the native *K. candel* plantation. The change in SOC $\delta^{13}C$ values with depth was similar in both plantations, indicating that the processes of C decay are similar in the native and exotic plantations. However, much larger quantities of mangrove-derived new C were stored at 16–60 cm in the *S. apetala* plantation than in the *K. candel* plantation based on a two-source isotope mixing model. Our study demonstrated that from the perspective of C storage, *S. apetala* was a better plantation species than *K. candel* in Southern China because the *S. apetala* plantation stored more labile, new C in its anaerobic sub-surface sediments. Additional sampling and long-term monitoring are needed to increase our understanding of SOC dynamics in these and other mangrove plantations.

## ACKNOWLEDGEMENTS

We thank Professor Li Zhi'an, Senior Engineer Zhang Qianmei, and Mr Zou Bi (the South China Botanical Garden), and Vice Director Xu Fanghong of the Administration of Zhanjiang Mangrove National Natural Reserve for their help in field sampling. We also thank Doctor Ding Ping of the Guangzhou Institute of Geochemistry, Chinese Academy of Sciences for help and assistance in field sampling and laboratory operation. This research was sponsored by funds of the Ministry of Science and Technology (No. 2007BAC28B04).

## REFERENCES

Agren, G.I., E. Bosatta, and J. Balesdent. 1996. Isotope discrimination during decomposition of organic matter: A theoretical analysis. *Soil Science Society of American Journal* 60:1121-1126.

Aller, R.C., and N.E. Blair. 2004. Early diagenetic remineralization of sedimentary organic C in the Gulf of Papua deltaic complex (Papua New Guinea): Net loss of terrestrial C and diagenetic fractionation of C isotopes. *Geochimica Et Cosmochimica Acta* 68:1815-1825.

Alongi, D.M. 2002. Present state and future of the world's mangrove forests. *Environmental Conservation* 29:331-349.

Balesdent, J., A. Mariotti, and D. Boisgontier. 1990. Effect of tillage on soil organic carbon mineralization estimated from $^{13}$C abundance in maize fields. *European Journal of Soil Science* 41(4):587-596.

Balesdent, J., C. Girardin, and A. Mariotti. 1993. Site-related delta-C-13 of tree leaves and soil organic-matter in a temperate forest. *Ecology* 74:1713-1721.

Barbier, E.B., E.W. Koch, B.R. Silliman, S.D. Hacker, E. Wolanski, J. Primavera, E.F. Granek, S. Polasky, S. Aswani, L.A. Cramer, D.M. Stoms, C.J. Kennedy, D. Bael, C.V. Kappel, G.M.E. Perillo, and D.J. Reed. 2008. Coastal ecosystem-based management with nonlinear ecological functions and values. *Science* 319:321-323.

Benfield, S.L., H.M. Guzman, and J.M. Mair. 2005. Temporal mangrove dynamics in relation to coastal development in Pacific Panama. *Journal of Environmental Mmanagement* 76(3):263-276.

Bouillon, S., A.V. Borges, E. Castaneda-Moya, K. Diele, T. Dittmar, N.C. Duke, S.Y. Lee, C. Marchand, J.J. Middelburg, V.H. Rivera-Monroy, T.J. Smith, and R.R. Twilley. 2008. Mangrove production and carbon sinks: a revision of global budget estimates. *Global Biogeochemical Cycles* 22:GB2013. doi:10.1029/2007GB003052.

Bouillon, S., and M.E. Bottcher. 2006. Stable isotopes in biogeosciences-Preface. *Organic Geochemistry* 37:1197-1199.

Boutton, T.W. 1996, Stable carbon isotope ratios of soil organic matter and their use as indicators of vegetation and climate change. In *Mass spectrometry of soils*, eds. T.W. Boutton, and S. Yamasaki, 311-339. Marcel Dekker. New York.

Chen Q.Q., C.D. shen, S.L. Peng, Y.M. Sun, W.X. Yi, Z.A. Li, and M.T. Jiang. 2002. Soil organic matter turnover in the subtropical mountainous region of South China. *Soil Science* 167(6): 401-415.

Collins, H.P., E.A. Paul, K. Paustain, and E.T. Elliott. 1997. Characterization of soil organic carbon relative to its stability and turnover. In The role of soil organic matter in agricultural systems and global change, eds. E.A. Paul, K. Paustain, E.T. Elliott, and C.V. Cole, 51-72. CRC Press Inc. New York.

Ehleringer, J.R., N. Buchmann, and L.B. Flanagan. 2000. Carbon isotope ratios in belowground carbon cycle processes. *Ecological Applications* 10:412-422.

Fujimoto, K. 2004. Below-ground carbon sequestration of mangrove forests in the Asia-Pacific region. In *Mangrove management and conservation: present and future*, eds. M. Vannucci, 138-146. United Nations University Press, Tokyo, Japan.

Gonneea, M.E., A. Paytan, and J.A. Herrera-Silveira. 2004. Tracing organic matter sources and carbon burial in mangrove sediments over the past 160 years. *Estuarine Coastal and Shelf Science* 61:211-227.

He, B.Y., T.H. Lai, H.Q. Fan, W.Q. Wang, and H.L. Zheng. 2007. Comparison of flooding-tolerance in four mangrove species in a diurnal tidal zone in the Beibu Gulf. *Estuarine Coastal and Shelf Science* 74:254-262.

Jennerjahn, T.C., and V. Ittekkot. 2002. Relevance of mangroves for the production and deposition of organic matter along tropical continental margins. *Naturwissenschaften* 89:23-30.

Kristensen, E., S. Bouillon, T. Dittmar, and C. Marchand. 2008. Organic carbon dynamics in mangrove ecosystems: A review. *Aquatic Botany* 89, 201-219.

Levin, I., and B. Kromer. 2004. The tropospheric ($CO_2$)-C-14 level in mid-latitudes of the Northern Hemisphere (1959-2003). *Radiocarbon* 46:1261-1272.

Levin, I., S. Hammer, B. Kromer, and F. Meinhardt. 2008. Radiocarbon observations in atmospheric $CO_2$: Determining fossil fuel $CO_2$ over Europe using Jungfraujoch observations as background. *Science of the Total Environment* 391:211-216.

Matsui, N., J. Suekuni, S. Havanond, A. Nishimiya, J. Yanai, and T. Kosaki. 2008. Determination of soil-related factors controlling initial mangrove (*Rhizophora apiculata* BL.) growth in an abandoned shrimp pond. *Soil Science and Plant Nutrition* 54:301-309.

Ren, H., H. Chen, Z.A. Li, and W.D. Han. 2009. Biomass accumulation and carbon storages of four different aged *Sonneratia apetala* plantations in Southern China. *Plant and Soil*. Doi:10.1007/s11104-009-0053-7.

Ren, H., S.G. Jian, H.F. Lu, Q.M Zhang, W.J. Shen, W.D. Han, Z.Y. Yin, and Q.F. Guo. 2008. Restoration of mangrove plantations and colonization by native species in Leizhou bay, South China. *Ecological Research* 23: 401-407.

Rethemeyer, J., P.M. Grootes, F. Bruhn, N. Andersen, M.J. Nadeau, C. Kramer, and G. Gleixner. 2004. Age heterogeneity of soil organic matter. *Nuclear Instruments and Methods in Physics Research B* 223-224:521-527.

Shaifullah, K.M., M. Mezbahuddin, M. Sujauddin, and S.M.S. Haque. 2008. Effects of coastal afforestation on some soil properties in Lakshmipur coast of Bangladesh. *Journal of Forestry Research (Harbin)*. doi: 10.1007/s11676-11008-10005-11678

Staddon, P.L. 2004. Carbon isotopes in functional soil ecology. *Trends in Ecology & Evolution* 19:148-154.

Stout, J.D., K.M. Goh, and T.A. Rafter. 1981. Chemistry and turnover of naturally occurring resistant organic compounds in soil. In *Soil biochemistry*, eds. E.A. Paul, and J.N. Ladd, 19-24. Marcel Dekker. New York.

Stuiver, M., and H.A. Polach. 1977. Discussion reporting of $^{14}C$ data. *Radiocarbon* 19: 355-363.

Trumbore, S. 2000. Age of soil organic matter and soil respiration: Radiocarbon constraints on belowground C dynamics. *Ecological Applications* 10:399-411.

Twilley, R.R., R.H. Chen, and T. Hargis.1992. Carbon sinks in mangroves and their implications to carbon budget of tropical coastal ecosystems. *Water Air and Soil Pollution* 64:265-288.

Vogel, J.S., D.E. Nelson, and J.R. Southon. 1987. C-14 background levels in an Accelerator Mass-Spectrometry system. *Radiocarbon* 29:323-333.

West, J.B., G.J. Bowen, T.E. Cerling, and J.R. Ehleringer. 2006. Stable isotopes as one of nature's ecological recorders. *Trends in Ecology & Evolution* 21:408-414.

Wynn, J.G. 2007. Carbon isotope fractionation during decomposition of organic matter in soils and paleosols: Implications for paleoecological interpretations of paleosols. *Palaeogeography Palaeoclimatology Palaeoecology* 251:437-448.

Wynn, J.G., J.W. Harden, and T.L. Fries. 2006. Stable carbon isotope depth profiles and soil organic carbon dynamics in the lower Mississippi Basin. *Geoderma* 131:89-109.

Yanagisawa, H., S. Koshimura, K. Goto, T. Miyagi, F. Imamura, A. Ruangrassamee, and C. Tanavud. 2009. The reduction effects of mangrove forest on a tsunami based on field surveys at Pakarang Cape, Thailand and numerical analysis. *Estuarine Coastal and Shelf Science* 81:27-37.

Zan, Q.J., Y.J. Wang, and W.B. Liao. 2001. Biomass and net productivity of *Sonneratia carseolaria-Sonneratia apetala* mangrove forest. *Journal of Wuhan Botanical Research* 15:391-397.

Zhang, J.P., C.D. Shen, H. Ren, and K.X. Liu. Quantification of soil organic carbon storage and turnover of high and middle tidal mangrove stands based on dual carbon isotopic measurements. *Ecological Modelling*, in press.

Zhang, J.P., H. Ren, W.J. Shen, S.G. Jian, and F.H. Xu. 2009. Community composition, species diversity and population biomass of the Gaoqiao mangrove forest in Southern China. In *International wetlands: Ecology, conservationa and restoration*, eds. J.R. Herrera, 177-189. Nova Science Publisher, Inc. New York.

In: Plantations
Editor: Hai Ren

ISBN: 978-1-62808-090-2


*Chapter 7*

# THE POTENTIAL FOR CARBON SEQUESTRATION IN CARBON DEPLETED AREAS OF THE BOREAL FOREST ECOZONE THROUGH AGROFORESTRY-BLOCK PLANTATION

***Silvia Lac*[1,*] *and Manuel Esteban Lucas-Borja*[2,†]**
[1]Forestry Engineer Consultant, University of Saskatchewan, Saskatchewan, Canada
[2]Department of Agroforestry Technology and Science and Genetics, School of Advanced Agricultural Engineering Castilla La Mancha University, Albacete, Spain

## ABSTRACT

Agroforestry systems have been considered an important tool in the mitigation for climate change. The conversion from forests to any other land-use is currently placing the forest sector as the second with greatest emission of carbon dioxide worldwide (larger than the transportation sector) and thus contributing to climate change. Actions to contain/reverse/minimize carbon losses from the conversion from forests to any other land-uses are furthermost important as can influence temperature rise in magnitude, velocity (gradual increase gives ecosystems greater chances to adapt, less mortality), and impacts (e.g., biodiversity, ecosystem functions and services). Where forest lands are in high demand and or have been permanently changed to other land-uses, agroforestry may be the only option available to restore to certain extent the original biome and recover the carbon stocks lost to the land use conversion that took place in the past (from forests to agriculture). Saskatchewan is the main agricultural province in Canada and is also the case where many farms have been developed on previous forest soils (Gray Luvisolic and Dark Gray Chernozemic/Luvisolic). The Boreal Forest is one of the world´s carbon richest ecosystems, especially in the soils. Carbon stock change or carbon sequestration potential was estimated for agroforestry systems type of wood fiber plantations adopted

* Silvia Lac. E-mail: lac.silvia@gmail.com.
† Manuel Esteban Lucas-Borja. Campus Universitario s/n, CP 02071, Albacete, Spain. E-mail: ManuelEsteban.lucas@uclm.es.

for Saskatchewan Boreal forest soils, in the nearby areas to wood processing facilities (converted into products). Carbon stocks were estimated for aboveground and belowground compartments, and in forest products. Research gaps were also identified to improve future estimations. Carbon sequestration estimations were carried out for future scenarios of approximately 500 years, for different species, adoption rates, carbon pools considered, and approaches to the forest products accounting system. The time-step used was the rotation length.

Above- and belowground C sequestration potential calculated in 10-ha farms converted to single species wood fiber plantation was higher for hybrid poplar (91.3 Mt C) than for trembling aspen (19.1 Mt C) and white spruce (17.8 Mt C). When the forest products pool) was considered (three pools), C sequestration potential was higher for both trembling aspen and white spruce (0.4 Mt C) than for hybrid poplar (-9 Mt C). The three- pools calculation was reproduced for hybrid poplar using the *production approach,* which tends to penalize exporting countries by accounting emissions for the forest products pool. This resulted in greater deficit/emission (-2.3 Mt C) due to Canada´s great exports. The most important research needs identified included growth curves, conversion and expansion factors, and soil organic carbon measurements. These data/measurements are required for each cultivated species, agroforestry design and silvicultural conditions (spacing, site quality, and stand management, etc).

**Keywords:** Climate change, carbon, carbon sequestration, carbon model, agroforestry, boreal forest

## Introduction: Climate Change and the Forestry Sector

Global warming is expected to cause both short-term and long-term effects, with earlier impacts closer to the poles where most warming is predicted and observed (Intergovernmental Panel on Climate Change, 2007), where carbon rich ecosystems and specially the boreal forest can play a critical role both in mitigating or aggravating the issue through positive feedback (the more temperature increase, the faster the decomposition rates and carbon dioxide emissions feeding further to the temperature increases through increase in greenhouse gas accumulation to the atmosphere).

All predictions indicate that temperature will increase, and increase more in the winter than in the summer, however, not all places will experience a steady increase in temperature, and locations with minimum and maximum temperature anomalies will change with time (Stewart et al., 1997; Singh and Wheaton, 1991). The increase in temperature is expected to lead to an increase in precipitation (Stewart et al., 1997; Singh and Wheaton, 1991). However the increase in precipitation would not compensate for the increased evapotranspiration and droughts would occur (Herrington et al., 1997). For the Prairies, the same authors reported that the drought of the late 1980s resulted in reduced volume growth, increased seedling mortality, and increased disturbances compared to the previous 50 years´ diseases (tent caterpillar and spruce budworm) and fires. The boreal forest vegetation growth and regeneration are predicted to be positively affected in some areas, while negatively affected in other areas (Stewart et al., 1997; Herrington et al., 1997).

As the geographical distribution of boreal forest species is controlled by temperature and precipitation, the boreal forest is also predicted to shift to new species composition as a result of climate change Singh and (Wheaton, 1991; International Institute for Applied Systems, Analysis Future climate change and carbon budgets, 2000).

In the next century, with suitable conditions, the boreal forest is migrating northwards into areas suited to sub-arctic and tundra ecosystems; and the climate at the southern edge of the boreal forest is changing towards conditions more suitable to temperate forests and grasslands. The boreal forest would then shrink considerably from its present size, as a result of its southern boundary migrating faster northwards than its northern boundary (Stewart et al., 1997).

This would result in a decreased forest carbon storage capacity. Wetlands and peatlands could also dry out as a result of climate change, losing their ability to maintain carbon stocks and releasing great amounts of carbon dioxide and methane into the atmosphere, thus contributing to further climate change/global warming (Singh and Wheaton, 1991). So far, the effects on the northern portion of the boreal forest are evident. A receding permafrost zone and a noticeable warming in permafrost temperatures have been observed in forest and tundra regions (Stewart et al., 1997). As the same authors report, there is evidence to the increase in vegetation productivity between 45 and 70 degrees north, improved seed production by boreal tree species along the northern boreal forest; increased disturbances related to fire compared to prior periods; and accelerated phenological development in spruce.

Despite most predictions agree on the consequences of climate change at the global scale, climatic parameters such as precipitation, temperature, and solar radiation intensity are still difficult to predict (International Institute for Applied Systems, Analysis Future climate change and carbon budgets, 2000; Waring and Running, 1998). Difficulties arise due to the current limitations of understanding and representing climatic changes and also in transferring global information to regional or local scales (Waring and Running, 1998). These, in turn, result in uncertainties about the timing, location, and rate of projected changes as well as the potential benefits of the climate change impacts (Stewart et al., 1997). Nevertheless, more and more observations and evidences have been collected and these to confirm much prediction ranges and all trends (Intergovernmental Panel on Climate Change, 2007).

The Canadian economy depends greatly on forestry. And the Canada's balance of trade has benefitted much in the past from the contribution of forest products exports (Canadian Forest Service, 1999). A loss of the current area due to the shrinkage of the boreal forest then would greatly affect the Canadian economy. Forest companies would be forced to move northwards and, if the boreal forest continues to shrink, the forestry sector may no longer be able to operate in this region sometime in the future. The market for forest products has shifted induced by the exports from lumber, wood pulp and printing paper to composite panels and prefabricated buildings (Canadian Forest Service, 1999). Saskatchewan, with only 6.9% of Canadian forestlands exports almost 2% of Canada′s exports as wood pulp, paper and paperboard, softwood lumber, and waferboard; and the main export markets for these products are the US, European Union, Japan, Central and South America (Canadian Forest Service, 1999). The market for export is growing for Saskatchewan, as well as for Canada and, according to Mistik Management (Mike Martel, personal communication) and Weyerhaeuser Corporation (Darryl Sande, personal communication), a shortage of wood is predicted within 20 years from now. According to predictions, the Meadow Lake, Prince Albert and Hudson Bay areas will see a reduction of the wood supply over the coming 5 to 10 years (Mak et al., 1999). Climate change forecasts are therefore highly incompatible to supply this growing market for forest products.

Adaptation and mitigation strategies include managing stands to reduce fires, prescribed burning, managing forest fuel, preserving and enhancing biodiversity, and managing forest landscapes that are becoming increasingly fragmented (Herrington, 1997). Forestry companies can replace natural species by planting better adapted species when the natural species die, or move to locations where timber becomes more abundant; however the adaptation of wildlife would be more difficult (Herrington, 1997).

Agroforestry can help to mitigate climate change directly through its carbon sequestration potential; or indirectly if through releasing the harvesting pressure over natural forests. They also include agricultural land use (potential to adopting agroforestry systems) that is in areas that are both accessible and highly productive in Canada (Mak et al., 1999). With only 6% of the forested area within agricultural rural Canada this supplies 19% of all round timber harvested in the country, 79% of maple products, 79% of firewood and 100% of Christmas trees. Furthermore difficulties in the forestry sector (economic downturns and diseases) can turn into opportunities to the new agroforestry sector. As a result, And the unprecedented combination of climate change and disturbances (e.g. flooding, drought, wildfire, insects, ocean acidification) and other global change drivers (e.g. land-use change, pollution, fragmentation of natural systems, overexploitation of resources) (Intergovernmental Panel on Climate Change, 2007), may yet bring additional problems together which can prove difficult to the Canadian forestry industry to overcome. And currently, the few mitigation options that can contribute to reducing risks (highest rate of carbon stock change) for the short term include stopping land-use conversion from primary forests to other land uses, and the restoration of forestlands previously converted to other land uses.

## METHODS TO ACCESS CARBON SEQUESTRATION IN THE FOREST COMPONENT OF AGROFORESTRY SYSTEMS

Two methods are currently accepted under the Kyoto Protocol to be used for carbon accounting in terrestrial ecosystems: the stock change and carbon flux approaches (Tipper and De Jong, 1998; Watson et al., 2000). In the stock change approach, the carbon stocks in vegetation tissues and in the soil are measured at different times, and the difference in the carbon stock is the net ecosystem productivity (Watson et al., 2000). As these authors report, carbon stock change receive a positive value when carbon stocks increase overtime (sequestration) and a negative value when carbon is released or lost from the ecosystem.

The carbon flux approach is based in periodic measurements of carbon fluxes into and out an ecosystem (Watson et al., 2000). As the same authors report, the carbon flux is usually based on chambers and the eddy correlation/covariance technique. Estimates of carbon sequestered with this method vary greatly due to climatic fluctuations and occasional insect outbreaks, and also are highly affected by inter-annual variations in climate (Kostela et al., 2000). Although this method is considered the most realistic, following forest carbon balances on a yearly basis, it is more difficult and costly (Kostela et al., 2000; Watson et al., 2000). Also, this method can be at best applied to local scales and short-term research (Kostela et al., 2000), as research stations capable to measure fluxes are currently not numerous enough to represent larger areas (Watson et al., 2000).

Although methods exist to measure stocks as well as the fluxes of carbon in all components of the terrestrial ecosystems (Watson et al., 2000), the stock change approach is a lower cost and easier method and is less influenced by short-term fluctuations in the carbon balance (Kostela et al., 2000; Watson et al., 2000) This method is normally applied for whole rotation periods, with data more readily available and therefore more representative of the ecosystems' heterogeneity (Watson et al., 2000). For this reason the stock change approach is used most extensively to predict changes to the carbon cycle.

As research stations or field data to predict changes in the carbon cycle for agroforestry ecosystems are currently not available, no other accounting method could be considered in this study other than the *stock change* approach. Therefore, the stock change approach was used for the forest component of the agroforestry ecosystem.

Within the stock change approach, the forest component of the agroforestry ecosystem can be further divided as follows: (1) the aboveground C component which includes stem, top, branches, foliage and stump; and (2) the belowground C component which includes roots and sometimes soils (Bonnor, 1985). For a particular forest ecosystem, the amount of biomass and therefore carbon allocated in aboveground components is primarily influenced by the soil capability, nutrient status, and available moisture (Fisher and Binkley, 1999). Better soils, nutrient status and moisture conditions will require trees less allocation of resources to root development and sooner development of trees aboveground biomass (and carbon) (Fisher and Binkley, 1999).

Different species of a given ecosystem may have particular allocation patterns, as a result of different characteristics, such as photoperiod, nutrient and moisture requirements, and root systems (Fisher and Binkley, 1999; Peterson and Peterson, 1992). Carbon allocation is also a function of tree age, which determines young trees to have higher proportion of roots in relation to the aboveground biomass, while the inverse occurs with mature trees (Fisher and Binkley, 1999; Kurz, 1992; Peterson and Peterson, 1992).

## Study Area

The study area included the rural areas of Meadow Lake, Prince Albert and Hudson Bay. These communities were chosen due to their heavy economic reliance on the forestry industry (Canadian Forest Service, 1999; Mak, 1999). These communities have two pulp mills and a paper mill, two OSB mills, a plywood mill, and over 130 sawmills (Brady and Appleby, 2000; Acton et al., 1998). These mills produce a wide range of marketable products, such as lumber, plywood, wood chips, OSB, pulp and paper (Brady and Appleby, 2000). Therefore a market for wood produced already exists, and regional deficits of softwoods are predicted for Prince Albert and Hudson Bay (Mak, 1999).

Another important aspect to consider selecting the study within the distance range which forestry companies, are willing to purchase and transport wood from.

From the carbon sequestration perspective, it is important that there is an incentive to harvest trees when they are highly productive (before declining in productivity and carbon sequestration potential), and that the carbon in the merchantable wood is be transferred to the products pool and therefore stored/sequestered for longer (specially in longer lived products such as for housing, but also in shorter lived products such as printing paper and recycling) before they are disposed and the carbon is returned to the atmosphere (through immediate release when burned for instance and or by decomposition where consumed by microorganisms which release carbon back through respiration) (Watson et al., 2000; Kurz et la., 1992; Canadian Climate Change Secretariat, 1999). The distances considered by the existing wood processing facilities to purchase and transport wood from consist in the following ranges: within a 150 km radius from the city of Meadow Lake (Mike Martel, personal communication); within 50 km south, 50 km north, 100 km east, and 150 km west from the city of Prince Albert; and within a 150 km radius from the city of Hudson Bay (Darryl Sande, personal communication). Another important aspect considered in selecting the study area is to work within the boundaries of the Boreal forest ecosystem. In such way, the benefits produced by the restoration to the biome and ecosystem´s functions and services (habitat, wildlife corridors, water quality, soil protection, etc.) can be maximized (Canadian Climate Change Secretariat, 1999). The Boreal forest is nevertheless a highly rich carbon biome (among the richest in the world). If carbon depletion is stopped and the lost C restored, this could potentially recover about 30% of carbon stocks lost in the first few years after the land-use conversion to agriculture (and other land uses), assuming that the maximum storage in native forest conditions can be fully restored and maximum storage does not surpass (Watson et al., 2000). Soils where these lands use changes (to agriculture) occurred (i.e., Gray Luvisolic and Dark Gray Chernozemic/Luvisolic) were included in this study. Thirty-five rural municipalities (RM's) were selected to take part in this study (figure 1). Each individual RM was included only if simultaneously met 2 criteria: (1) the RM area had minimum of 50% cover of forest soils (Gray Luvisolic and Dark Gray Chernozemic/Luvisolic) and (2) RM was within the distance range from wood processing facilities (information compiled from questionnaires sent to the main facilities). Maps were used in GIS from the Saskatchewan Land Resource Centre/University of Saskatchewan and agricultural land-use data for RM´s was compiled from Statistics Canada (Statistics Canada, 1996). Ecozones and Ecoregions maps were provided by the Saskatchewan Land Resource Centre/University of Saskatchewan. The analysis indicates that 7,206 farms (383.68 ha in average) have the potential to be converted to agroforestry systems, where each farm had an average area. The methods and assumptions are further described in (Lac, 2002).

The study area was characterized in Acton et al. (1998). Its elevation ranges from 500 m in its eastern part to 600 m in its western part, with generally a gradual break from plains to uplands. It has a humid continental climate at lower elevations and in the southern part of the study area, and sub-arctic climate at higher elevations and in its northern part, with temperatures generally decreasing from south to north and from east to west. Total precipitation is variable and low, increasing from southwest to northeast the study area. The area has long and cold to very cold and snowy winters and short, warm and moist summers. There is a gradual increase in wildlife productivity and diversity from north to south in the study area, as the vegetation diversity increases. Aspen forests and their understory vegetation provide habitat for many wildlife, such as beaver, deer, snowshoe, white tailed rabbit, etc. The main human activities in the study area described by the same authors are forestry, agriculture, hunting, trapping outdoor recreation, tourism, oil and gas exploration and production, and mining. Forestry operations focus on pulp and paper production, with some panelboard and saw mill operations, and less than 9% of the study area is within some form of park or protected area, the majority of which are Wildlife Habitat Protection Act lands (Acton et al., 1998).

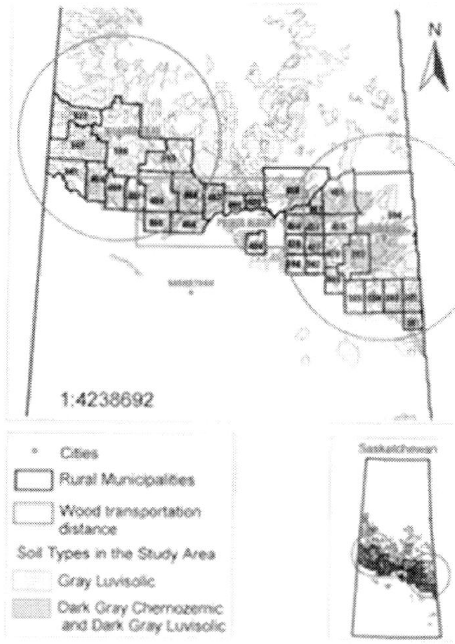


Figure 1. Study area, area available for the adoption of agroforestry systems in Saskatchewan.

# FUTURE SCENARIOS RESULTS

Future scenarios of approximately 500 years were produced (Table 1). The time scale was selected by the research team as the most reliable since using past records to estimate future conditions is inherently limited.

Two native species (trembling aspen and white spruce) were considered to maximize benefits (biodiversity, wildlife habitat and corridors, etc.) and according to best ecozones and ecorregion maps and soil types inference for vegetation; using GIS and further described in Lac (2002). A fast growing species was considered to maximize production and shorter term income to farmers.

Thus, hybrid poplar varieties Walker and Assiniboine were included in this study as they show best performance currently in the study area, according to the PFRA Shelterbelt Centre and their publications, further described in the author´s M.Sc. thesis (Lac, 2002) which this article is largely a summary. Due to the lack of information to different types of agroforestry systems in the study area, block-arrangements/wood fiber plantations were considered to resemble natural stands from which data could be used for next best substitute. A complete description for methods, conversion and expansion factors used and assumptions is available in Lac (2002).

Hybrid poplar growth curves (models for trees growth in the forestry sector, expressed in volume per hectare yearly and for commercial use to specific forestry products) adapted curves (Peterson et al., 1999), available for the Prairies (pulpwood) were used, assuming an intermediate quality site, 3x3m tree spacing, no mortality, and medium intensity management- site preparation, some cultivation and weed control.

**Table 1. Future scenarios considered for carbon sequestration in wood fiber plantations**

| | |
|---|---|
| Hybrid poplar | Scenario 1: All farms currently without agroforestry systems convert 10 ha to hybrid poplar wood fiber plantations. The projection was done for 495 years, included 2 carbon pools (aboveground and belowground), and rotation periods of 15 years [19]. Result: 91.3 Mt C (sequestration). |
| | Scenario 2: All farms currently without agroforestry systems convert 10 ha to hybrid poplar wood fiber plantations. The projection was done for 495 years, included 3 carbon pools (aboveground, belowground, and forest products), and rotation periods of 15 years. Result: - 9 Mt C (emission). |
| | Scenario 3: All farms currently without agroforestry systems convert 10 ha to hybrid poplar wood fiber plantations. The projection was done for 495 years, included 3 carbon pools (aboveground, belowground, and forest products), and rotation periods of 15 years. In this scenario, all sequestration in the forest products pool was accounted for Canada (*production* approach). Result: -2.3 Mt C (emission). |
| | Scenario 4: 360 farms convert 10 ha to hybrid poplar wood fiber plantations over 15 years (average adoption of 24 farms per year), until all farms currently without agroforestry systems are converted. The projection was done for 495 years, included 2 carbon pools (aboveground and belowground carbon stocks), and rotation periods of 15 years. Result: 65.5 Mt C (sequestration). |
| Trembling aspen | Scenario 5: All farms currently without agroforestry systems convert 10 ha to trembling aspen wood fiber plantations. The projection was done for 490 years, included 2 carbon pools (aboveground and belowground), and rotation periods of 70 years. Result: 19.1 Mt C (sequestration). |
| | Scenario 6: All farms currently without agroforestry systems convert 10 ha to trembling aspen wood fiber plantations. The projection was done for 490 years, included 3 carbon pools (aboveground, belowground, and the forest products), and rotation periods of 70 years. Result: 0.4 Mt C (sequestration). |
| | Scenario 7: 1,680 farms convert 10 ha to trembling aspen wood fiber plantations over 70 years (average adoption of 24 farms per year), until al farms currently without agroforestry systems are converted. The projection was done for 495 years, included 2 pools aboveground and belowground), and rotations of 70 years. Result: 15 Mt C (sequestration). |
| White Spruce | Scenario 8: All farms currently without agroforestry systems convert 10 ha to white spruce wood fiber plantations. The projection was done for 490 years, included 2 carbon pools (aboveground and belowground carbon stocks), and rotation periods of 70 years. Result: 17.8 Mt C (sequestration). |
| | Scenario 9: All farms currently without agroforestry systems convert 10 ha to white spruce wood fiber plantations at once. The projection was done for 490 years, included 3 carbon pools (aboveground, belowground, and the forest products), and rotation periods of 70 years. Result: 0.4 Mt C (sequestration). |
| | Scenario 10: 1,680 farms convert 10 ha to white spruce wood fiber plantations over 70 years (average adoption of 24 farms per year), until al farms currently without agroforestry systems are converted. The projection was done for 495 years, included 2 carbon pools (aboveground and belowground carbon stocks), and rotations of 70 years. Result: 13.9 Mt C (sequestration). |

The marketable wood volume was converted to biomass using a wood basic density/ conversion factor of 0.358 t/m$^3$ (Shroeder, 1996) an expansion factor of 1.454 was applied to include all aboveground non-merchantable components (Peterson et al., 1999), and all aboveground biomass was converted to carbon using a conversion factor of 0.42 (Horwath et al., 1994). Soil carbon sequestration values are adopted from north central US (Hansen, 1993).

For trembling aspen, the state level forest government agency (Saskatchewan Environment- Phil Loseth, personal communication) provided with growth curves from natural stands (Golder Associates Ltd. Report on Construction of Yield Curves for Inventory Zones C20, C30, C40, and C50 in Saskatchewan, 2001). The provincial growth curves for trembling aspen are a compilation of data from various local inventories, which were part of the Volume Sampling Program carried out between 1970 and 1990. The median of these growth curves (C20, C50 and C60) was taken, which produced one growth curve for each of the four crown density classes (i.e., crown cover of >10% $\leq$30%, >30% $\leq$55%, >55% $\leq$80%, and >80%).

Second, the average between the two growth curves for the intermediate crown density classes was taken and assumed representative of the study area. Marketable volume was converted to biomass using a wood basic density/conversion factor of 0.408 t/m$^3$ (Singh, 1987), an expansion factor of 1.18 was applied to include all aboveground non-merchantable components (Singh, 1987), and a conversion factor of 0,502 was applied to convert all aboveground biomass to carbon (Tillman, 1981).

For white spruce, growth curves were also provided from Saskatchewan Environment for natural stands (Golder Associates Ltd. Report on Construction of Yield Curves for Inventory Zones C20, C30, C40, and C50 in Saskatchewan, 2001), and data was processed in the same manner as for white spruce whereas intermediate crown density classes was used and assumed representative. The marketable volume was converted to biomass using a factor (wood basic density) of 0.364 t/m$^3$ (Wang and Micko, 1984), an expansion factor of 1.20 was applied to include all aboveground non-merchantable components [28], and a conversion factor of 0.527 was applied to convert biomass to carbon (Tillman, 1981).

The belowground carbon stocks for trembling aspen and white spruce were calculated using values compiled for the carbon for the entire soil profile (Siltanen, 1997), because often forestry methods used result in great soil disturbance and mixing of in the entire soil profile. The best available soil organic carbon database which also permitted simultaneous queries for soil zones, species (at mature age), ecoclimatic provinces or provinces, and also soil organic carbon measurements for the entire soil profile, was the "Soil Profile and Organic Carbon Data Base for Canadian Forest and Tundra Mineral Soils" (Siltanen, 1997).

The assumption therefore was that all carbon lost to the land-use conversion to agriculture (30%) is fully recovered within the first rotation (70 years) to the maximum potential for carbon storage, although few sources are available for long term records to confirm this assumption (Siltanen, 1997).

Nevertheless it is expected that the maximum potential for carbon storage in agricultural soils is unlikely to exceed that of their native condition (Yanai, et al., 2000; Nagle, 1990; Janzen et al., 1998; Bruce et al., 1999). Values for soil carbon are assumed to include roots as a conservative approach to avoid double accounting issues.

Due to difficulties in distinguishing the sources of soil organic carbon when sampling, adding estimations of carbon stocks in roots biomass to estimations of carbon stocks in the soil organic matter and mineral soil horizons is likely to cause double accounting for the belowground carbon pool according to re-known soil scientists in the research team (Dan Pennock and Marie Boehm), and therefore one of these should be chosen.

However, this can underestimate the C in the soil compartment. This is not addressed directly (in other accounting methods such as the Carbon Budget Model for the Canadian Forest Service (Kurz, 1992; Bruce et al., 1999).

For the forest products pool, it was possible to estimate future use of the species, in consultation with the wood processing facilities and expert's opinion. Using the *stock change* approach, the exports were estimated, together with the selection of an end disposal for products and tree parts resulting from the industrial process. Half-lives (Skog, 1999) (time that takes for half carbon to be released) were applied for each product when disposed to decompose, and carbon was accounted as instant release when burning is used. Very long carbon sequestration periods are considered in cases such as sawdist piles, whereas partial decomposition is negligible (Price et al., 1996). Hybrid poplar end use is mostly to oriented strand board- OSB; trembling aspen end use is mostly OSB and pulp and paper; and white spruce end-use is to lumber, panels, and paper. The *production approach* was also used for comparison (where all carbon sequestration/emission is accounted from the ecosystem). Further details for accounting methods are available in (Lac, 2002).

Results indicate that the highest carbon sequestration potential will be reached for future scenarios considering single species established on gray soils (Gray Luvisolic and Dark Gray Chernozemic/Luvisolic) of Saskatchewan (10 ha per farm), and accounting for 2 compartments only (i.e., including aboveground and belowground C and excluding the forest products compartment from the accounting). Total ecosystem carbon sequestration will be 91.3 Mt C for hybrid poplar (scenario 1), 19.1 Mt C for trembling aspen (scenario 5), and 17.8 Mt C for white spruce wood fiber plantations (scenario 8).

Under these scenarios, hybrid poplar had the highest carbon sequestration potential, followed by trembling aspen and white spruce. This result can be largely explained by the high rate of accumulation in the aboveground carbon pool of hybrid poplar, which compensates for low levels of sequestration in the belowground carbon pool.

Future 500 years projections also including forest products (stock change approach) indicate a total carbon sequestration of -9 Mt C for hybrid poplar (scenario 2), and 0.4 Mt C for both trembling aspen and white spruce single species wood fiber plantations (scenarios 6, 9). Although the only difference between scenarios 2, 6, and 9 and scenarios 1, 5, and 8 was the inclusion of forest products, results diverged from those previously calculated. They showed the largest carbon sequestration for trembling aspen, followed by white spruce and hybrid poplar wood fiber plantations.

This diverging result can be explained by a sum of factors, such as the end use of each species in products, the number of rotations in these simulations for each species, and the storage capacity in the carbon pools considered. The lower C sequestration in scenarios including the forest products pool can be partially attributed to the method used for accounting (stock change method). This method results in higher emissions in the producing country (Canada), and higher sequestration in the consuming country. The scenario 3 was similar to scenario 2, with the exception of using the *production* approach for the products pool (i.e., carbon sequestration and emissions are accounted for the producing country). This resulted in a total carbon sequestration (after 500 years) of -2.3 Mt C. As expected, scenario 3 resulted in lower carbon sequestration than scenario 2.

## Discussions and Conclusions

Results from this study suggest that including forest products pool shows distinct results from the non-inclusion of this important pool, and would result in a different species selection for the future simulations on wood fiber plantations and within the considered time length. Wood fiber plantations with higher carbon sequestration potential (scenario 1) had the potential to offset 5.6 years of Saskatchewan carbon dioxide emissions, i.e., 1.2% of total emissions in 495 years, assuming Saskatchewan emissions continue at same rate (Pembina Institute for Appropriate Development, 2000).

Likewise, a similar scenario would result in 1.3 times higher total marketable volume at the end of each rotation in the context of the Saskatchewan annual allowable cut (Canadian Forest Service, 1999).

Through applying conversion and expansion factors that have been extensively used for Canada (Lemprière and Booth, 1998; Robinson et al., 1999) to the growth curves considered in this study, the aboveground carbon sequestration at harvesting age was found to be overestimated for hybrid poplar by 43%, for trembling aspen by 29%, and for white spruce 36%; as compared to results from this study.

Finally, the quality of estimates is so good as the quality of the data sources available. A higher amount of data and information is likely to allow improving the quality of these estimates.

## Research Needs

The most important research gaps resulting from this analysis are:

- Specific growth curves for each species, compiled for each soil and ecological zones. Within them, compiled for each agroforestry design and growth conditions, such as spacing, site quality, and stand management;
- Soil organic carbon measurements for the entire soil profile (due to the mixing at soil preparation for planting trees) from planted stands in agricultural or former agricultural soils, for each species, for each agroforestry design, and growth conditions (such as spacing, site quality, and stand management); and, within those, could be compiled by either soil zone or other ecological framework;
- Conversion and expansion factors for each species, agroforestry design, and growth conditions (such as spacing, site quality, and stand management); and, within those, could be compiled by either soil zone or other ecological framework;
- More field data collection, preferably from long-term plots (or else from chronosequential studies) for each agroforestry design and growth conditions (such as spacing, site quality, and stand management);

Furthermore, the standardization of sampling, measurement and laboratory techniques and technologies would greatly enhance the chances of field data to be compared and compiled.

## REFERENCES


Acton, D. F., Padbury, G. A. and Stushnoff, C. T. (1998). *The Ecoregions of Saskatchewan*. Saskatchewan Environment and Resource Management. Regina, Saskatchewan: Canadian Plains Research Centre/University of Regina.

Bonnor, G. M. *Inventory of Forest Biomass in Canada* (1985). Chalk River, Ontario: Petawawa National Forestry Institute/Canadian Forest Service.

Brady, A. and Appleby, A. (2000). *Deforestation: Lack of Regeneration in Saskatchewan Forests.* Technical Report 1.

Bruce, J. P., Frome, M., Haites, E., Janzen, H., Lal, R., and Paustian, K. (1999). Carbon sequestration in soils. *Journal of Soil and Water Conservation* 54:382-389.

Canadian Climate Change Secretariat (1999). Sinks Table Option Paper, Land-Use, Land-Use Change and Forestry in Canada and the Kyoto Protocol (National Climate Change Process).

Canadian Forest Service (1999). The State of Canada's Forests 1998-1999. Ottawa, Ontario: Canadian Forest Service.

Fisher, R. F. and Binkley, D. (1999). *Ecology and Management of Forest Soils*. Third Edition. New York, New York: John Wiley and Sons Inc.

Golder Associates Ltd. Report on Construction of Yield Curves for Inventory Zones C20, C30, C40, and C50 in Saskatchewan (2001). Unpublished contract report. Prince Albert, Saskatchewan: Saskatchewan Environment/Forest Ecosystems Branch.

Guy, R. D. and Benowicz, A. (1998). Can Afforestation Contribute to a Reduction in Canada's net CO2 Emissions? A Report prepared for the CPPA.

Hansen, E. A. (1993). Soil carbon sequestration beneath hybrid poplar plantations in the north central United States. *Biomass and Bioenergy* 5:431-436.

Herrington, R., Johnson, B. and Hunter, B. (1997). *Responding to Global Climate Change in the Prairies.* Volume III of the Canada Country Study: Climate Change Impacts and Adaptation. Adaptation and Impacts section, Atmospheric and Hydrologic Sciences Division, Environment Canada.

Horwath, W. R., Pregitzer, K. S. and Paul, E. A. (1994). 14C allocation in tree-soil systems. *Tree Physiology* 14:1163-1176.

Intergovernmental Panel on Climate Change (2007). *Climate Change: Impacts, Adaptation and Vulnerability.* Contribution of Working Group II to the Fourth Assessment Report of the IPCC. Cambridge, United Kingdom: Cambridge University Press.

International Institute for Applied Systems Analysis Future climate change and carbon budgets (2000). Available at <http://www.iiasa.ac.at/Research/FOR/papers/medea/10_future_climate_change.htm> Last accessed 09 February 2013.

Janzen, H. H., Campbell, C. A., Gregorich, E. G., and Ellert, B. H. (1998). Soil carbon dynamics in Canadian agroecosystems. Ch.5 in: Rattan, L., Kimble, J. M., Follett, R. F., and Stewart, B. A. *Advances in Soil Science: Soil Processes and the Carbon Cycle*. Boca Raton, Florida: CRC Press.

Kostela, J., Nygren, P., Berninger, F., and Luukkanen, O. (2000). Implications of the Kyoto Protocol for Tropical Forest Management and Land Use: Prospects and Pitfalls. *Tropical Forestry Reports 22.* Hakapaini Oy, Helsinki: Department of Forest Ecology/ University of Helsinki.

Kurz, W. A., Apps, M. J., Webb, T. M., and McNamee, P. J. (1992). *The Carbon Budget Model of the Canadian Forest Sector: Phase I.* Edmonton, Alberta: Northern Forestry Centre/Forestry Canada. Inf. Rep. NOR-X-326.

Lac, S. (2002). Carbon Sequestration in the Boreal Forest Plain Ecozone Through the Conversion from Agriculture to Agroforestry. Master of Science thesis. Saskatoon, Saskatchewan: Division of Environmental Engineering/University of Saskatchewan.

Lemprière, T. and Booth, D. (1998). Preliminary Estimates of Carbon Stock Changes in 2008-2012 Resulting from Reforestation, Afforestation and Deforestation Activity in Canada since 1990. Internal Draft 27 February 1998. Ottawa, Ontario: Canadian Forest Service/Natural Resources Canada.

Mak, K., Lavoie, A. and Grundberg, B. (1999). *Evaluating the Potential and Opportunities for Agroforestry in the Province of Saskatchewan.* Regina, Saskatchewan: Saskatchewan Agriculture and Food.

Nagle, G. S. (1990). *Technical Background Paper: Trees for Canada Program.* Victoria, British Columbia: Nawitka Renewable Resource Consultants Ltd.

Natural Resources Canada, (2012). Carbon Budget Model. Natural Resources Canada. Available at <http://cfs.nrcan.gc.ca/pages/94> Last accessed 09 February 2013.

Pembina Institute for Appropriate Development (2000). Provincial Government Performance on Climate Change: 2000. Available at <http://www.pembina.org/> Last accessed 09 February 2013.

Peterson, E. B., Bonnor, G. M., Robinson, G. C., and Peterson, N. M. (1999). *Carbon sequestration aspects of an afforestation program in Canada's Prairie Provinces.* Submitted to Joint Forest Sector Table/Sinks Table (National Climate Change Process).

Peterson, E. B. and Peterson, N. M. (1992). *Ecology, Management, and Use of Aspen and Balsam Poplar in the Prairie Provinces, Canada.* Special Report 1, Forestry Canada Northwest Region, Northern Forestry Centre. Victoria, British Columbia: Western Ecological Services Ltd.

Price, D. T., Mair, R. M., Kurz, W. A., and Apps, J. (1996). Effects of forest management, harvesting and wood processing on ecosystem carbon dynamics: a boreal case study. Chapter 23, 279-292 pp. (1996). In: Apps, M. J. and Price, D. T. (Eds.). Forest Ecosystems, Forest Management and the Global Carbon Cycle. *NATO ASI series*, Series 1, No. 40. New York, New York.

Robinson, G. C., Peterson, E. B., Smith, S. M., and Nagle, G. S. (1999). *Estimating the Carbon Sequestration Associated with Reforestation in Western Canada.* Report by Nawitka Renewable Resource Consultants for the Joint Forest Sector and (Sinks Table National Climate Change Process).

Shroeder, W. R. (1996). *Determination of Poplar Wood Quality and Land Suitability. Project No. 96000472, Final Report.* Indian Head, Saskatchewan: PFRA Shelterbelt Centre.

Siltanen, R. M., Apps, M. J., Zoltai, S. C., Mair, R. M., and Strong, W. L. (1997). A Soil Profile and Organic Carbon Data Base for Canadian Forest and Tundra Mineral Soils. *Natural Resources Canada.* Edmonton, Alberta: Northern Forestry Centre/Canadian Forest Service.

Singh, T. (1982). Biomass Equations for ten Major Tree Species of the Prairie Provinces. *Inf. Environment Canada, Rep. NOR-X-242.* Edmonton, Alberta: Northern Forestry Centre/ Canadian Forest Service.

Singh, T. (1987). Wood density variations in thirteen Canadian tree species. *Wood Fiber Science* 19:363-369.

Singh, T. and Wheaton, E. E. (1991). Boreal forest sensitivity to global warming: implications for forest management in Western Interior Canada. *The Forestry Chronicle* 67:342-348.

Skog, K. E. and Nicholson, G. A. (1999). Carbon cycling through wood products: the role of wood and paper products in carbon sequestration. *Forest Products Journal* 48: 75-83 (1998). Cited in: Canadian Climate Change Secretariat. National Climate Change Process, National Sinks Table Foundation Paper, Final Report.

Statistics Canada (1996). 1996 Census of Agriculture CD-ROM release 2.1. Saskatoon, Saskatchewan: Data Library Services/University of Saskatchewan.

Stewart, R. B., Wheaton, E. and Spittlehouse, D. (1997). Climate change: implications for the boreal forest. Pp. 22-24. In: *Air and Water Waste Management Association Symposium*. Implications of Climate Change: What Do We Know? Conference proceedings. Calgary, Alberta.

Tillman, D. A., Rossi, A. J. and Kitto, W. D. (1981). *Wood Combustion: Principles, Processes and Economics.* Orlando, Florida: Academic Press.

Tipper, R. and De Jong, B. (1998). Quantification and regulation of carbon offsets from forestry: comparison of alternative methodologies with special reference to Chiapas, Mexico. *Commonwealth Forestry Review* 77:219-228. Cited in: Kostela, J., Nygren, P., Berninger, F., and Luukkanen, O. (2000). Implications of the Kyoto Protocol for Tropical Forest Management and Land Use: Prospects and Pitfalls. *Tropical Forestry Reports* 22. Hakapaini Oy, Helsinki: Department of Forest Ecology/ University of Helsinki.

Wang, I. C. and Micko, M. M. (1984). Wood quality of white spruce from north central Alberta. *Canadian Journal of Forest Research* 14:181-185.

Waring, R. H. and Running, S. W. (1998). *Forest Ecosystems Analysis at Multiple Scales*. Second Edition. San Diego, California: Academic Press.

Watson, R. T., Noble, I. R., Bolin, B., Ravindranath, N. H., Verardo, D. J., and Dokken, D. J. (2000). IPCC Special Report on Land Use, Land-Use Change and Forestry. *A Special Report of the Intergovernmental Panel on Climate Change*. New York, New York: Cambridge University Press.

Yanai, R. D., Arthur, M. A., Siccama, T. G., and Federer, C. A. (2000). Challenges of measuring forest floor organic matter dynamics: repeated measures from a chronosequence. *Forest Ecology and Management* 138:273-283.

In: Plantations
Editor: Hai Ren

ISBN: 978-1-62808-090-2


*Chapter 8*

# ROLES OF ARBUSCULAR MYCORRHIZAL ASSOCIATION IN PLANT NUTRITION AND GROWTH OF TROPICAL FORESTRY AND AGROFORESTRY IN DEGRADED SOIL RECLAMATION

***Nelson W. Osorio*** [*] ***and Juan D. León***
Universidad Nacional de Colombia, Colombia

## ABSTRACT

Plant productivity is severely constrained by the inadequate soil nutrient supply, particularly in highly weathered tropical soils. One of the most relevant strategies of plants to grow in these soils is the formation of arbuscular mycorrhizal association with soil fungi.

The plant supplies carbonaceous compounds to the fungus, while the fungus provides nutrients, particularly diffusion-limited ones such as phosphate (Pi), $Cu^{2+}$, and $Zn^{2+}$. Roots can absorb available Pi from distances not exceeding a few millimeters (mm) away from their surface, while mycorrhizal hyphae can extend to several centimeters (cm) from the root surface, exploring a greater volume of soil. Mycorrhizal hyphae have a higher affinity for absorbing Pi than roots. However, plant species exhibit different degrees of mycorrhizal dependency (MD) to produce maximum growth at a given level of soil nutrient availability. The MD of several tropical plant species used in forestry and agro-forestry has been determined in relation to plant root characteristics and environmental conditions.

This chapter explores the importance of the mycorrhizal association to plant nutrition and growth of seedlings in the establishment of forestry/agro-forestry ecosystems and vegetation restoration in degraded lands.

**Keywords**: Mycorrhizae, mycorrhizal dependency, soil phosphate, tropical agro-forestry, degraded lands

---

[*] Corresponding Author. Calle 59 A No. 63-20, 050034, Medellín, Colombia, Email: nwosorio@unal.edu.co and jdleon@unal.edu.co.

## INTRODUCTION

Plant productivity is severely constrained in soils from the tropical humid area due to strong soil acidity (pH <5.0), high levels of exchangeable and soluble aluminum (Al) and low nutrient bioavailability (Brady and Weil, 2002). Al ions not only cause toxicity in plants but also reduce soil microbial activity (Lindsay, 2001). Given the intense precipitation in some tropical areas (>2000 mm $yr^{-1}$), nutrients such as nitrogen, potassium, calcium, and magnesium have been leached out from the soils leaving them depleted (Uphoff et al., 2006). On the other hand, plant nutrients such as phosphate (Pi), sulfate, and boron are converted into plant-unavailable forms by soil constituents (Havlin et al., 1999). For this reason, soil soluble Pi concentration is commonly low in most tropical soils (0.001-0.01 mg $L^{-1}$) (Fox, 1979; Barber, 1995). The behavior of Pi in tropical soils has received much attention because its deficiency is quite common and this limits plant performance in agriculture and forestry ecosystems (Turner et al., 2006).

Plants exhibit different strategies to grow in Pi deficient soils that include: (i) development of an elongated root system with fine roots and abundance of root hairs; (ii) exudation of phosphatase enzymes capable of releasing Pi from organic compounds, (iii) release of organic acids that dissolve Pi compounds, and (iv) formation of a symbiotic association with soil fungi called mycorrhiza (Sylvia, 1999; Smith, 2002; Smith et al., 2003). Mycorrhiza association occurs in the roots of most plants, however, there are different kinds of mycorrhizae (Harley and Smith, 1983) and two types are commonly formed in forests and plantation: arbuscular-mycorrhiza (AM) and ectomycorrhiza (ECM) (Smith and Read 1997; Sylvia, 1999). The former mycorrhiza is formed by fungi of Glomeromycota (Schübler et al., 2001), while the latter is formed by fungi of Ascomycota and Basidiomycota. The structure, physiology and ecology are variable between these two types of associations (Smith and Read, 1997). The arbuscular mycorrhiza is globally widespread in different types of plants (herbaceous, trees, crops, etc.), while the ECM is more restrictive to conifer trees and other species of forestry plants. Although little is known about mycorrhizal symbiosis for a large number of tree species, it has been widely accepted that this association is not only important for plants and trees survival in nutrient-poor soil (Read, 1991), but also plays a key role in nutrient cycling and retention of nutrients in the soil (Medina and Cuevas, 1993; Rilling et al., 2001). The arbuscular micorrizal fungi (AMF) provide also protection against drought stress and pathogens pressure on the roots (Dehne, 1982; Harley and Smith, 1983; Langley and Hungate, 2003) in addition to impacting on the composition of plant communities (Allen et al., 1995; Van der Heijden et al., 1998; Kottke, 2002).

In this chapter, our aim is to discuss the importance of AMF as a key element in the growth and Pi nutrition status of plant species used in tropical forestry, agroforestry, and vegetation restoration programs of degraded lands.

### Abundance and Functional Aspects of Arbuscular Micorrizal Fungi in Natural Forests and Plantations

The mycorrhizal association is important for the growth of tropical trees, its dynamics in the establishment of plantations and forest restoration has been acknowledged (Janos 1980a,

1980b, 1988; Huante et al., 1993). Tropical forests rich in plant species of different families are dominated by trees capable of forming AMF, whereas temperate and boreal forests, eucalyptus forests and dipterocarp forests are dominated by a few genera or species of trees on which the ectomycorrhizal association is established. However, ECM forming trees have also been found in patches or large spaces between trees that form AM in the humid tropics (Farsi and Fontana 1962; Hogberg and Nylund, 1981; Hogberg, 1982; Onguene and Kuyper, 2001).

Perhaps one of the most important aspects of mycorrhizae in tropical ecosystems is related to nutrient recycling. Particularly, mycorrhizal associations represent for tropical tree species a key strategy to overcome the shortage of soil nutrients (Read, 1991), especially for Pi. This is one of the most serious limitations to plant development and productivity in the humid tropics. Studies on terrestrial ecosystems, have determined the abundance of spores forming mycorrhizae in soils, given their importance in the nutrient supply to the host plant in exchange for carbohydrates (Trappe, 1987).

The study of biogeochemical cycling of Pi in Andean forests of Colombia was conducted by Barreto (2005) who determined the abundance of AMF spores (Brundrett et al., 1996) in samples of the A-horizon of Andisols (Fulvudands and Hapludands, US Soil Taxonomy System; Buol et al., 1997). The aim was to determine the differences between native mature forests of oak (*Quercus humboldtii*) and mature plantation of cypress (*Cupressus lusitanica*) (> 50 years-old). Eight different AMF spore morphotypes were detected, but two were most common; higher abundance was recorded in cypress plantations (131 spores per 100 g) than in oak forests (79 spores per 100 g). This difference may be associated with a higher amount of fine roots found in the first 30-cm soil depth of cypress plantations (14.8 Mg $ha^{-1}$) than in oak forest (4.2 Mg $ha^{-1}$) (Barreto and León, 2005). Also, this may reflect the differences in nutrient supply in each ecosystem from litter decay; in cypress plantation the annual decomposition rate was k= 0.37, while in oak forests was k= 1.02 (León, 2007). In cypress plantations in the same region, Alvarado (1988) found in all collected soil samples spores of the AMF such as *Glomus* sp. and *Entrophospora* sp.; likely, the abundance of AMF spores was associated with the increase in the density of fine roots and decrease in soil acidity (Picone, 2002).

Some authors have suggested that Pi resorption just before leaf abscission may have an adaptive significance for plants in soil Pi-deficient soil (Vitousek and Sanford, 1986). In the case of cypress, León et al. (2009) determined lower Pi re-uptake (37%) than in the oak (43%). Furthermore, in accordance with Medina et al. (1990), the values of the ratio Pi: N in mature leaves (0.03) indicated a low availability of soil Pi to oak, while the cypress would be slightly above the threshold value (0.05). Hence, the difference in AMF spores may be associated with Pi recycling and availability in these ecosystems.

Reports on abundance of AMF spores from different tropical forests are numerous. For instance, in forests of Nicaragua and Costa Rica, Picone (2002) found 110-770 AMF spores per 100 g of dry soil, while in tropical rainforests of Cameroon, Musoko et al. (1994) found approximately 250 AMF spores per 100 g. Wilson et al. (1992) found in Ivorian forests 100-300 AMF spores per 100 g of dry soil. Louis and Lim (1987) reported between 100 and 500 spores per 100 g in Singapore forests. Allen et al. (1998) reported 100-280 and 30-90 AMF spores per 100 g of soil in red and yellow tropical deciduous forest of Mexico, respectively.

On the other hand, plantations have also been used to remediate degraded soils; one of the plant species most used for this purpose has been *Acacia mangium*, a fast growing tree, native from Australia, widely planted in Asia and Africa (Lee and Nguyen, 1991; Ren and Yu, 2008). An advantage of *A. mangium* is its tolerance to degraded soils conditions and capability to establish association with soil microorganisms such as nitrogen-fixing bacteria (*Bradyrhizobium* group) (Galiana et al., 1994), AMF (Ba et al., 1996), and ECM (Anino 1992; Duponnois and Ba, 1999).

Local experience with plantations of *A. mangium* (11 years old) established on degraded land by alluvial mining in Colombia (León et al., 2008) allow us to determine the presence of AMF spores in its rhizosphere (19000 AMF spores per 100 g of soil) and AM mycelia inside the roots. This value is comparable to that reported by Johnson and Wedin (1997) for humid forests of Costa Rica (12000 spores/100 g). Biogeochemical cycling studies in these plantations (Castellanos-Barliza and León, 2010; León et al., 2010) indicated an extreme shortage of soil Pi supply characterized by low soil Bray II-Pi (<3 mg $kg^{-1}$), foliar Pi content (0.04%), high leaf litter N:Pi ratio (124) and low return of Pi in the litter fall (0.5 kg $ha^{-1}$ $yr^{-1}$) (Swift et al., 1979; Aerts, 1997). The results indicated abundance of AMF spore, despite records of high degree of human disturbance of that soil (Stahl et al., 1988; Jaramillo et al., 2004). The development of mycorrhizae observed was a clear indication that the plant-mycorrhizal fungus association was formed in plantations of *A. mangium* as well as other associated plants such as *Vismia* sp., *Piper* sp. and *Ceiba pentandra* and several grass species.

Despite the abundance, there was low diversity with a clear dominance of a morphotype that later was tentatively identified as *G. microaggregatum* (Medina et al., 1999). Likely, its dominance is due to its adaptability to unfavorable soil conditions (Taylor, 2000), such as poor soil structure, extreme acidity, and Hg presence. It needs to be determined whether the presence of AMF spores is the result of its ability to persist despite the mining activities or due to invasion of mycorrhizal propagules (e.g., spores and hyphae) from surrounding land by runnoff. Indeed, field observations at these sites suggest that both processes may occur (León et al., 2008). In any case, the presence of these spores and the mycorrhizal colonization clearly indicate the ability of these fungi to infect *A. mangium* roots and presumably promote plant growth (as shown later in this chapter based on the results of Daza and Osorio, 2011). Presumably, the benefits to plant growth and performance of *A. mangium* in these soils are also the result of $N_2$-fixing bacteria found associated with the roots, which offers a competitive advantage to the plant in those environments (Andrade et al., 2000), and facilitating nutrient recycling process towards the support of ecosystem productivity (fine litter fall 3.10 t $ha^{-1}$ $yr^{-1}$) (Castellanos-Barliza and León, 2011).

## Soil Phosphate Deficiency as a Constraint for Plant Growth

Soil Pi is constituted by organic and inorganic compounds of variable nature and complexity (Havlin et al., 1999). Soil inorganic Pi can be in different pools that are at chemical equilibria: (i) primary minerals mainly apatites [$Ca_5(PO_4)(OH,F)$]; (ii) secondary minerals forming precipitates of iron-, aluminum-, and calcium-phosphates; (iii) strongly or weakly adsorbed Pi on the surface of soil minerals, which depends on the type mineral; (iv) soluble in the soil solution as $H_2PO_4^-$ in acid soils or $HPO_4^{2-}$ in alkaline soils. When plant roots and soil microbes absorb Pi from the soil solution, this breaks the equilibria among these Pi-pools and triggers a series of dissolving or desorbing reactions from the solid-phase that slowly replenishes the concentration of soluble Pi in the soil solution (Lindsay, 2001). On the other hand, soil organic Pi is forming compounds of variable nature (some of them unknown) that are decomposed by plant and microbial enzymes (generically called *phosphatases*) that breaks ester-links and slowly release Pi ions into the soil solution (Stevenson, 1986).

The low bioavailability of Pi in tropical soils is caused by the chemical stability of soil Pi compounds and the strength which Pi is retained on minerals surfaces. As a consequence, most of Pi in these soils is unavailable and only a little portion called labile-Pi ($\leq 1.0\%$ of the total-Pi) can be easily absorbed by plant roots (Table 1).

This is particularly evident in highly weathered soils of the tropics (Oxisols and Ultisols) and in volcanic-ash soils (Andisols) (Sanchez and Uehara, 1980; Parfitt, 1989; Shoji et al., 1993). Sanchez and Logan (1992) estimated that 1018 billion hectareas (ha) in the tropical zone have high Pi retention capacity; 659 million are in the Neotropics. In agroecosystems, the only significant source of Pi to plants roots is from the addition of Pi inputs (fertilizers or organic amendments).

The long-term production without the application of Pi progressively will reduce the soil Pi availability and exhaust the natural reserves in the soil (Smeck, 1985; Clark and Paul, 1989; Barber, 1995; Havlin et al., 1999). Soil Pi losses are represented by the removal in cropping products and erosion, leaching is negligible in most soils (Smeck, 1985).

The immediate source of Pi for plant roots is the soil solution, which usually contains a low soluble Pi concentration (Pi: 0.001-0.01 mg $L^{-1}$) (Fox, 1979; Barber, 1995). When Pi is removed from the soil solution by roots, soil microbes and/or by Pi fixation reactions, a gradient of Pi concentration is created between the solid phase and the soil solution around the roots. Adsorbed Pi on the soil solid surfaces must desorb in order to replenish Pi in the soil solution (Do Carmo Harta and Torrent, 2007).

Hence, Pi diffuses from the solid phase, where it is more concentrated, to the soil solution around the root surface where its concentration is continuously being depleted. However, the rate of Pi diffusion is quite slow ($10^{-12}$-$10^{-15}$ $m^2$ $s^{-1}$) (Schachtman et al., 1998) which limits the Pi supply and creates a depletion zone of Pi around the roots of approximately 1-2 mm (Barber, 1995). Pi ions beyond the zone of depletion cannot be accessed by the root surface (Barber, 1995).

Phosphate ions that slowly diffuse into the soil solution originate mainly from Pi weakly adsorbed on soil colloids and from those freshly precipitated (Sanchez, 1976; Stevenson, 1986; Lindsay, 2001).

Soil Pi supply depends on the Pi buffering capacity of soils (Pypers et al., 2006), which can be estimated from the relationship between the concentration of Pi in soil solution and the quantity of labile Pi in the solid phase (Barber, 1995; Holford, 1997).

Soil Pi sorption capacity and soluble Pi fertilizer requirement can be estimated through the use of isotherm of sorption (Hue and Fox, 2010). Briefly, this consisted of the use of a little aliquot of soil (2-3 g) that received grading amounts of Pi (0-2000 mg $kg^{-1}$) in plastic centrifuge tubes. The most common source of Pi for this procedure is $KH_2PO_4$ dissolved in 0.01 M $CaCl_2$.

Then, the tubes are frequently shaken during six days at room temperature (25°C). After that, the solution is separated by centrifugation 5000 rpm (2500xg) for 15 min and the soluble Pi concentration measured by the molybdate-phosphate method (Murphy and Riley, 1962). The data results are presented as a figure and regression model fitted the relationship between adsorbed and soluble Pi (Figure 1).

**Table 1. Fractionation of inorganic soil phosphate (Pi) compounds in tropical soils of Colombia (COL) and Hawai'i (HI) (USA). Source: Osorio (2012)**

| Soil | Labile-Pi[a] (%) | Al-Pi (%) | Fe-Pi (%) | Ca-Pi (%) |
|---|---|---|---|---|
| *Lowly weathered:* | | | | |
| Vertisol (Lualualei,HI) | 0.5 | 16.7 | 27.7 | 55.0 |
| Mollisol (Neira, COL) | 0.6 | 35.4 | 11.1 | 52.9 |
| | | | | |
| *Highly weathered:* | | | | |
| Oxisol (Molokai, HI) | 0.03 | 32.6 | 58.2 | 9.1 |
| Oxisol (Wahiawa, HI) | 0.5 | 24.0 | 67.8 | 7.7 |
| Oxisol (Carimagua, COL) | 0.3 | 10.3 | 87.6 | 1.8 |
| Ultisol (Caucasia, COL) | 0.2 | 8.8 | 90.9 | 0.0 |
| | | | | |
| *Volcanic ash soils:* | | | | |
| Andisol (Guarne, COL) | 0.4 | 71.3 | 2.4 | 25.9 |
| Andisol (La Selva, COL) | 0.4 | 71.0 | 9.2 | 19.4 |
| Andisol (Naranjal, COL) | 1.0 | 79.7 | 7.5 | 11.8 |
| Andisol (Tantalus, HI) | 0.3 | 16.5 | 69.0 | 14.2 |

Fractionation of mineral was made following the procedure of Kuo (1996). [a] Labile-Pi = soluble and weakly adsorbed.

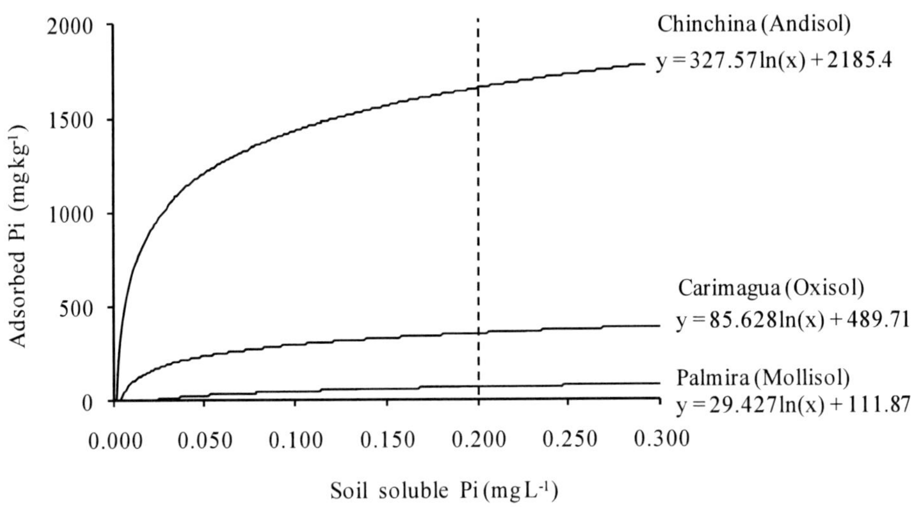


Source: Osorio (2012).

Figure 1. Pi sorption isotherms of three tropical soils from Colombia. The dashed line-intercept indicates the amount required in each soil to achieve a soil soluble Pi concentration of 0.2 mg $L^{-1}$.

## Arbuscular Mycorrhizal Association as an Alternative to Enhance Plant Growth in Tropical Soils

Arbuscular mycorrhizal fungi (AMF) form fine hyphae that grow inside and around the cells of the root cortex and form arbuscules. These are fine hyphae highly branched and exchange materials between the plant host and the fungus (Figure 2). The plant supplies carbonaceous compounds to the fungus, while the fungus provides nutrients, particularly diffusion-limited ones such as Pi, $Cu^{2+}$, and $Zn^{2+}$ (Lynch and Ho, 2005; Hamel, 2004; Habte and Manjunath, 1991; Barber, 1995; Marschner, 1995).

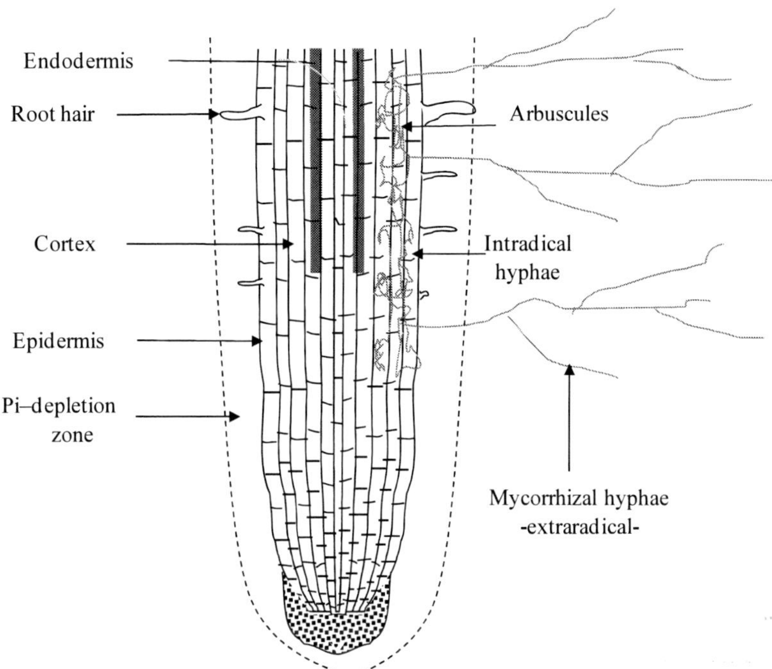


Source: Osorio (2012).

Figure 2. Diagram showing on the right side the arbuscular mycorrhizal association in the root tissues and the mycorrhiza-free tissue on the left.

AMF only can absorb nutrients from the soil solution and then transport them to the roots to which they associate (Gerdemann, 1975; Tinker and Gildon, 1983; Bolan, 1991; Habte and Fox, 1993; Newsham et al., 1994; Cardoso et al., 2006). The most appreciated effect of AMF on plant performance is the enhancement of Pi, Cu, and Zn uptake (Pacovsky, 1986; Manjunath and Habte, 1988). This is particularly relevant in plant species whose root system is morphological and/or physiologically inefficient in plant Pi uptake due to lack or scarcity of root hairs (Koide and Schreiner, 1992; Manjunath and Habte, 1991). For this reason, the influence of AMF on plant growth is associated in most cases to the amelioration of the Pi-nutritional status (Bolan, 1991) as well as the balance uptake of other nutrients (Cu, Zn, and Mn).

AMF increase plant Pi uptake by several mechanisms that include enhanced rhizosphere microbial activity (Bago and Azcon-Aguilar, 1997), increases in root phosphatase activity (Gavito and Miller, 1998; Radersma and Grierson, 2004; McCully, 1999; Hetrick, 1991), excretion of chelating agents (Gohre and Paszkowski, 2006), and particularly a higher exploration of the soil (Jacobsen et al., 1992, 1994). In Pi-deficient soils the plant nutrient requirement exceeds the rate at which Pi diffuses to the root surface, which produces a Pi depletion zone of 1-2 mm that surrounds the root system. On the other hand, AMF can vigorous extend their perennating hyphae up to 10–12 cm from the root surface, which is beyond of the Pi depletion zone. In this way, the mycorrhizal hyphae explore a greater volume of soil for P uptake than the root alone (Hattingh et al., 1973; Mosse, 1981; Hayman, 1983; Jacobsen et al., 1994; Jeffries et al., 2003). Some authors have estimated that the AMF hyphae can explore 100-1000 more soil volume than the mycorrhizal-free roots (Sieverding, 1991). Also, the fact that the hyphae have a small diameter than roots can penetrate through soil micropores that are not available for root systems (Bolan, 1991).

In addition, mycorrhizal hyphae have a higher affinity for absorbing Pi than roots (Howeler et al., 1981). Schachtman et al. (1998) reported that the hyphae of *Gigaspora margarita* had an affinity constant for Pi (Km) of 2.5 μM (Pi: 0.077 mg $L^{-1}$), while many plants usually exhibited a Km of 6-44 μM (Pi: 0.19-1.36 mg $L^{-1}$), particularly those highly dependent on mycorrhizal association (Nye and Tinker, 1977; Barber, 1995). In addition, Barber (1995) affirmed that due to the small radius of the mycorrhizal hypha (1-3 μm) there is no Pi depletion zone around the hypha.

This allows the mycorrhizal hypha to take up Pi more effectively due to a higher and more constant Pi concentration. However, Li et al. (1991) found a very narrow Pi depletion zone around the mycorrhizal hypha. In contrast, roots with a greater radius (150 μm) generate a zone of depletion of at least 1 mm, resulting in low Pi concentration around the root surface. Smith and Read (1997) reported P influx in mycorrhizal roots of 3 to 5-fold higher than non-mycorrhizal roots (10-11 mol $m^{-1}$ $s^{-1}$).

For the reasons mentioned above, the mycorrhizal association can reduce the requirement of soluble Pi concentration of plants. In this way the rate of Pi fertilizers can be significantly reduced (up to 40 times) as reported by Habte and Osorio (2001) for *Leucaena leucocephala*. This represents not only an economical advantage but also an environmental benefit because soil with high levels of soluble Pi can pollute surface waterways when soil particles are eroded and deposited in them causing eutrophication (Culley et al., 1983; Sharpley et al., 1992).

When plants rely on the mycorrhizal association and low Pi fertilization rather than heavy Pi fertilization, the risks of water pollution are reduced (Miyasaka and Habte, 2001). Therefore, AMF can be included in soil nutrient management programs oriented to reduce environmental pollution (Habte and Osorio, 2001).

AMF have also been studied for their contribution to plant health because they can reduce the negative impact of plant pathogens such as *Phytophtora* spp., *Pythium* spp., *Fusarium* spp. and nematodes, among others (Linderman, 1988; Cooper and Grandison, 1986; Hooker et al., 1994; Newsham et al., 1994; Trotta et al., 1996).

Similarly, AMF can improve the stability of soil aggregates, which is highly relevant for the bioremediation of degraded soils (eroded, compacted, and mined) (Bethlenfalvay, 1992; Wright and Upadhyaya, 1996, 1998; Rilling et al., 2001). It has been demonstrated that AMF can also improve plant tolerance to water stress, salinity, soil acidity, and heavy metal toxicity (Bethlenfalvay et al., 1988).

## Mycorrhizal Dependency of Tropical Plants Used in Forestry and Agroforestry

Despite multiple benefits that plants can derive from the mycorrhizal association, not all plants depend on AMF. In fact, plant species exhibit different degrees of mycorrhizal dependency (MD) (Plenchette et al., 1983). Mycorrhizal dependency is a measure of the degree to which a plant species relies on mycorrhizal association for nutrient uptake and growth (Habte and Osorio, 2001).

The earliest definition of this concept was made by Gerdemann (1975) who stated that the dependency of plant species on mycorrhizal condition is a function of soil fertility. This definition has since been modified to make it more operational by replacing the imprecise term "soil fertility" with "soil solution Pi concentration" (Habte and Manjunath, 1991). It is well established that plant species vary in their response to AMF inoculation (Plenchette et al., 1983; Pope et al., 1983; Hetrick et al., 1988; Khalil et al., 1999). Most of the variation has been related to different ability of plant species to take up Pi at very low soil-Pi concentrations in the absence of the mycorrhizal association (Baon et al., 1993). Habte and Manjunath (1991) found that the MD of several plant species was determined by root characteristics of the host species (root length, root density, root surface area, incidence and length of hair roots), which determine their capacity to explore and absorb Pi from the soil solution. Other authors found that the seed size and seed Pi reserve also control the value of MD (Jaramillo and Osorio, 2005).

According to Plenchette et al. (1983) the MD should be calculated as the percentage of the difference between the dry matter of inoculated and uninoculated plants over the dry matter of inoculated plants:

MD = [(Inoculated plant dry matter - Uninoculated plant dry matter) ÷ Inoculated plant dry matter] x 100

On the other hand, when soil soluble Pi concentration is < 0.02 mg $L^{-1}$, most plant species will respond significantly to mycorrhizal colonization (Figure 3). As soil soluble Pi increased from 0.02 to 0.1–0.2 mg $L^{-1}$, the dependency of plants on AMF for Pi uptake diminishes progressively (Figure 4). Thus, at 0.2 mg $L^{-1}$ only very highly mycorrhizal-dependent species respond significantly to mycorrhizal colonization. Since MD is significantly affected by soil solution Pi level, Habte and Manjunath (1991) proposed that MD should be estimated at different levels of available Pi, particularly at the soil solution Pi of 0.02 mg $L^{-1}$. Based on the value of MD five categories have been proposed to classify plant species (Table 2). In Table 3 some plant species used in tropical forestry and agroforestry are presented according to their MD category. Some examples of contrasting MD are presented in Figure 3 with *Nageia rospigliosii* (romeron-pine) and *Calophyllum brasiliense* (barcino) inoculated or not with *G. aggregatum* at three levels of soil soluble Pi.

**Table 2. Categories of mycorrhizal dependence of plant species measured at a soluble Pi level of 0.02 mg L-1. Source: Habte and Manjunath (1991)**

| Category | MD (%) | Further considerations |
|---|---|---|
| Independent | 0 | Plant species do not form the mycorrhizal association |
| Marginal | 0-25 | - |
| Moderate | 25-50 | - |
| High | 50-75 | Plant dry matter does not increase by AMF inoculation at 0.2 mg $L^{-1}$ |
| Very high | >75 | Plants dry matter increases by AMF inoculation at 0.2 mg $L^{-1}$ |

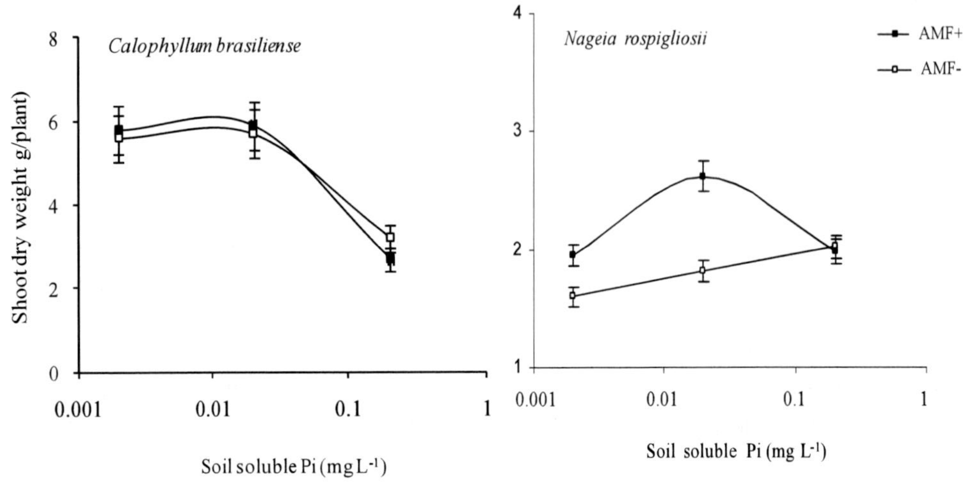


Sources: Sierra (2006) and Diez (2006).

Figure 3. Shoot dry weight of seedlings of *Calophyllum brasiliense* (independent) and *Nageia rospigliosii* (moderately dependent) as a function of *G. aggregatum* inoculation at three levels of soil soluble Pi.

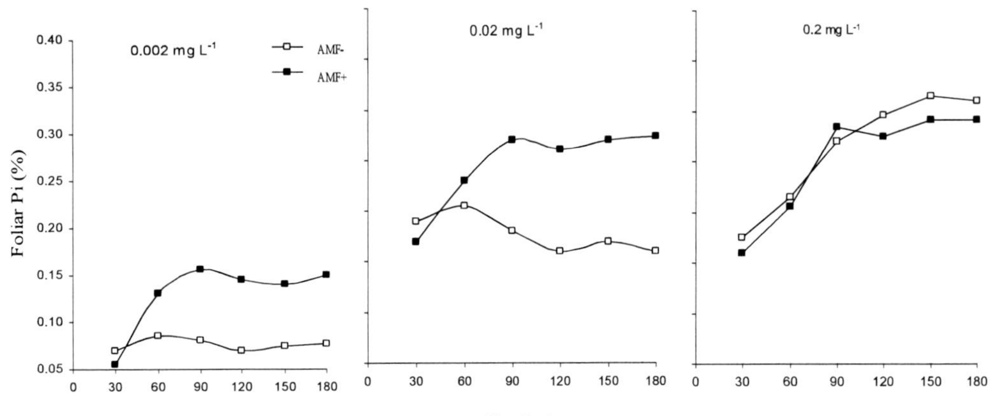


Source: Diez et al., 2008.

Figure 4. Foliar Pi concentration of *Nageia rospigliosii* either uninoculated (AMF-) or inoculated with *G. aggregatum* (AMF+) at three levels of soil solution Pi.

## Mycorrhizal Effectiveness in Tropical Soils

Soils differ highly in AMF diversity, population structure, and capacity to enhance plant performance (Kottke, 2002; Langley and Hungate, 2003). Although AMF are present in most soils, they may be absent if the soils have been degraded, sterilized, severely deforested, frequently burnt, Pi-over-fertilized, mono-cultured with non-mycorrhizal plant species, maintained for long periods of time in fallow (with non-mycorrhizal plant species) and intensively exposed to fungicides (Habte, 1989; Kabir et al., 1999). The lack of low density of AMF propagules may be a limiting factor for crop plant productivity, land restoration, and forestry if the target plant species are mycorrhizal dependent and the soil is soluble Pi-deficient (Habte and Osorio, 2001).

**Table 3. Mycorrhizal dependency of tropical plants used in forestry and agroforestry**

| MD category | Plan species | Reference |
|---|---|---|
| Independent | *Calophyllum brasiliense* Cambers | Sierra et al., 2012 |
| | *Platyciamus regnellii* Benth., *Ormosia aborea* (Vell.), *Platypodium elegans* Vog., *Macheria stipitatum* (DC), *Myroxylon peruiferum* L.f., *Hymenaea courbaril* L., *Dendropanax cuneatum* (DC), *Ceiba speciosa* (St.Hil), *Tabebuia roseo-alba* (Rid.). | Siqueira and Saggin-Junior, 2001 |
| Marginal | *Sesbania formosa, S. pachycarpa, S. sesban* *Cassia reticulata* Willd, *Chloris gayana* Kunth. | Habte and Manjunath, 1991 Habte and Turk, 1991 |
| Moderate | *Acacia mangium* Willd., *Gliricidia sepium* (Jacq.) | Miyasaka et al., 2003 |
| | *Leucaena retusa, Sesbania grandiflora* | Habte and Manjunath, 1991 |
| | *Acacia koa* (Gray) | Miyasaka et al., 1993 |
| | *Laurel* (*Ocotea* sp.) *Chamaesenna colombiana* (Britton and Killip) | Sierra et al., 2009 Sierra, 2006 |
| | *Nageia rospigliosii* (Pilger) | Diez et al., 2008 |
| High | *Sophora crhysophylla* | Miyasaka et al., 1993 |
| | *Cassia siamea* | Habte, 1995 |
| | *Leucaena diversifolia, Leucaena trichodes* | Habte and Manjunath, 1991 |
| | *Sesbania tormentosa* | Gemma et al., 2002 |
| | *Albizia ferruginea, Allium* cepa L., *Azadirachta indica* A. Juss, *Cajanus cajan, Enterolobium cyclocarpum* Jcq., *Paraserianthes falcataria* (L.), *Sauropus androgynus* (L.) | Habte, unpublished |
| | *Aspidosperma parvifolium* A.DC., *Solanum granuloso, Lithraea molleoides* (Vell.), *Trema micrantha* (L.) | Siqueira and Saggin-Junior, 2001 |
| Very high | *Luehea grandiflora* Mart., *Senna spectabilis* (A.DC.), *Croton floribundus* Spreng, *Tibouchina granulosa* Cogn., *Cecropia pachystachya* Trec., *Cordia trichotoma* (Vell.), *Senna macranthera* (Collad), *Cedrela fissilis* Vell., *Caesalpinia ferrea* Mart., *Myrsine umbellata* Mart., *Tabebuia impetiginosa* (Mart.), *Sapindus saponaria* L., *Tabebuia serratifolia* (Vahl), *Copaifera lagsdorffii* Desf. | Siqueira and Saggin-Junior, 2001 |
| | *Leucaena leucocephala* | Habte and Manjunath, 1991 |
| | *Bidens sandvicencis* | Gemma et al., 2002 |

Knowing the effectiveness of AMF in a soil is important because the extent to which plant species respond to inoculation of soils with known AMF fungi will depend, among other factors, on the host species, the Pi status of the soil, and the infectivity and effectiveness of indigenous AMF populations (Habte and Osorio, 2001). A reliable method of determining the effectiveness of indigenous AMF fungi contributes significantly to the use of soil as well as known AMF inocula with predictable outcomes (Habte et al., 1987).

Commonly the presence of mycorrhizal propagules in soils is quantified by the number of AMF-spores, which is unfortunately not completely reliable (Osorio and Habte, 2001). AMF spores can be unviable or exhibit a period of latency (Abbot and Robson, 1984; Allen et al., 1995). For this reason, the presence of AMF spores is not a good indicator of AMF effectiveness (Brundrett et al., 1996). The mycorrhizal effectiveness is directly related with the amount of infective mycorrhizal propagules, not only viable spores but also extraradical hypahe, and infected root fragments present in the soil (Brundrett et al., 1996).

The method used to measure mycorrhizal effectivenes of indigenous soil AMF consists of the inoculation with aliquots of the soil (25 g) into a substrate conducive for optimal mycorrhizal activity (soluble Pi level= 0.02 mg $L^{-1}$, pH 6.0), which is then planted with seedlings of a plant species that exhibit high mycorrhizal dependency (e.g., *Leucaena leucocephala*). Since plant growth and Pi uptake consistently increase as a result of AMF infection, one of the best ways to determine the symbiotic effectiveness of soil AMF is to monitor the Pi status of host plants as the symbiosis develops (Habte and Osorio, 2001). To this aim the pinnule technique developed by Habte et al. (1987) is a rapid, nondestructive, and precise technique for monitoring the development of symbiotic effectiveness in the arbuscular mycorrhizal association.

In the next paragraphs we described two experiments carried out to determine mycorrhizal effectiveness of indigenous AMF in (i) soils degraded by alluvial mining and intensive cropping, and (ii) soils subjected to different uses such as forestry, fallow, crops and grassland.

## Mycorrhizal Effectiveness of Degraded Soils

This investigation was conducted under greenhouse conditions at the Universidad Nacional de Colombia (6° 15′ N, 75° 35′ W and 1495 m altitude). The substrate was a B-horizon from clay-loam Ultisol which was air-dried, sieved at 4 mm, autoclaved (120°C, 0.1 MPa, 2h), and then limed with $CaCO_3$ to reach a soil pH of 6.0. Later the soil fertilized with $KH_2PO_4$ (350 mg of P $kg^{-1}$) to obtain soil soluble Pi concentration of 0.02 mg $L^{-1}$, which is optimal for mycorrhizal activity. To this purpose a Pi sorption isotherm was previously conducted as explained before (Fox and Kamprath, 1970).

Then, 600g of the substrate were transferred into plastic pots and later inoculated separately with either 15 g of a degraded soil by alluvial mining, 15 g of a degraded soil from an oil-palm plantation (*Elaeis guineensis*, 30 year-old), or 15 g of an effective mycorrhizal inoculum composed by 17 infective propagules of *Glomus aggregatum* per g (Table 4). An uninoculated control was included for comparison purposes. After that, the substrate was planted with germinated seeds of *Leucaena leucocephala*, previously scarified with $H_2SO_4$ for 20 min. Plants were grown for 49 days and frequently watered to maintain 60% of the maximal water holding capacity. A Pi-free Hoagland solution was added weekly.

Foliar Pi content was monitored as a function of time from the fourth pinnule counting from the base in youngest pinna fully expanded as explained by Habte et al. (1987), Aziz and Habte (1987), and Habte and Fox (1993) (Figure 5). Mycorrhizal colonization of plant roots were measured by the grid-line method (Giovannetti and Mosse, 1980) after clearing with KOH (Phillips and Hayman, 1970) and staining with fucsin acid (Kormanik et al., 1980). The experimental design was completely randomized with three treatments, each treatment had four replicates. Data were subjected to ANOVA and Duncan tests ($P \leq 0.05$).

The results indicated that leucaena plants grown in the substrate inoculated with the degraded soils did not develop the mycorrhizal association. Consequently, they exhibited poor growth accompanied with visual symptoms of Pi deficiency, i.e. chlorosis, defoliation (Smith et al., 1992), which was also observed in uninoculated control plants. By contrast, plants inoculated with *G. aggregatum* exhibited a high mycorrhizal colonization (74%) and as a result vigorous growth and pinnule Pi contents significantly higher after the 28$^{th}$ day (Figure 6). At harvest (49$^{th}$ day), uninoculated control plants and those inoculated with the degraded soils has a mean pinnule Pi content of 1.5 μg/pinnule, while those inoculated with *G. aggregatum* had 8.9 μg/pinnule that was significantly higher ($P \leq 0.001$) and represents an increase of 6-fold.

The shoot dry weight and Pi content of *Leucaena* grown in the substrate inoculated with *G. aggregatum* were significantly higher (by 1.96 and 6.40-fold, respectively) than those plants in either uninoculated or inoculated with degraded soils (Table 5). The latter treatments did not have significant differences among them.

**Table 4. Soil and crude inoculum of *G. aggregatum* used as source of indigenous mycorrhizal inoculums**

| Inoculum source | Mycorrhizal propagules/g[a] | pH (w, 1:1) | SOM[b] (g kg$^{-1}$) | Ca[c] | Mg | K | Soluble Pi[d] (mg L$^{-1}$) |
|---|---|---|---|---|---|---|---|
| | | | | (cmol$_c$ kg$^{-1}$) | | | |
| Oil palm plantation-soil | 0 | 4.6 | 50 | 1.0 | 0.3 | 0.16 | 0.011 |
| Alluvial mined-soil | 2 | 5.1 | 14 | 1.1 | 0.5 | 0.11 | 0.008 |
| *G. aggregatum* | 17 | 6.2 | 20 | 2.4 | 1.0 | 0.25 | 0.020 |

[a] determined by the most probable number technique (Porter, 1979); [b] Soil organic matter determined by Walkley and Black method; [c] Ca, Mg, and K were extracted by 1 *M* ammonium acetate; [d] determined in 0.01 *M* $CaCl_2$ soil extract 1:10.

Figure 5. View of the youngest –mature pinna of a seedling of leucaena, the arrow shows the location of the pinnule to be collected. Original photo.

**Table 5. Leucaena shoot dry weight and shoot Pi content as a function of inoculation with two degraded soils and *G. aggregatum*. Means with different letter are significant different (Duncan test, $P \leq 0.05$). Source: Jaramillo et al. (2004)**

| Treatment | Shoot dry weight (g/plant) | Shoot Pi content (mg/plant) |
|---|---|---|
| Uninoculated | 0.59 c | 0.43 b |
| Oil palm plantation-soil | 0.60 c | 0.60 b |
| Alluvial mined-soil | 0.66 b | 0.73 b |
| *G. aggregatum* | 1.20 a | 3.76 a |

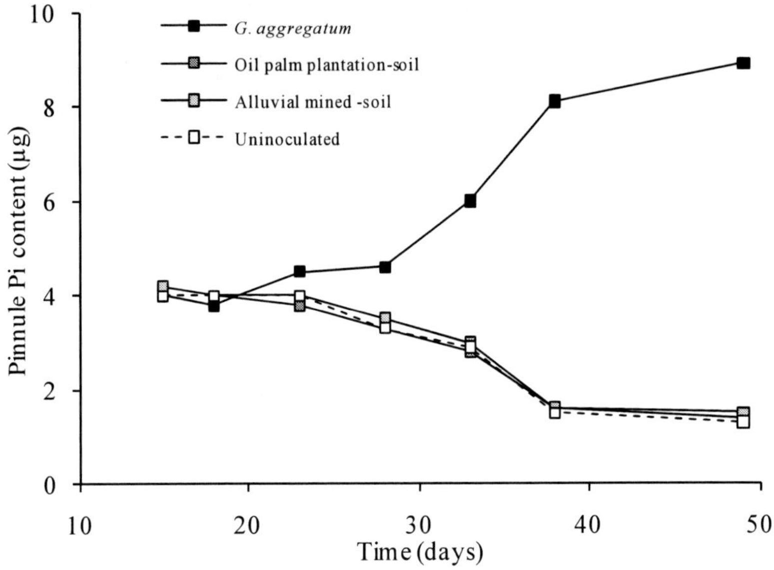


Source: Jaramillo et al. (2004).

Figure 6. Leucaena pinnule Pi content as a function of inoculation with two degraded soils and *G. aggregatum* over time after transplanting. Each point is the mean of four replicates.

## Mycorrhizal Effectiveness of Soils with Different Uses (Forestry Plantations, Fallow, Crops, and Grassland)

A procedure similar to above mentioned was used. In this case the soils used as source of indigenous AMF were collected from the A horizon (0-20 cm) under different plant coverage in the region of the Eastern of Antioquia (Table 6). Soil samples were identified according to plant coverage (CP= conifer plantation; BS= secondary forest; FL: fern-land; HF= high fallow; CB= common bean; PT= potato crop; GL= grassland of kikuyu grass).

Mycorrhizal inoculation was carried out separately with 20 g of each soil samples per pot (900 g) and were mixed thoroughly with the substrate. A crude inoculum of *Glomus aggregatum* (20 g/pot) was used as a positive control, as well as an uninoculated negative control that received washing of the crude inoculum filtered by Whatman No.1 filter paper. The number of indigenous mycorrhizal infective propagules in each soil was detected following the procedure proposed by Porter (1979).

**Table 6. Location, altitude, dominant plant species, and soil chemical properties of the soils used as source of indigenous AMF**

| Soil/coverage | Location | Altitude (m) | Dominant plant species | pH w,1:1 | SOM[a] (%) | Al[b] | Ca[c] | Mg | K | Pi[d] | Fe[e] |
|---|---|---|---|---|---|---|---|---|---|---|---|
| | | | | | | ----------$cmol_c$ $kg^{-1}$---------- | | | | mg $kg^{-1}$ | |
| Conifer plantation (CP) | 6°15′38′′N, 75°27′19′′W | 2390 | *Cupressus lusitanica, Pinus patula* | 4.4 | 28 | 5.7 | 0.2 | 0.3 | 0.41 | 3 | 139 |
| Secondary forest (SF) | 6°06′41′′N, 75°32′46′′W | 2500 | *Hedyosmum bomplandianum, Billia rosea, Clethra fagifolia, Viburnum anabaptista, Alchornea verticillata, Schefflera* sp., *Hedyosmum* sp., *Turpinia heterophylla, Miconia* sp., *Myrsine coriaceae* | 3.2 | 52 | 11.3 | 0.4 | 0.8 | 0.75 | 23 | 399 |
| Fern-land (FL) | 6°16′04′′N, 75°27′53′′W | 2400 | *Pteridium aquilinum* (fern) | 4.4 | 37 | 10.2 | 1.3 | 0.5 | 0.52 | 7 | 247 |
| High fallow (HF) | 6°06′41′′N, 75°32′46′′W | 2500 | *Miconia* sp., *Brunellia subsessilis, Solanum* sp., *Palicourea angustifolia, Tibouchina lepidota, Meriania nobilis. Cecropia angustifolia, Vismia* sp., *Clusia* sp. | 3.3 | 68 | 15.0 | 0.4 | 1.1 | 0.93 | 40 | 235 |
| Common bean (CB) | 6°11′43′′N, 75°22′05′′W | 2135 | *Phaseolus vulgaris* | 5.7 | 15 | - | 11.7 | 0.9 | 1.32 | 10 | 66 |
| Potato crop (PT) | 6°11′43′′N, 75°22′05′′W | 2135 | *Solanum tuberosum* | 5.8 | 12 | - | 13.8 | 0.6 | 0.33 | 11 | 73 |
| Kikuyu grass (GL) | 6°11′43′′N, 75°22′05′′W | 2135 | *Pennisetum clandestinum* | 5.9 | 14 | - | 11.7 | 1.5 | 0.38 | 1 | 65 |

Source: Sierra (2006).

[a] Soil organic matter contente determined by Walkley and Black method; [b] Al extracted by 1 *M* KCl; [c] Ca, Mg, and K were extracted by 1 *M* ammonium acetate; [d] Pi extracted by the Bray-II method; [e] extracted by Olsen-EDTA.

*Leucaena leucocephala* was used as indicator plant and was growing for 90 days under greenhouse conditions and treated as explained before. Similar variables (pinnule-Pi content, shoot dry weight, and mycorrhizal colonization) were considered in this study and data analyzed in the same way.

Pinnule Pi content of *Leucaena* was significantly affected by treatments, in uninoculated substrates this was decreasing over time until the level of 0.63 μg of Pi per pinnule at harvest (Figure 7). Similar behavior was observed in plants inoculated with soils taken from the conifer plantation, the secondary forest, the fern-land, and the high fallow and their pinnule Pi contents were statistically comparable to that of uninoculated plants.

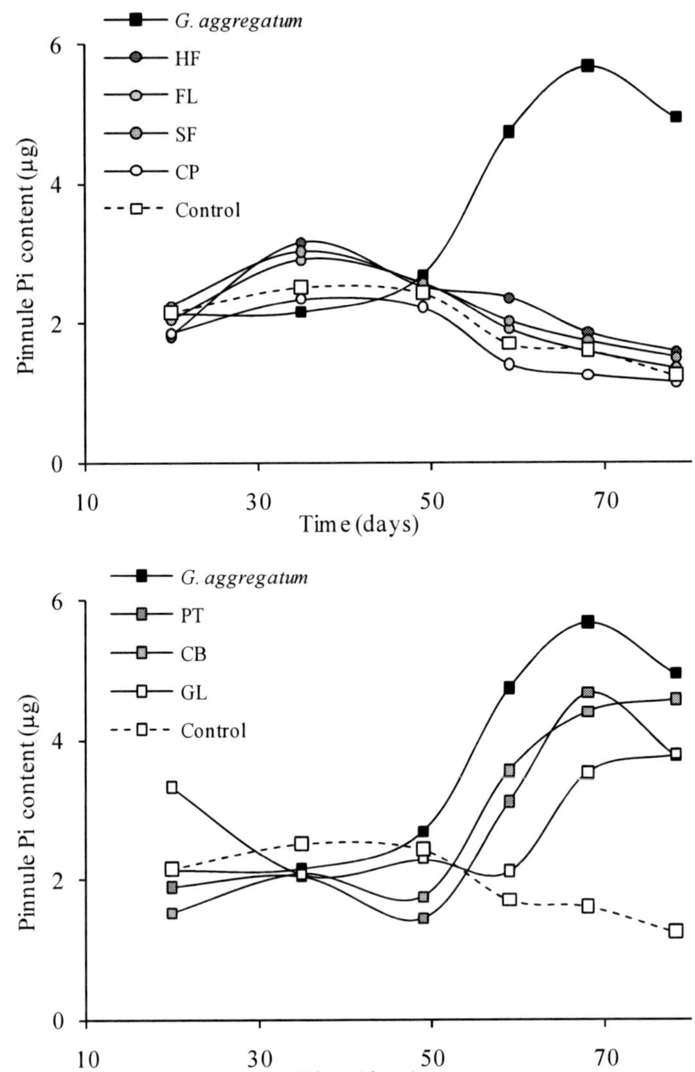


Source: Osorio et al. (2008).

Figure 7. Leucaena pinnule Pi content as a function of inoculation with soils with different plant coverage and variable number of propagules of indigenous AMF over time.

These results suggest an ineffectiveness of the indigenous AMF population in these soils. The most probable number technique allow us to detect that these soils had a low number of AMF propagules (Table 6), which fluctuate between 100 to 650 per kg of soil, which developed a low AMF colonization in the roots (3%).

Consistently, the plant growth was significantly constrained by the low infection in the roots; the values of shoot dry weights were statistically similar to those of the uninoculated plants (Table 6). By contrast, a higher mycorrhizal effectiveness was found in the soils with crops and grass as illustrated in Figure 4. The behavior of pinnule Pi content with these treatments followed overtime the pattern of plants inoculated with G. *aggregatum*. This was associated to high number of propagules of indigenous AMF (3950-27000 per g of soil), higher values of mycorrizal colonization (10-25%), and, consequently, higher shoot dry weights; 80-94 % of the relative dry growth (Table 7).

The low amount of indigenous AMF propagules is consistent with soils with extreme acidity (pH<5.0), high levels of exchangeable Al (> 2 cmlc $kg^{-1}$), and low content of nutrients, particularly Ca and Mg (Table 6). Such conditions may impair plant performance and, consequently, reduce the mycorrhizal activity (infectivity, effectiveness, spore production) or perhaps it has direct effects on these fungi (Hayman 1982; Sieverding, 1991; Soedarjo and Habte, 1995).

Conversely, the high mycorrhizal effectiveness found in the soils cultivated with crops and grass was observed with higher values of soil pH (5.7-5.9), lack of Al, and high levels of Ca and Mg. These conditions are the outcome of historic liming and fertilization programs in these soils. Likely, the enhanced soil fertility has improved plant performance and thus favoring the nutrition, multiplication and activity of AMF. Note that the soil cultivated with potato crops and common bean had even more AMF propagules than the inoculum of *G. aggregatum*.

**Table 7. Leucaena shoot dry weight (SDW) and its mycorrhizal colonization as a function of inoculation with soils containing different number of propagules of indigenous AMF. Source: Sierra (2006)**

| Soil/coverage | Number of propagules of indigenous AMF $kg^{-1}$ | Mycorrhizal colonization (%) | SDW (g/plant) | Relative SDW (%) |
|---|---|---|---|---|
| Control | 0 | 0 | 0.58 f | 39 |
| SF | 100 | 3 | 0.93 de | 63 |
| CP | 100 | 3 | 0.79 e | 53 |
| HF | 200 | 3 | 1.07 cd | 73 |
| FL | 650 | 3 | 0.78 e | 53 |
| GL | 3950 | 10 | 1.39 ab | 94 |
| PT | 24500 | 19 | 1.18 bc | 80 |
| CB | 27000 | 25 | 1.39 ab | 94 |
| G. *aggregatum* | 8400 | 24 | 1.48 a | 100 |

## Plant Response to Mycorrhizal Inoculation in Soils with Variable Number of Indigenous AMF

The results obtained in the previous section encouraged us to evaluate the plant response to mycorrhizal inoculation in soils that exhibited contrasting mycorrhizal effectiveness (variable number of indigenous AMF propagules). To this purpose and in separate experiments we selected soils with low and high effectiveness. In the first experiment, plastic pots were filled out with 2 kg of a degraded soil by alluvial mining from Taraza, Colombia. Previously, the soil was air-dried and sieved at 4 mm. Soil chemical properties were pH (water, 1:2) 5.9, O.M. 0.7%, Bray-Pi 13 mg kg$^{-1}$; Ca, Mg and K 8.3, 4.6, and 0.04 cmol$_c$ kg$^{-1}$, respectively.

The potted soil was amended with Huila rock phosphate at the rate of 300 mg of Pi kg$^{-1}$ of soil. Later, the soil was either uninoculated or inoculated with an AMF inoculum that contained 500 spores of *G. microaggregatum*, an indigenous fungus found in this soil that was multiplied for this study in corn and brachiaria-grass as host plants. The soil was planted with germinated seeds of *L. leucocephala*, our indicator plant of reference. Plants were grown for 90 days under greenhouse conditions, watered if required to maintain 50-60% of the maximal water holding capacity.

At the end of the period of growth the plant height, shoot and root dry weight, and shoot Pi content were measured as explained above. Data were subjected to ANOVA and LSD for mean separation ($P \leq 0.05$).

The results indicated that the mycorrhizal inoculation significantly increased plant height by 92%, shoot and root dry weight by 120 and 156 %, respectively, and shoot Pi uptake by 5-times (Figure 8). For these reasons, AMF inoculation is a highly recommended practice to promote plant in degraded soils subjected to bioremediation projects.

A second experiment was carried out to evaluate the plant response to mycorrhizal inoculation in soils with different levels of indigenous AMF propagules: 100, 3850 and 4300 per g (Table 6 and 7) (Paternina, 2006).

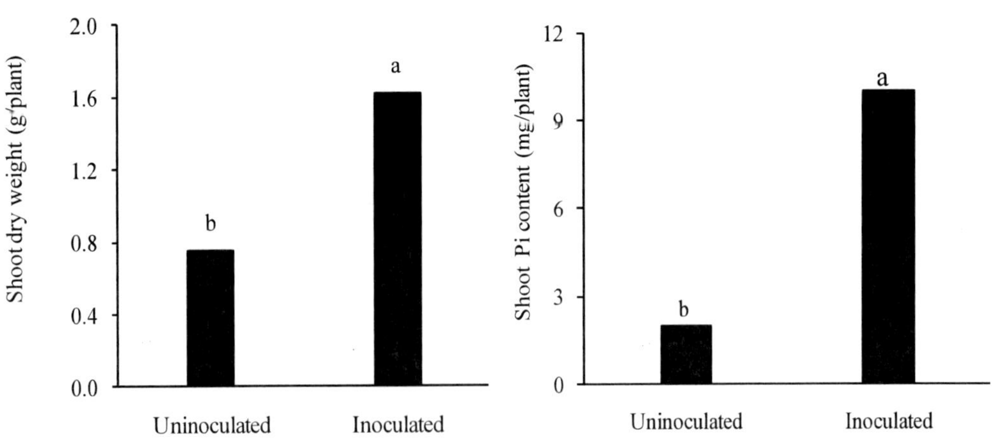


Source: Daza and Osorio (2011).

Figure 8. Response of leucaena in plant growth and Pi uptake to *G. fasciculatum* inoculation in a degraded soil by alluvial mining. Different letters on the columns indicate significant differences (LSD test, $P \leq 0.05$).

To this aim, leucaena plants were grown under greenhouse conditions in three potted soils (900 g per pot) either uninoculated or inoculated with the mycorrhizal fungus *G. fasciculatum* (25 g $kg^{-1}$) that contained 43000 AMF propagules per kg. We measured the plant response in terms of the relative plant Pi uptake and growth with and without the inoculation.

The results also showed that the magnitude of the response to the mycorrhizal inoculation decreased as the number of indigenous propagules of AMF in the soil increased (Figure 9). When the soil had 100 AMF propagules per kg the response was highly significant ($P<0.01$) and nil when the soil had 4300 propagules per kg. These results confirmed the hypothesis derived from the experiment conducted by Sierra (2006) (Figure 7 and Table 7). In this way, plant derived benefits from the AMF inoculation if the target plant is mycorrhiza-dependent, the soil had a low level of available Pi, and the soil had a low number of indigenous AMF propagules.

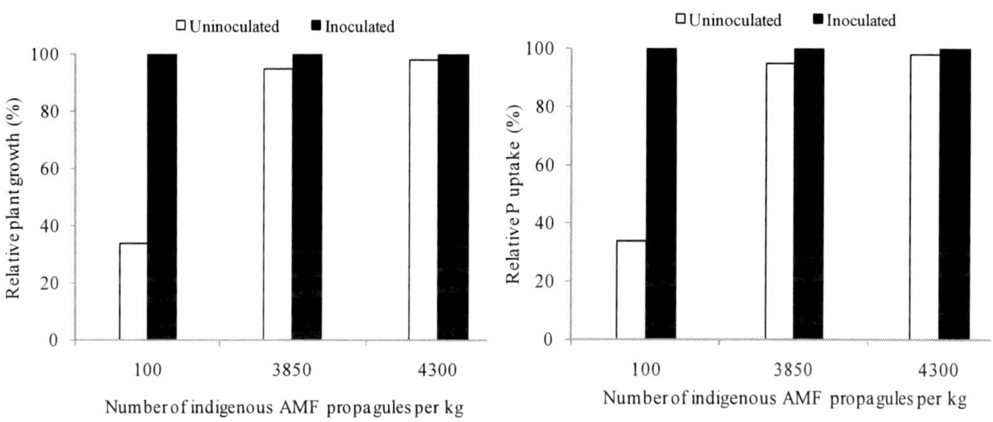


Source: Osorio et al. (2008).

Figure 9. Relative response of leucaena in plant growth and Pi uptake to *G. fasciculatum* inoculation in three soils with different levels of indigenous AMF propagules.

## Conclusion

Our study reaffirms the importance of mycorrhizal association in the establishment of plant species in tropical soils. The roles that AMF play in plant growth and Pi uptake are so relevant that their use through biofertlization is feasible and cost-effectively practice in tropical cropping, agroforestry, and forestry systems. Mycorrhiza-forming plants can grow with only a fraction of the Pi required for growth compared to mycorrhizal-free plants. Maximal benefits of AMF inoculation can be obtained with inoculum of good quality and when plant and soil conditions are favorable for AMF functioning. The use of AMF is recommendable if: (i) the target plant species exhibits mycorrhizal dependency, (ii) the soil soluble Pi level is low, and (iii) the effectiveness of soil indigenous AMF is low. The potential role of mycorrhiza at the ecosystem-level is significant; particularly in the reestablishment of soil Pi cycling in degraded soils as well as in forestry and agroforestry ecosystems.

## Acknowledgments

We thank the Universidad Nacional de Colombia for financial support of the Project "Restoration of lands in a process of desertification with Neem (*Azadirachta indica*) plantations in Western Antioquia" (project No. 9388).

## References

Abbot, L.K., Robson, A.D., (1984). The effect of VA mycorrhizae in plant growth. In: C.L. Powell, and D.J. Bagyaraj (Eds.), *VA mycorrhiza* (pp.113-130). Boca Raton, FL: CRC Press.

Aerts, R., (1997). Climate, leaf litter chemistry and leaf litter decomposition in terrestrial ecosystems: a triangular relationship. *Oikos* 79, 439-449.

Allen, E.B, Allen, M.F., Helm, D.J., Trappe, J.M., Molina, R., Rincón, E., (1995). Patterns and regulation of mycorrhizal plant and fungal diversity. *Plant Soil* 170, 47-62.

Allen, E.B., Rincon, E., Allen, M.F., Perez-Jimenez, A., Huante, P., (1998). Disturbance and seasonal dynamics of mycorrhizae in a tropical deciduous forest in Mexico. *Biotropica* 30, 261-274.

Alvarado, B., (1988). Determinación de micorrizas en *Cupressus sp*. In: F.H. Orozco (Ed.), *Investigaciones sobre micorrizas en Colombia* (2nd edition, pp. 111-117). Medellín, Colombia: Sociedad Colombiana de la Ciencia del Suelo.

Andrade, A.G., Costa, G.S., Faria, S.M., (2000). Deposition and decomposition of litter of four years old *Mimosa caesalpinifolia*, *Acacia mangium* and *Acacia holosericaea* in a Planosol. *Rev. Bras.Cienc. Solo* 24, 777-785.

Anino, E.O., (1992). Natural ecto-mycorrhiza of *Acacia mangium*. *Nitrogen fixing tree Research Reports* 10, 96-97.

Aziz, T., Habte, M., (1987). Determining vesicular-arbuscular mycorrhizal effectiveness by monitoring P status of leaf disks. *Can. J. Microbiol.* 33, 1097–1101.

Ba, A.M., Dalpé, Y., Guissou, T., (1996). Les Glomales d´ *Acacia holosericea* y *Acacia mangium*. *Bois et Forests des Tropiques* 250, 5-18.

Bago, B., Azcon-Aguilar, C., (1997). Changes in the rhizosphere pH induced by arbuscular mycorrhiza formation in onion (*Allium cepa*). *Z. Pflanz. Bodenkunde* 160, 333–339.

Baon, J.B., Smith, S.E., Alston, A.M., (1993). Mycorrhizal response of barley cultivars differing in P efficiency. *Plant Soil* 157, 97–105.

Barber, S.A., (1995). *Soil nutrient bioavailability: a mechanistic approach*. New York, NY: John Wiley and Sons.

Barreto, L.H., (2005). *Estudio exploratorio del compartimento edáfico de ecosistemas forestales altoandinos de Piedras Blancas, Antioquia*. Tesis Ingeniería Forestal, Universidad Nacional de Colombia, Medellín.

Barreto, L.H., León, J.D., (2005). Masa de raíces finas en ecosistemas forestales (*Quercus humboldtii*, *Pinus patula* y *Cupressus lusitanica*) de Piedras Blancas, Antioquia. *Rev. Fac. Nac. Agron Med.* 58, 2907-2929.

Bethlenfalvay, G.J., (1992). Mycorrhiza and crop productivity. In: G.J. Bethlenfalvay, R.G. Linderman (Eds.), *Mycorrhizae in sustainable agriculture* (pp.1-27). Madison, WI: ASA, CSSA, SSSA.

Bethlenfalvay, G.J., Cantrell, I.C., Mihara, K.L., Schreiner, R.P., (1998). Relationship between soil aggregation and mycorrhiza as influenced by soil biota and nitrogen. *Biol. Fertil. Soils* 28, 356– 363.

Bolan, N.S., (1991). A critical review on the role of mycorrhizal fungi in the uptake of phosphorus by plants. *Plant Soil* 134,189–207.

Brady, N.C., Weil, R.R., (2002). *The nature and properties of soils*. Upper Saddle River, NJ: Prentice Hall.

Brundrett, M., Bougher, N., Dell, B., Grove, T., Malajczuk, N., (1996). *Working with mycorrhizas in forestry and agriculture*. Canberra, Australia: Australian Centre for International Agricultural Research.

Buol, S., Hole, F.D., McCraken, R.J., Southard, R.J., (1997). *Soil Genesis and Classification*. Ames, IA: Iowa State University Press.

Cardoso, I., Boddington, C.L., Janssen, B.H., Oenema, O., Kuyper, T., (2006). Differential acces to phosphorus pool of an Oxisol by mycorrhizal and non-mycorrhizal maize. *Commun. Soil Sci. Plan.* 37, 1537-1552.

Castellanos-Barliza, J., León, J.D., (2010). Caída de hojarasca y dinámica de nutrientes en plantaciones de *Acacia mangium* (Mimosaceae) de Antioquia, Colombia. *Acta Biol. Colomb.* 15, 289-308.

Castellanos-Barliza, J., León, J.D., (2011). Descomposición de hojarasca y liberación de nutrientes en plantaciones de *Acacia mangium* (Fabales: Mimosaceae) establecidas en suelos degradados de Colombia. *Rev. Biol. Trop.* 59, 113-128.

Clark, F.E., Paul, E.A., (1989). *Soil Microbiology and Biochemistry*. San Diego, CA: Academic Press.

Cooper, K.M., Grandison, G.S., (1986). Interaction of vesicular-arbuscular mycorrhizal fungi and root knot nematode on cultivars of tomato and white clover susceptible to *Meloidogyne hapla*. *Annals Appl. Biol.* 108, 555–565.

Culley, J.L.B., Bolton E.F., Bernyk B., (1983). Suspended solids and phosphorus loads from a clay soil; I. Plot studies. *J. Environ. Qual.* 12, 493–503.

Daza, P.C., Osorio, N.W., (2011). Plant growth and phosphorus uptake promotion in leucena by a mycorrhizal fungus in leucaena grown in a degraded soil by alluvial mining. Suelos Ecuat. 41, 144-149.

Dehne, H.W., (1982). Interaction between vesicular-arbuscular mycorrhizal fungi and plant pathogens. *Phytopathology* 72, 1115-1119.

Diez, M.C., (2006). Dependencia micorrizal de pino romerón, *Retrophyllum rospigliosii* (Pilg.) C.N. Page (Podocarpaceae) bajo diferentes regímenes lúminicos. Tesis M.Sc., Universidad Nacional de Colombia, Medellín.

Diez, M.C., Osorio, N.W., Moreno, F.H., (2008). Evaluation of mycorrhizal dependency of romeron pine (*Nageia rospigliosii* Pilger) under contrasting light conditions. *Rev. Fac. Nac. Agron. Med.* 61, 4554-4563.

Do Carmo Harta, M., Torrent, J., (2007). Phosphorus desorption kinetics in relation to phosphorus forms and sorption properties of Portuguese acid soils. *Soil Sci.* 172, 631-638.

Duponnois, R., Ba, A.M., (1999). Growth stimulation of *Acacia mangium* Willd by *Pisolithus* sp. in some Senegal soils. *Forest Ecol.Manag.* 119: 209-215.

Farsi, B., Fontana, A., (1962). Micorrize ectotrofiche di *Brachystegia laurentii* e di alcune altre di Caesalpiniaceae minori del Congo. *Allionis* 8, 21-131.

Fox, R.L., (1979). Comparative responses of field grown crops to phosphate concentrations in soil solutions. In: H. Munsell, and R. Staples (Eds.), *Stress Physiology in Crop Plants* (pp. 81-106). New York, NY: John Wiley and Sons.

Fox, R.L., Kamprath, E.J., (1970). Phosphate solution isotherms for evaluating the phosphate requirements of soils. *Soil Sci. Soc. Am. Pro.* 34,902–907.

Galiana, A., Prin, Y., Mallet, B., Gnahoua, G.M., Poitel, M., Diem, H.G., (1994). Inoculation of *Acacia mangium* with alginate beads containing selected *Bradyrhizobium* strains under field conditions: Long-term effect on plant growth and persistence of the introduced strains in soil. *Appl. Environ. Microbiol.*60, 3974-3980.

Gavito, M.F., Miller, M.H., (1998). Early phosphorus nutrition, mycorrhizal development, dry matter partitioning and yield of maize. *Plant Soil* 199, 177–186.

Gemma, J.N., Koske R.E., Habte, M., (2002). Mycorrhizal dependency of some endemic and endangered Hawaiian plant species. *Am. J. Bot.* 89, 337-345.

Gerdemann, J.W., (1975). Vesicular-arbuscular mycorrhiza. In: J.G. Torrey, and D.T. Clarkson (Eds.), *The development and functions of roots* (pp. 575-591). London, UK: Academic Press.

Giovannetti, M, Mosse, B., (1980). An evaluation of techniques for measuring vesicular-arbuscular mycorrhizal infection in roots. *New Phytol.* 84, 489–500.

Gohre, V., Paszkowski, U., (2006). Contribution of the arbuscular mycorrhizal symbiosis to heavy metal phytoremediation. *Planta* 223, 1115–1122.

Habte, M., (1989). Impact of simulated erosion on the abundance and activity of indigenous vesicular arbuscular mycorrhizal endophytes in an Oxisol. *Biol. Fert. Soils* 7, 164–167.

Habte, M., (1995). Dependency of *Cassia siamea* on vesicular arbuscular mycorrhizal fungi. *J. Plant Nutr.* 18, 2191-2198.

Habte, M., Fox, R.L., (1993). Effectiveness of VAM fungi in nonsterile soils before and after optimization of P in soil solution. *Plant Soil* 151, 219–226.

Habte, M., Fox, R.L., Huang, R.S., (1987). Determining vesicular-arbuscular mycorrhizal effectiveness by monitoring P status of subleaflets of an indicator plant. *Commun. Soil Sci. Plan.*18, 1403–1420.

Habte, M., Manjunath, A., (1991). Categories of vesicular-arbuscular mycorrhizal dependency of host species. *Mycorrhiza* 1, 3-12.

Habte, M., Osorio. N.W., (2001). *Arbuscular mycorrhizas: producing and applying arbuscular mycorrhizal inoculum*. Honolulu, HI: University of Hawaii.

Habte, M., Turk, D., (1991). Response of two species of *Cassia* and *Gliricidia sepium* to vesicular arbuscular infection. *Commun. Soil Sci. Plan.* 22, 1861–1872.

Hamel, C., (2004). Impact of arbuscular mycorrhizal fungi on N and P cycling in the root zone. *Can. J. Soil Sci.* 84, 383-395.

Harley, J.L., Smith, S.E., (1983). *Mycorrhizal symbiosis*. New York, NY: Academic Press.

Hattingh, M.J., Gray, L.E., Gerdemann, J.W., (1973). Uptake and translocation of $^{32}$P-labeled phosphate to onion roots by endomycorrhizal fungi. *Soil Sci.* 116, 383-387.

Havlin, J., Beaton, J., Tisdale, S.L., Nelson W., (1999). *Soil Fertility and Fertilizers: an introduction to nutrient management*. Upper Saddle River, NJ: Prentice Hall.

Hayman, D.S., (1982). Influence of soils and fertility on the activity and survival of arbuscular mycorrhizal fungi. *Phytopathology* 72, 1119–1125.

Hayman, D.S., (1983). The physiology of vesiculararbuscular endomycorrhizal symbiosis. *Can. J. Bot.* 61, 944–963.

Hetrick, B.A.D., (1991). Mycorrhizas and root architecture. *Experientia* 47, 355-362.

Hetrick, B.A.D., Kitt, D.G., Wilson, G.T., (1988). Mycorrhizal dependence and growth habits of warm-season and cool-season tall grass prairie plants. *Can. J. Bot.* 66, 1376–1380.

Hogberg, P., (1982). Mycorrhizal association in some woodland and forest trees and shrubs in Tanzania. *New Phytol.* 92, 407-415.

Hogberg, P., Nylund, J., (1981). Ectomycorrhizae in coastal Biombo woodland of Tanzania. *Plant Soil* 63, 283-289.

Holford, I.C.R., (1997). Soil phosphorus: its measurement, and its uptake by plants. *Aust. J. Soil Res.* 35, 227-39.

Hooker, J.E., Jaizme-Vegas, M., Atkinson, D., (1994). Biocontrol of plant pathogens using arbuscular mycorrhizal fungi. In: S. Gianinazzi, and H. Schuepp (Eds.), *Impacts of arbuscular mycorrhizal on sustainable agriculture and natural ecosystems* (pp. 191-199). Basel, Switzerland: Birkhauser Verlag.

Howeler, R.H., Edwards, D.G., Asher, C.J., (1981). Applications of the flowing solution culture techniques to study involving mycorrhizas. *Plant Soil* 59, 179–183.

Huante, P., Rincon, E., Allen, E.B., (1993). Effect of vesicular-arbuscular mycorrhizae on seedling growth of four tree from the deciduous tropical forest in Mexico. *Mycorrhiza* 2, 141-145.

Hue, N.V., Fox, R., (2010). Predicting plant phosphorus requirements for Hawaii soils using a combination of phosphorus sorption isotherms and chemical extraction methods. *Commun. Soil Sci. Plan.* 41, 133–143.

Jacobsen, I., Abbott, L.K., Robson, A.D., (1992). External hyphae of vesicular-arbuscular mycorrhizal fungi associated with *Trifolium subterraneum* L. *New Phytol.* 120, 371-380.

Jakobsen, I., Jones, E.J., Larse, J., (1994). Hyphal phosphorus transport, a keystone to mycorrhizal enhancement of plant growth. In: S. Gianinazzi, and S. Schuepp (Eds.), *Impacts of arbuscular mycorrhizal on sustainable agriculture and natural ecosystems* (pp. 133-146). Basel, Switzerland: Birkhauser Verlag.

Janos, D.P., (1980a). *Mycorrhizae influence tropical succession. Biotropica* 12, 56-64.

Janos, D.P., (1980b). Vesicular-arbuscular mycorrhizae affect lowland tropical rainforest plant growth. *Ecology* 61, 151-162.

Janos, D.P., (1988). Mycorrhiza applications in tropical forestry: are temperate-zone approaches appropriate? In: *Trees and mycorrhiza, Proceedings of the Asian Seminar* (pp. 133-188). Kuala Lumpur, Malaysia: Malaysian Forest Research Institute.

Jaramillo, S., Osorio, N.W., (2005). Factors that determine the mycorrhizal dependence of plants. *Suelos Ecuat.* 35, 34-40.

Jaramillo, S.P., Silva J.M., Osorio, N.W., (2004). Potential simbiotic and mycorrhizal effectiveness of three soils subjected to different uses. *Rev. Fac. Nac. Agron. Med.* 57, 2203-2214.

Jeffries, P., Gianinazzi, S., Perotto, S., Turnau, K., Barea, J.M., (2003). The contribution of arbuscular mycorrhizal fungi in sustainable maintenance of plant health and soil fertility. *Biol. Fert. Soils* 37, 1-16.

Johnson, N.C., Wedin, D.A., (1997). Soil carbon, nutrients and mycorrhizae during conversion of dry tropical forest to grassland. *Ecol. Appl.* 7, 171-182.

Kabir, Z., O'Halloran, I.P., Hamel, C., (1999). Combined effects of soil disturbance and fallowing on plant and fungal components of mycorrhizal corn (*Zea mays* L.). *Soil Biol. Biochem.* 31, 307–314.

Khalil, S., Loynachan, T.E., Tabatabai, M.A., (1999). Plant determinants of mycorrhizal dependency in soybean. *Agron. J.* 91, 135–141.

Koide, R.T., Schreiner, R.P., (1992). Regulation of the vesicular-arbuscular mycorrhizal symbiosis. *Ann. Rev. Plant Physiol. Plant Mol. Biol.* 43, 557–581.

Kormanik, P.P., McGraw, A.C., Shultz, R.C., (1980). Procedures and equipment for staining large numbers of plant roots for endomycorrhizal assay. *Can. J. Microbiol.* 26, 536–538.

Kottke, I., (2002). Mycorrhizae-rhizosphere determinants of plant comunities. In: Y. Waisel, A. Eshel, and U. Kakkafi (Eds.), *Plant roots: the hidden half* (3rd ed., pp. 919-932). New York, NY: Marcel Dekker.

Kuo, S., (1996). Phosphorus. In: D.L. Sparks (Ed.), *Methods of Soil Analysis: Chemical Methods* (pp. 869-919). Madison, WI: Soil Sci. Soc. Am. and Am. Soc. Agron.

Langley, J.A., Hungate, B.A., (2003). Mycorrhizal control on belowground litter quality. *Ecology* 84, 2302-2312.

Lee, D.K., Nguyen, H.N., (1991). Growth of some Acacia species in Vietman. In: J.W. Turnbull (Ed.), *Australian Acacias in developing countries* (pp. 173-176). Canberra, Australia: Australian Centre for International Agricultural Research.

León, J.D., (2007). Contribución al conocimiento del ciclo de nutrientes en bosques montanos naturales de *Quercus humboldtii* y reforestados (*Pinus patula* y *Cupressus lusitanica*) de la región de Piedras Blancas, Antioquia (Colombia). Tesis Doctoral, Universidad de Salamanca, Salamanca, España.

León, J.D., González, M.I., Gallardo, J.F., (2009). Retranslocación y eficiencia en el uso de nutrientes en bosques del centro de Antioquia. *Colombia Forestal* 12, 119-140.

León, J.D., Osorio, N.W., Castellanos, J., Osorio, L.F., (2010). Recuperación de suelos degradados por minería aluvial de oro con plantaciones de Acacia en la región del Bajo Cauca (Antioquia, Colombia). *Suelos Ecuat.* 50: 62-67.

León, J.D., Osorio, N.W., Díez, M.C., Castellanos, J., Osorio, L.F., Marín, N., (2008). Grupos funcionales de microorganismos en suelos degradados por minería de aluvión plantados con *Acacia mangium*. *Suelos Ecuat.* 38, 75-80.

Li, X., Ecckhard, G., Marschner, H. (1991). Phosphorus depletion and pH decrease at the root-soil and hypha-soil interfaces of VA mycorrhizal white clover fertilized with ammonium. *New Phytol.* 119, 397-404.

Linderman, R.G., (1988). Mycorrhizal interaction with the rhizosphere microflora: the mycorhizosphere effect. *Phytopathology* 78, 366-371.

Lindsay, W.L., (2001). *Chemical equilibria in soils*. Caldwell, NJ: The Blackburn Press.

Louis, I., Lim, G., (1987). Spore density and root colonization of vesicular-arbuscular mycorrhizal in tropical soil. *T. Brit. Mycol. Soc.* 88, 207-212.

Lynch, J.P., Ho, M.D., (2005). Rhizoeconomics: carbon costs of phosphorus acquisition. *Plant Soil* 269, 45-56.

Manjunath, A., Habte, M., (1988). The development of vesicular-arbuscular mycorrhizal infection and the uptake of immobile nutrients in *Leucaena leucocephala*. *Plant Soil* 106, 97–103.

Manjunath, A., Habte, M., (1991). Root morphological characteristics of host species having distinct mycorrhizal dependency. *Can. J. Bot.* 69, 671–676.

Marschner, H., (1995). *Mineral nutrition of higher plants*. London, UK: Academic Press.

McCully, M., (1999). Roots in soil: unearthing the complexities of roots and their rhizospheres. *Ann. Rev. Plant Physiol. Plant Mol. Biol.* 50, 695-718.

Medina, E., Cuevas, E., (1993). Mineral nutrition: humid tropical forests. *Prog. Bot.* 55, 115-129.

Medina, E., García, V., Cuevas, E., (1990). Sclerophylly and oligotrophic environments: relationships between leaf structure, mineral nutrient content, and drought resistance in tropical rain forest of the upper Rio Negro region. *Biotropica* 22, 51-64.

Medina, M., (1999). Caracterización molecular por RAPDs de cepas nativas de hongos endomicorrícicos de suelos de una zona de minería en el Bajo Cauca Antioqueño. Tesis Maestría de Microbiología, Universidad Nacional de Colombia, Bogotá.

Miyasaka, S.C., Habte, M., (2001). Plant mechanisms and mycorrhizal symbioses to increase phosphorus uptake efficiency. *Commun. Soil Sci. Plan.* 32, 1101-1147.

Miyasaka, S.C., Habte, M., Friday, J.B., Johnson, E.V., (2003). *Manual on arbuscular mycorrhizal fungus production and inoculation techniques*. Honolulu, HI: University of Hawaii. (Soil and crop management No. 5).

Miyasaka, S.C., Habte, M., Matsuyama, D.T., (1993). Mycorrhizal dependency of two Hawaiian endemic tree species: koa and mamane. *J. Plant Nutr.* 16, 1339-1356.

Mosse, B., (1981). Vesicular-arbuscular mycorrhiza research for tropical agriculture. Honolulu, HI: University of Hawaii (Research Bulletin No. 194).

Murphy, J., Riley, J.P., (1962). A modified single solution method for the determination of phosphate in natural waters. *Anal. Chim. Acta* 27, 31–36.

Musoko,M., Last, F.T., Mason, P.A., (1994). Populations of spores of vesicular-arbuscular mycorrhizal fungi in undisturbed soils of secondary semideciduous moist tropical forest in Cameroon. *Forest Ecol. Manag.* 63, 359-377.

Newsham, K.K., Fitter, A.H., Watkinson, A.R., (1994). Root pathogenic and arbuscular mycorrhizal fungi determine fecundity of asymptomatic plants in the field. *J. Ecol.* 82, 805–814.

Nye, P.H., Tinker, P.B., (1977). Solute movement in the soil-root system. Oxford, UK: Blackwell Scientific Publications.

Onguene, N.A., Kuyper, T.W., (2001). Micorrhizal association in the rain forest of south Cameroon. *Forest Ecol. Manag.* 140, 227-287.

Osorio, N.W., (2012). *Manejo de nutrientes en suelos del trópico*. Medellín, Colombia: Universidad Nacional de Colombia.

Osorio, N.W., Diez, M.C., Sierra, J., Paternina, L., (2008). Consideraciones ecológicas sobre la asociación micorrizal en suelos de la región altoandina. In: J.D. León (Ed.) *Ecología de bosques andinos: experiencias de investigación* (pp. 179-196). Medellín, Colombia: Universidad Nacional de Colombia.

Pacovsky, R.S., (1986). Micronutrient uptake and distribution in mycorrhizal or phosphorus-fertilized soybeans. *Plant Soil* 95, 379–388.

Parfitt, R.L., (1989). Phosphate reactions with natural allophane, ferrihydrite and goethite. *J. Soil Sci.* 40, 359-369.

Paternina, L., (2006). Effect of arbuscular mycorrhizal inoculation on soils with different mycorrhizal symbiotic potentials by monitoring leucaena seedling growth and phosphorus status. Thesis, Universidad Nacional de Colombia, Medellin.

Phillips, J.M., Hayman, D.S., (1970). Improved procedures for clearing and staining parasitic and vesicular-arbuscular mycorrhizal fungi for rapid assessment of infection. *T. Brit. Mycol. Soc.* 55, 158–161.

Picone, C., (2002). Diversity and abundance of arbuscular-mycorrhizal fungus spores in tropical forest and pasture. *Biotropica* 32, 734-750.

Plenchette, C., Fortin, J.A., Furlan, V., (1983). Growth response off several plant species to mycorrhiza in a soil of moderate P fertility I. Mycorrhizal dependency under field conditions. *Plant Soil* 70, 191–209.

Pope, P.E., Chaney, W.R., Rhodes, J.D., Woodhead, S.H., (1983). The mycorrhizal dependency of four hardwood tree species. *Can. J. Bot.* 61, 412–417.

Porter, W., (1979). The "Most Probable Number" method for ennumerating infective propagules of vesicular arbuscular micorrizal fungi in soil. *Aust. J. Soil Res.* 17, 515-519.

Pypers, P., Delrue, J., Diels, J., Smolders, E., Merckx, R., (2006). Phosphorus intensity determines short term P uptake by pigeon pea (*Cajanus cajan* L.) grown in soils with different P buffering capacity. *Plant Soil* 284, 217-227.

Radersma, S., Grierson, P., (2004). Phosphorus mobilization in agroforestry: organic anions, phosphatase activity and phosphorus fractions in the rhizophere. *Plant Soil* 259, 209-219.

Read, D.J., (1991). *Mycorrhizas in ecosystems. Experientia* 47, 376-391.

Ren, H., Yu, Z., (2008). Biomass changes of an *Acacia mangium* plantation in southern China. *J. Trop. For. Sci.* 20, 105-110.

Rilling, M.C., Wright, S.F., Nichols, K.A., Schmidt, W.F., Torn, M.S., (2001). Large contribution of arbuscular mycorrhizal fungi to soil carbon pools in tropical forest soils. *Plant Soil* 233, 167-177.

Sanchez, P., (1976). *Properties and Management of Soils in the Tropics*. New York, NY: John Wiley and Sons.

Sanchez, P., Logan, T., (1992). Myths and Science about the chemistry and fertility of soils in the tropics. In: R. Lal and P. Sanchez (Eds.), *Myths and Science of Soils of the Tropics* (pp. 35-46). Madison, WI: Soil Sci. Soc. Am.

Sanchez, P., Uehara, G., (1980). Management considerations for acid soils with high phosphorus fixation capacity. In: F.E. Khasawneh (Ed.), *The Role of Phosphorus in Agriculture* (pp. 471-514). Madison, WI: Soil Sci. Soc. Am.

Schachtman, D.P., Reid, R., Ayling, S.M., (1998). Phosphorus uptake by plants: from soil to cell. *Plant Physiol.* 116, 447-453.

Schübler, A., Schwarzott, D., Walker, C., (2001). A new fungal phylum, the Glomeromycota: phylogeny and evolution. *Mycol. Res.* 105, 1413-1421.

Sharpley, A.N., Smith, S.J., Jones, O.R., Berg, W.A., Coleman, G.A., (1992). The transport of bioavailable phosphorus in agricultural run-off. *J. Environ. Qual.* 21, 30–35.

Shoji, S., Nanzyo, M., Dahlgren, R.A., (1993). Volcanic ash soils-genesis, properties, and utilization. Amsterdam, Netherlands: Elsevier Publishing.

Sierra, J.A., (2006). Micorrhizal dependency of some forestry trees and mycorrhizal effectiveness in highland-soils from Eastern Antioquia. Thesis M.Sc., Universidad Nacional de Colombia, Medellin.

Sierra, J.A., Castro, D., Osorio, N.W., (2009). Mycorrhizal dependency of laurel. *Rev. Colomb. For.* 12, 17-24.

Sierra, J.A., Castro, D., Osorio, W., (2012). Mycorrhizal dependence of barcino (Clusiaceae: *Calophyllum brasiliense* Cambess). *Actualidades Biológicas* 34, 199-206.

Sieverding, E., (1991). *Vesicular-arbuscular mycorrhiza management in tropical agroecosystems*. Eschborn, Germany: Deutsche Gesellschaft Technische Zusammenabeit (GTZ).

Siqueira, J.O., Saggin-Júnior, O.J., (2001). Dependency on arbuscular mycorrhizal fungi and responsiveness of some Brazilian native woody species. *Mycorrhiza* 11, 245–255.

Smeck, N.E., (1985). Phosphorus dynamics in soils and landscapes. *Geoderma* 36, 185-199.

Smith, F.W., (2002). The phosphate uptake mechanism. *Plant Soil*, 245:105-114.

Smith, F.W., Mudge, S.R., Rae, A.L., Glassop, D., (2003). Phosphate transport in plants. *Plant Soil* 248, 71-83.

Smith, F.W., VandenBerg, P.J., Gonzalez, A., Andrew, C.S., Pieters, W.H.J., (1992). *Foliar symptoms of nutrient disorders in the tropical shrub legume Leucaena leucocephala.* St. Lucia, Queensland, Australia: CSIRO, Division of tropical crops and pastures (Technical paper No.32).

Smith, S.E., Read, D.J., (1997). *Mycorrhizal symbiosis*. San Diego, CA: Academic Press.

Soedarjo, M., Habte, M., (1995). Mycorrhizal and nonmycorrhizal host growth in response to changes in pH and P concentration in a manganiferous Oxisol. *Mycorrhiza* 5, 337–345.

Stahl, P.D., Williams, S.E., Christensesn, M., (1988). Efficacy of native vesicular-arbuscular mycorrhizal fungi after severe soil disturbance. *New Phytol.* 110, 347–354.

Stevenson, F.J., (1986). *Cycles of soil.* New York, NY: John Wiley and Sons.

Swift, M.J., Heal, O.W., Anderson, J.M., (1979). *Decomposition in terrestrial ecosystems.* Oxford, UK: Blackwell scientific publications.

Sylvia, D.M., (1999). Mycorrhizal symbiosis. In: D.M. Sylvia, J. Furhrmann, P. Hartel and D. Zuberer (Eds.), *Principles and applications of soil microbiology* (pp. 408–426). Upper Saddle River, NJ: Prentice Hall.

Taylor, D.L., (2000). A new dawn: the ecological genetics of mycorrhizal fungi. *New Phytol.* 147, 236-239.

Tinker, P.B., Gildon, A., (1983). Mycorrhizal fungi and ion uptake. In: D.A. Rob and W.S. Pierpoint (Eds.), *Metals and micronutrients, uptake and utilization by plants* (pp. 21–32). London, UK: Academic Press.

Trappe, J.T., (1987). Phylogenetic and ecologic aspects of mycotrophy in the angiosperms from an evolutionary stand-point. In: G.R. Safir (Ed.), *Ecophysiology of VA mycorrhizal plants* (pp. 5-25). Boca Raton, Florida: CRC Press.

Trotta, A., Varese, G.C., Gnavi, E., Fusconi, A., Sampo, S., Berta, G., (1996). Interaction between the soil-borne root pathogen *Phytophthora nicotianae* var. parasitica and the arbuscular mycorrhizal fungus *Glomus mosseae* in tomato plants. *Plant Soil* 185, 199–209.

Turner, B.L., Frossard, E., Oberson, A., (2006). Enhancing phosphorus availability in low-fertility soils. In: N. Uphoff (Ed.) *Biological approaches to sustainable soil systems* (pp. 191-206). Boca Raton, FL: CRC Press.

Uphoff, N., Ball, A.S., Fernandes, E.C.M., Herren, H., Husson, O., Palm, C., Pretty, J., Sanginga, N., Thies, J. E., (2006). Understanding the functioning and management of soil systems. In: N. Uphoff (Ed.), *Biological approaches to sustainable soil systems* (pp. 3-13). Boca Raton, FL: CRC Press.

Van Der Heijden, M.G., Klironomos, J.N., Ursic, M., Mouroglis, P., Strettwolf-Engel, R., Boller, T., Wiemken, A., Sanders, I.R., (1998). Mycorrhizal fangal diversity determines plant biodiversity, ecosystem variability, and productivity. *Nature* 396, 69-72.

Vitousek, P.M., Sanford, R.L., (1986). Nutrient cycling in moist tropical forest. *Annu. Rev. Ecol. Syst.* 17, 137-167.

Wilson, J., Ingleby, K., Mason, P.A., Ibrahim, K., Lawson, G.J., (1992). Long-term changes in vesicular-arbuscular mycorrhizal spore populations in *Terminalia* plantations in Cote d'Ivore. In: D.J. Read, D.H. Lewis, A.H. Fitter and I.J. Alexander (Eds.), *Mycorrhizas in ecosystems* (pp. 268-275). Wallingford, UK: CAB International.

Wright, S.F., Upadhyaya, A., (1996). Extraction of an abundant and unusual protein from soil and comparsion of hyphal protein of arbuscular mycorrhizal fungi. *Soil Sci.* 161, 575–586.

Wright, S.F., Upadhyaya, A., (1998). A survey of soils for aggregate stability and glomalin, a glycoprotein produced by hyphae of arbuscular mycorrhizal fungi. *Plant Soil* 198, 97–107.

In: Plantations
Editor: Hai Ren
ISBN: 978-1-62808-090-2


*Chapter 9*

# ALLUVIAL GOLD-MINING DEGRADED SOILS RECLAMATION USING ACACIA MANGIUM PLANTATIONS: AN EVALUATION FROM BIOGEOCHEMISTRY

***Juan D. León,[1] Jeiner Castellanos,[1] Maria Casamitjana,[2] Nelson W. Osorio[1] and Juan C. Loaiza[1,3,*]***

[1]Universidad Nacional de Colombia, Departmento de Ciencias Forestales y Departamento de Biociencias, Columbia
[2]Universitat de Girona, Geosciencies Department, Girona, Spain
[3]Centre Tecnologic Forestal de Catalunya, Solsona, Spain

## ABSTRACT

A study was carried out to evaluate the use of *Acacia mangium* plantations in reactivating biogeochemical cycles of soil nutrients in a soil degraded by alluvial gold-mining. Fine litterfall production was 10.4 Mg ha$^{-1}$ yr$^{-1}$ and it was dominated by the leaf fraction 5.7 Mg ha$^{-1}$ (55%). The highest nutrient return found corresponded to nitrogen (N) (69 kg ha$^{-1}$) and the lowest to phosphorus (P) (0.5 kg ha$^{-1}$).

The results of a litter-bag study indicated that the annual decomposition constant ($k$) ranged from 1.25 to 1.80. Chemical composition of the leaf litter (*e.g.,* N and P concentrations, C/N and N/P ratios, and phenols content) and rainfall play an important role in the decay process, which was characterized by a slow release of N, Ca, and Mg from the leaf litter, while P was immobilized. The C return to soil from the decomposition of the leaf litterfall oscillated between 2.0 and 2.4 Mg ha$^{-1}$, which was lower than expected, likely because the mesh size of the litter bags restricted our observation only for the leaf fraction. The *A. mangium* plantations showed a great potential in the reclamation of these degraded soils due to the high values of fine litter production and nutrient return, which contributed to the reactivation of biogeochemical cycles.

[*] Corresponding autor: Juan C. Loaiza. Universidad Nacional de Colombia, Departmento de Biociencias, Calle 59A No. 63-20, Office 14-208, 050034, Medellín, Colombia. E-mail: jc.loaza@ctfc.es.

Therefore, the soil under this plantation showed significant increases in chemical and physical properties respect to unplanted soil: nitrogen content (0.24 vs. 0.5%), available P (2.48 vs. 6.53 mg $kg^{-1}$), cation exchange capacity (6.12 vs. 11.19 $cmol_c$ $kg^{-1}$), organic matter content (6.14 vs. 18.66%, and aggregate stability (73.0 vs. 85.35%). *A. mangium* plantations have a high potential to remediate soils degraded by alluvial gold-mining.

**Keywords:** Soil bioremediation, biogeochemical fluxes, soil physics, *Acacia mangium*, soil degradation

## INTRODUCTION

In tropical forests the fine litterfall production is the main process that determines the potential return of organic matter and nutrients onto the soil, which support plant development and soil biota (Proctor 1983, Spain 1984, Vitousek and Sanford 1986, Landsberg and Gower 1997; Norgrove and Hauser 2000, Kumar and Agrawal 2001). However, nutrient recycling is achieved when the leaf litterfall is decomposed by soil biota, a key process in forestry systems that determines soil fertility and forest primary productivity (Wang *et al.* 2008). If the nutrients are quickly released, they could be lost by leaching or volatilization (Palma *et al.* 1998, Schlesinger 2000). If the decomposition occurs slowly, the nutrient supply to plant roots will be insufficient, thus limiting plant growth and development (Jordan 1985, Swift and Anderson 1989, and Bubb *et al.* 1998, Montagnini and Jordan 2002). For these reasons, the rates at which litter decomposition and subsequent nutrient release occur constitute key factors of the ecosystem functioning.

In the case of degraded lands by superficial mining activities, the lost of plant coverage on the soil surface disrupt these biogeochemical cycles (León *et al.* 2008). Land reclamation of these soils may be achieved by establishing forestry species, which must chosen based on their ability to adapt to extreme and restrictive soil conditions. Several studies have demonstrated that forestry plantations play an important role in the improvement of soil fertility, due to their capacity to induce ecological and physicochemical changes in the soil (Lugo *et al.* 1990, Parrota 1992, Lugo 1997, Garten 2002, Singh *et al.* 2002).

Species of the *Acacia* genus such as *A. albida, A. Senegal* and *A. mangium* are commonly employed for forestry and land reclamation in the tropics (Laclau *et al.* 2000). *A. mangium* grows quickly and has a high capacity to adapt to nutrient-poor acidic soils, due to its capacity to establish symbiotic associations with $N_2$ fixing bacteria (Ribet and Drevon 1996, Higa and Higa 2000, Garay *et al.* 2004) and mychorrizal fungi (Lim 1988, Duponnois and Ba 1999). In extremely nutrient-poor soil *A. mangium* has exhibited an outstanding growth rate higher than other plant species such as *Eucalyptus* sp. and *Gmelina arborea.*

In Colombia, at the beginning of the 1990s, *A. mangium* was succesfully introduced in the Bajo Cauca Antioqueño region (Departments of Cordoba and Antioquia, Colombia) for forestry use and land restoration (Torres and del Valle 2007, León *et al.* 2008). Although, several plantations were established and currently they cover more than 700 ha, little is known about the functioning of these restored ecosystems.

The main purpose of this chapter was to evaluate the nutrient dynamics in the fine litterfall and its decomposition in established *A. mangium* plantations in soils degraded by alluvial mining.

# MATERIALS AND METHODS

## Site Description

The study took place on 11-year-old *A. mangium* plantations over the course of one year. The plantations were established by the regional environmental authority (Corporación Autónoma Regional del Centro de Antioquia, CORANTIOQUIA) to restore degraded lands by alluvial-gold- mining in the Bajo Cauca region, northwest of Colombia (Figure 1).

The study was performed in the last foothills of the central Andean range, in the sourrondings of Rio Rayo (Tarazá; 07°31'44"N, 75°21'04"W) and Jardin (Cáceres; 07°45'30"N, 75°14'26"W) villages. In Tarazá and Cáceres the annual average temperature and the annual average rainfall is 27 and 28°C and 3133 and 2771 mm, respectively (IGAC 2007). The soil erosion produced by mining activities has been so severe that the first soil horizons were completely removed; the surface residual material lacks soil organic matter, natural seed bank, and soil biota (Osorio 2000).

Due to the strong surface compaction that occurs after the end of mining activities, the land was ploughed with a subsoiler-till (at 60-70 cm depth) before plantation establishment, thus favoring plant root development. Aside from the serious physical limitations of this land has for an adequate plant growth, the remaining soil was very acidic, poor in macronutrients, rich in iron content, and with high levels of mercury (Hg) that limited agronomical production activities (León *et al.* 2008) (Table 1).

On this soil an *A. mangium* plantation was established, but it did not undergo a comprehensive plan of forest management. The stand had a mean diameter at breast height (DBH) of 16.9 cm, mean tree height of 17.1 m, and basal area of 22 $m^2$ $ha^{-1}$ (Table 1).

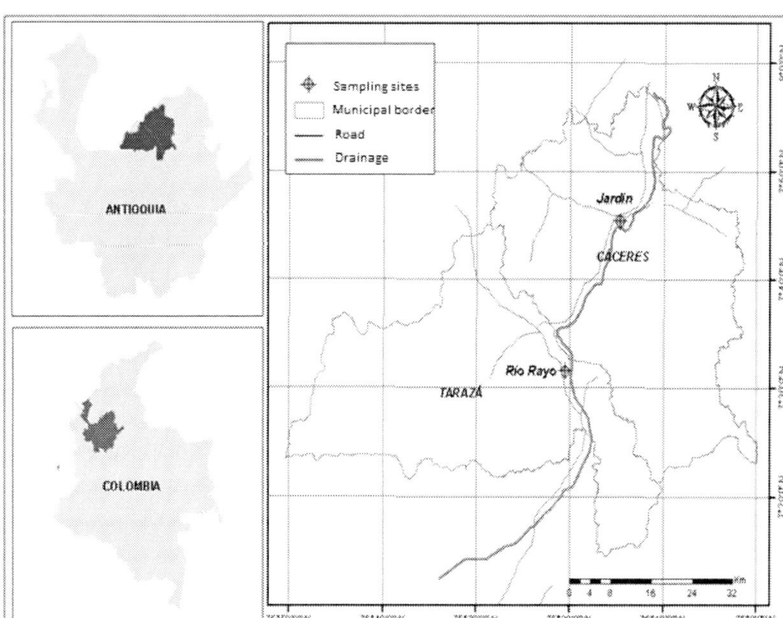


Figure 1. Study area in Bajo Cauca, Colombia.

**Table 1. Stand characteristics of *A. mangium* and soils characteristics into the studied zones in the Bajo Cauca Antioqueño, Colombia**

| Structural Stand parameters | | | | | | |
|---|---|---|---|---|---|---|
| Site | N (trees $ha^{-1}$) | DBH (cm) | G ($m^2$ $ha^{-1}$) | V ($m^3$ $ha^{-1}$) | B (t $ha^{-1}$) | H (m) |
| S1 | 1133 | 19.21 | 32.25 | 241.13 | 204.61 | 21.44 |
| S2 | 1026 | 14.29 | 16.92 | 84.68 | 82.63 | 12.79 |
| N1 | 826 | 20.48 | 29.10 | 211.30 | 185.81 | 20.83 |
| N2 | 760 | 13.67 | 12.30 | 62.28 | 58.77 | 13.28 |
| Mean | 936 | 16.91 | 22.64 | 149.85 | 132.96 | 17.08 |

| Soil chemical parameters | | | | | | | | | | | | | | |
|---|---|---|---|---|---|---|---|---|---|---|---|---|---|---|
| Sites | pH | OM | N | C | C/N | Ca | Mg | K | Al | Cu | Fe | Mn | Zn | P |
| | | -----------%--------- | | | | ---cmol(+) $kg^{-1}$--- | | | ------------------mg $kg^{-1}$-------------------- | | | | | |
| S1 | 4.5 | 6.22 | 0.69 | 3.60 | 5.22 | 5.65 | 3.28 | 0.84 | 0.70 | 1.42 | 34.48 | 41.14 | 2.45 | 4.67 |
| S2 | 4.5 | 6.06 | 0.40 | 3.51 | 8.78 | 5.49 | 3.23 | 0.85 | 0.67 | 1.56 | 34.32 | 41.27 | 2.89 | 4.04 |
| N1 | 4.4 | 5.36 | 0.40 | 3.10 | 7.77 | 6.41 | 3.07 | 0.77 | 0.67 | 1.28 | 28.77 | 41.60 | 2.14 | 7.35 |
| N2 | 4.6 | 9.34 | 0.52 | 5.41 | 10.41 | 8.22 | 3.56 | 0.68 | 0.63 | 2.40 | 33.00 | 42.07 | 5.99 | 8.38 |
| Mean | 4.5 | 6.75 | 0.50 | 3.91 | 8.05 | 6.44 | 3.29 | 0.78 | 0.67 | 1.67 | 32.64 | 41.52 | 3.37 | 6.11 |

N: density G: Base area, DBH: diameter at breast height (1.3 m), H: high, V: volume of wood, B: biomass, pH: determined with a soil/water suspesion 1:2. S1: Plots with good tree development: Subsoiling, S2: Plots with general poor tree development: Subsoiling, N1: Plots with general good tree development: No-till, N2: Plots with general poor tree development: No-till (Universidad Nacional de Colombia - CORANTIOQUIA 2008, unpublished data).

## Return of Organic Matter and Nutrients Via Leaf Litterfall

The plots were classified according to: (i) the general state of the plantation: good plant development (G1) and poor plant development (G2), and (ii) the tillage treatments received at the time of plantation establishment: 'subsoiling' and 'no-till'. The treatments consisted of four groups: in G1 (subsoiling -S1- and no-till –N1-) and in G2 (subsoiling -S2- and no-till -N2-). All plots were at the same substrate (alluvial-mining residual soil) and the same environmental and silvicultural conditions. In the plantation, 12 circular plots were established with an area of 250 $m^2$, allowing three replicates per treatment (S1, N1, S2, N2). On each, nine plot traps at 1 m height from the soil surface were installed to monitor the fine litter production. The traps used consisted of a circular receptacle of 0.5 $m^2$ on fine mesh, used to avoid the loss of fine material (Ramírez *et al.* 2007). The collection of material was performed twice each month for a year (March 2007 to March 2008). Rainfall was registered using rain gauges close to the plantations, which were monitored weekly on each site.

## Fine Litterfall, Leaf Litter Decomposition and Nutrient Release

The litter bags technique was employed with a plastic mesh of 2 mm aperture size, which allowed us to avoid material loss by fragmentation and at the same time to retain some important undecomposed materials (Wieder and Lang 1982, Sundarapandian and Swamy 1999). On each study site, 18 bags (20 x 20 cm size) containing 10 g of oven-dried leaf litter were distributed in three groups, with six bags per tree (a total of 54 sample bags). To evaluate the first factor (subsoiling), different sites were established with the same climatic conditions, but different soil pre-treatments (JS: Jardin subsoiling; JNS: Jardin no-till). To evaluate the impact of rainfall, the study was divided into sites with different pluviometric registers (RS: Río Rayo subsoiling; JS: Jardín subsoiling).

Each month, three bags were removed, with a total of nine bags per sampling. The bags were opened and the litter materials were air dried, brushed gently to remove soil particles, oven-dried at 65°C, and weighed. The residual dry matter (RDM) was estimated at each sampling date.

## Soil Sampling

The soil samples were taken at a 20 cm depth on each study site in order to analyze some soil physical and chemical properties. Subsamples were taken in a zig-zag pattern, making a diametric trajectory in a northerly direction (Osorio and Casatmijana 2011). Samples of undisturbed soil were taken as well, in order to analyze aggregate stability. Six sites were selected as control plots where *A. mangium* was unplanted and without interventions. The samples were conserved hermetically and cooled at 5°C in the field until they were processed in the laboratory of biogeochemistry of the Universidad Nacional de Colombia at Medellin.

## Laboratory Analysis

The collected material was separated into several components: leaves of *A. mangium* (L), leaves of other species (OL), woody fraction (W), reproductive material (RM), and other rests (OR). The litter bag material was extracted and cleaned manually to eliminate roots, invertebrates, and soil. Subsequently, the material was dried at 65°C until reaching constant weight.The leaves were milled and heated to 600°C for elemental analysis (Ca, Mg and K content per dry dissolution and atomic absorption spectrophotometry); leaf P content was determined using acid dissolution and visible spectrophotometry; N content by micro-Kjeldahl (Benton et al. 1991); C content was determined by dry combustion by means of an analyzer LECO CNS-2000; lignin content was determined by acid detergent and $KMnO_4$ (Horwitz 2003, Tejada 1992). Phenols were extracted by means of methanol over the course of five days, and were analyzed with the colorimetric Folin-Ciocalteu assay (Singleton and Rossi 1965). Results were expressed as mg of Gallic acid per 100 g of extract.

Soil samples were air dried and sieved ($\leq$ 2 mm). Soil pH was measured in a soil: water suspension (1:2 ratio) with a pH electrode (Beckman 601). Labile soil P was determined according to the method of Kalra and Maynard (1991) using spectrophotometry (Baush and Lamb Milton Roy 601) at 660-nm wavelength. Exchangeable Ca, Mg, and K were extracted with neutral 1 *M* ammonium acetate and measured with atomic absorption spectrophotometry. Total N was measured using the micro-Kjeldahl method.

Additionally, the mean weight diameter (MWD) of soil aggregates and the soil aggregate stability (AE) were determined by the Yoder method.

## Data Analysis

Nutrient return per treatment was calculated based on fine litter production values and elemental concentrations. The residual amount of nutrients at each sampling was calculated as the product of the RDM and the elemental concentration, and then divided by 100. The residual dry matter (RDM) evolution or remaining dry matter in the litter bags and the nutrient release dynamics were compared among sites through a one-way ANOVA, checking the statistical normality and the data homocedasticity beforehand.

When the treatment had significant effects, the mean separation was made using the Tukey test ($P \leq 0.05$).

To determine the association between RDM and other factors such as total rainfall, subsoil treatment, and substrate quality, Pearson correlation coefficients were obtained ($P \leq 0.05$). To describe RDM evolution, the negative exponential simple model (nonlinear), was adjusted according to Equation 1 (Olson 1963).

$$Xt / Xo = e^{-kt} \quad (1)$$

where, $Xt$: dry weight of the remaining material at time t (g), $Xo$: initial dry weight (g), $k$: decomposition constant, $t$: time spent (days).

From this constant, we calculated the mean life or necessary time to reach leaf litter decomposition of 50% ($t_{0.5}$) and 99% ($t_{0.99}$): $t_{0.5}$= -0.693/$k$ and $t_{0.99}$= -4.605/$k$ (Olson 1963, Arunachalam and Singh 2002).

The indicators used to choose the model were: the determination coefficient ($R^2$), the Durbin-Watson statistic, and the sum of squares of error. The Mann-Whitney test was used between two groups ($P \leq 0.05$) to check the soil parameter values between unplanted plots (control) and plantations (Wilcoxon). The statistical analysis was performed with Statgraphics Centurion XV (StatPoint Technologies, Inc).

# RESULTS

## Fine Litter Production

Annual fine litterfall was 10350 kg $ha^{-1}$ $yr^{-1}$, the monthly amount lessening during the rainy months. Among treatments the greatest values were represented by S1 and N1. Fine litterfall was dominated by leaf fraction (5543 kg $ha^{-1}$ $yr^{-1}$) and reproductive materials (2462 kg $ha^{-1}$ $yr^{-1}$). The values registered in OR and OL fractions were 577.7 and 153.3 kg $ha^{-1}$ $yr^{-1}$, respectively. The annual fine litter production did not show significant differences among treatments ($P$=0.21) (Figure 2).

On the other hand, significant correlations ($P \leq 0.05$) between total rainfall and fine litterfall were not detected in any treatment. The variation coefficients were quite low in all the fractions under all treatments, with the exception of the OL fraction (Table 2).

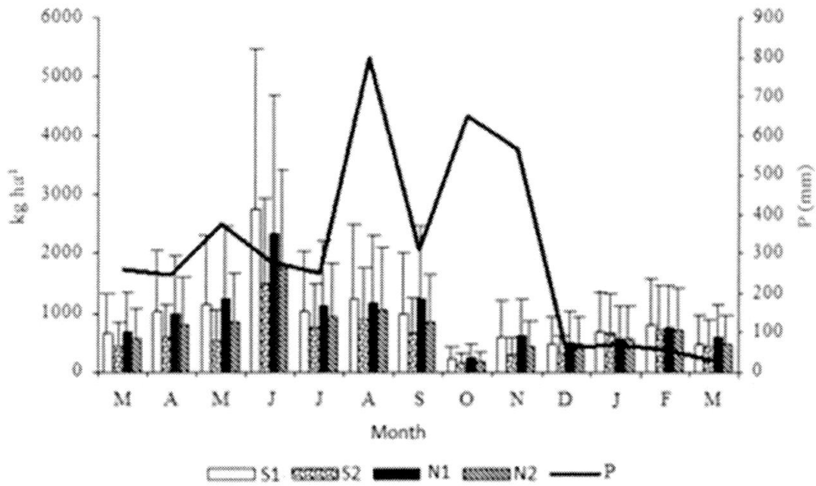


Figure 2. Mean values of monthly rainfall (P) and fine litterfall in *A. mangium* plantations in the Bajo Cauca region, Colombia. Bars represent standard deviation.

**Table 2. Annual production of fine litter (kg ha$^{-1}$), separated by fractions and treatments, in *A. mangium* plantations in Bajo Cauca, Colombia**

| Treatments | Fractions | | | | | |
|---|---|---|---|---|---|---|
| | L | OL | RM | W | OR | Total |
| S1 | 6454.98 (2.17) | 30.29 (9.92) | 2790.11 (6.37) | 2229.26 (6.73) | 596.75 (5.03) | 12101.39 (3.11) |
| S2 | 4520.36 (2.62) | 0.57 (20.75) | 1811.07 (5.38) | 1048.85 (6.40) | 446.30 (6.40) | 7827.15 (2.75) |
| N1 | 5957.14 (2.03) | 571.97 (5.00) | 2871.94 (5.43) | 1893.76 (5.62) | 672.66 (4.78) | 11967.47 (2.40) |
| N2 | 5239.83 (2.43) | 10.18 (12.23) | 2373.07 (4.59) | 1277.88 (6.20) | 595.02 (5.32) | 9495.97 (2.43) |
| Mean | 5543.08 | 153.25 | 2461.55 | 1612.44 | 577.68 | 10348.00 |

Values between brackets represent the variation coefficient (%). L: *A. mangium* leaves; OL: Other species leaves, W: Woody material, RM: Reproductive material, OR: Other rests.

## Potential Nutrient Return from Leaf Litterfall

Leaf litterfall of *A. mangium* showed high N content and low P content (Table 3). The greatest N and P values were registered with the N1 and S1 treatments (N: 1.32% and 0.012%).

The lowest values of both nutrients were found under the S2 treatment (N: 1.16% and P: 0.007%). P concentrations were found to be significantly different ($P$=0.0001) among the S1, N1 and S2, N2 treatments. The variation coefficient of P concentration was greater in the subsoiling sites (S1 and S2), than in the non-subsoiling sites (N1, N2). N concentration showed relatively low variation coefficients among treatments (Table 3). Ca, Mg and K concentrations decreased considerably beginning in September and increased gradually until February; this trend was due to the transition between rainy periods and drier periods (Figure 3). N and P concentrations showed consistent trends under all treatments; however, some peaks in concentration were observed in N during September, coinciding with a decrease in rain (Figure 2).

P concentration peaked during June and September, months which also corresponded to a drier weather; the increase was most evident under the S1 treatment. The highest potential nutrient return via leaf litterfall were observed with N (69.1 kg $ha^{-1}$ $yr^{-1}$) and Ca (30.0 kg $ha^{-1}$ $yr^{-1}$), and the lowest with P (0.6 kg $ha^{-1}yr^{-1}$).

Non-significant differences in nutrient return among treatments were observed in the following nutrients: N, Ca, Mg and K. In contrast, significant differences were observed in P return among the S2, S1 and N1 treatments.

**Table 3. Mean values of leaf nutrient concentration (%± SD) and annual nutrient return via leaf litterfall (kg $ha^{-1}$ $yr^{-1}$± SD) in *A. mangium* plantations in Bajo Cauca region (Colombia)**

| Nutrient concentration (%) | | | | | |
|---|---|---|---|---|---|
| Treatments | ---------P-------- | --------N--------- | -------Ca------ | -------Mg------- | --------K-------- |
| S1 | 0.012±0.005 b (40.86) | 1.245±0.148ab (11.86) | 0.511±0.152a (29.68) | 0.131±0.031a (23.54) | 0.147±0.054a (36.78) |
| S2 | 0.007±0.002a (23.35) | 1.158±0.095a (8.22) | 0.493±0.152a (30.83) | 0.118±0.027a (23.01) | 0.141±0.075a (53.10) |
| N1 | 0.011±0.002b (16.95) | 1.316±0.131b (9.94) | 0.489±0.138a (28.23) | 0.123±0.030a (24.40) | 0.203±0.075a (36.85) |
| N2 | 0.009±0.001a (16.82) | 1.249±0.121ab (9.72) | 0.525±0.167a (31.78) | 0.121±0.026a (21.36) | 0.164±0.059a (35.89) |
| Mean | 0.010 | 1.242 | 0.505 | 0.123 | 0.164 |

| Nutrient return (kg $ha^{-1}$ $yr^{-1}$) | | | | | |
|---|---|---|---|---|---|
| Treatments | --------P--------- | ---------N------- | --------Ca----- | -------Mg----- | ---------K------- |
| S1 | 0.8±0.036ac (4.76) | 80.5±2.790a (3.46) | 35.1±1.623a (4.62) | 8.9±0.360a (4.06) | 9.9±0.418a (4.21) |
| S2 | 0.3±0.012b (3.82) | 52.2±1.576a (3.02) | 23.9±1.121a (4.69) | 5.6±0.213a (3.81) | 7.0±0.411a (5.89) |
| N1 | 0.7±0.025acd (3.80) | 78.0±2.416a (3.09) | 31.4±1.486a (4.74) | 7.7±0.319a (4.16) | 12.8±0.548a (4.29) |
| N2 | 0.5±0.015bd (3.32) | 65.8±2.031a (3.08) | 29.7±1.444a (4.86) | 6.6±0.244a (3.72) | 9.3±0.421a (4.51) |
| Mean | 0.5 | 69.1 | 30.0 | 7.2 | 9.7 |

Variance coefficient (%) shown in brackets. SD: Standard deviation. For each nutrient, means followed with different letters are significantly different ($P \leq 0.05$). S1: Plots with generally high tree development: Subsoiling, S2: Plots with generally low tree development: Subsoiling, N1: Plots with generally high tree development: No-till, N2: Plots with generally low tree development: No-till.

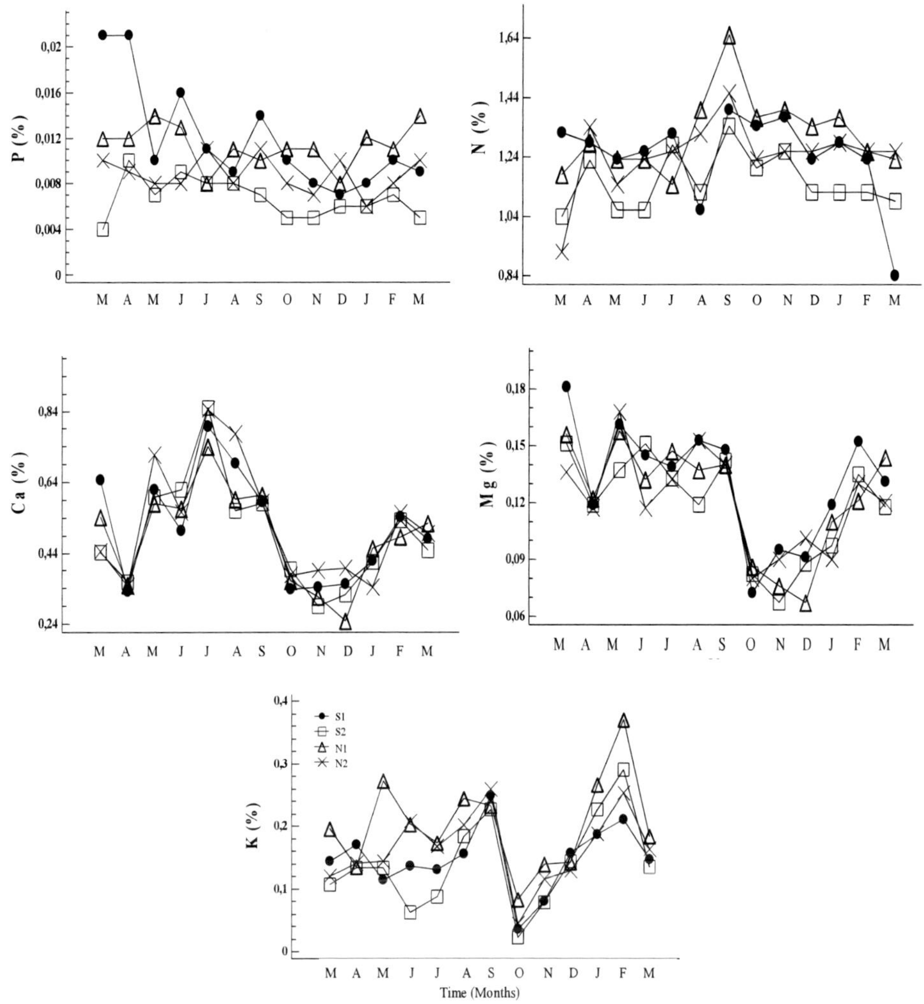


Figure 3. Nutrient concentration of leaf litter deposited in litter bags in *A. mangium* plantations in Bajo Cauca region (Colombia). S1: Plots with generally high tree development: Subsoiling, S2: Plots with generally low tree development: Subsoiling, N1: Plots with generally high tree development: No-till, N2: Plots with generally low tree development: No-till.

## Leaf Litter Decomposition

In all sites, the tendency was that litter bag material lost weight over time (Figure 4.). With reference to the initial material in the litter bags, the residual dry matter (RDM) at the end of the study was found 0.55 in JS, 0.41 in JNS, and 0.42 in RS.

The annual decomposition rates of the simple exponential model (Table 4) followed the decreasing sequence RS ($k$=1.80)>JNS ($k$=1.72)>JS ($k$=1.24).

**Table 4. Regression models adjusted to residual dry matter at different times of decomposition in *A. mangium* plantations in the Bajo Cauca region (Colombia)**

| Site | Model | $t_{0.5}$ (years) | $t_{0.99}$ (years) | *k* (1/year) | $R^2$ (%) | SSE | D-W |
|---|---|---|---|---|---|---|---|
| JS | $10x\ e^{(-0.0034127*t)}$ | 0.56 | 3.70 | 1.25 | 89.44 | 3.95 | 2.19 |
| JNS | $10xe^{(-0.00471566*t)}$ | 0.40 | 2.70 | 1.72 | 86.82 | 8.29 | 1.07 |
| RS | $10xe^{(-0.00493899*t)}$ | 0.38 | 2.60 | 1.80 | 96.20 | 2.10 | 2.71 |

Xo: Initial dry weight (10 g), *k*: annual decomposition rate, $R^2$: determination coefficient, SSE: Sum of squared error, D-W: Durbin-Watson statistic, $t_{0.5}$ = necessary time to decompose 50% of the litter, $t_{0.99}$ = necessary time to decompose 99% of the litter. JS: Jardín subsoiling, JNS: Jardín no-till, RS: Río Rayo subsoiling.

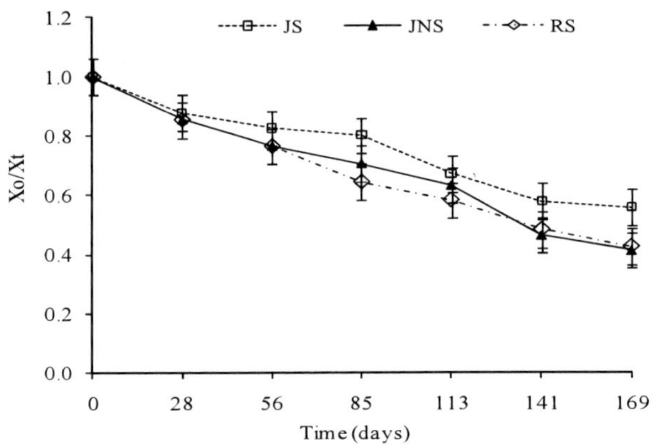


Figure 4. Residual dry matter (Xo/Xt) of leaf litter in the three sites in *A. mangium* plantations in Bajo Cauca region, Colombia. JS: Jardín subsoiling, JNS: Jardín no-till, RS: Río Rayo subsoiling. Bars indicate a 95% confidence interval.

## Nutrient Release from Leaf Litter

The most and least abundant nutrients in RDM were, respectively, N and P. In all the studied sites, increasing P concentration over time was found to be the predominant pattern. At the end of this study, N and P concentrations increased 25-52% and 150-325%, respectively (Table 5). N release was slow on the three sites until day 113, and became faster at the end of the study (Figure 5). Ca and Mg clearly showed trends similar to N in JNS and RS. Regarding P, the immobilization and increase of RDM were found to be dominant trends, with the increase at the end of the study being (169 day) 150% higher in JNS and RS. K and Mg were released throughout the study in JS and RS, but not in JNS.

## Quality Evolution of Leaf Litter

Parameters such C/N, N/P, and total phenol content showed a strong decrease from the beginning to the end of the study (37.1-23.1, 169.0- 69.7, 67.1-11.0%, respectively), with similar tendencies in JNS and RS (Table 6).

**Table 5. Nutrient concentration dynamics in the RDM of leaf litter in the sites studied in *A. mangium* plantations in Bajo Cauca region (Colombia) (values in percentage)**

| Time (days) | Sites | | | | | | | | | | | | | | |
|---|---|---|---|---|---|---|---|---|---|---|---|---|---|---|---|
| | JS | | | | | JNS | | | | | RS | | | | |
| | P (*a*) | N (*a*) | Ca (*a*) | Mg (*a*) | K (*a*) | P (*b*) | N (*a*) | Ca (*a*) | Mg (*a*) | K (*b*) | P (*b*) | N (*a*) | Ca (*a*) | Mg (*a*) | K (*a*) |
| 0 | 0.008 | 1.347 | 0.398 | 0.161 | 0.083 | 0.008 | 1.347 | 0.398 | 0.161 | 0.083 | 0.008 | 1.347 | 0.398 | 0.161 | 0.083 |
| 28 | 0.009 | 1.297 | 0.388 | 0.170 | 0.084 | 0.020 | 1.410 | 0.386 | 0.194 | 0.120 | 0.015 | 1.310 | 0.417 | 0.178 | 0.075 |
| 56 | 0.012 | 1.251 | 0.426 | 0.183 | 0.091 | 0.023 | 1.503 | 0.467 | 0.188 | 0.136 | 0.023 | 1.492 | 0.475 | 0.203 | 0.074 |
| 85 | 0.013 | 1.288 | 0.484 | 0.196 | 0.080 | 0.022 | 1.568 | 0.434 | 0.180 | 0.079 | 0.030 | 1.568 | 0.559 | 0.229 | 0.089 |
| 113 | 0.018 | 1.568 | 0.558 | 0.207 | 0.085 | 0.027 | 1.680 | 0.600 | 0.225 | 0.116 | 0.032 | 1.876 | 0.536 | 0.208 | 0.086 |
| 141 | 0.020 | 1.480 | 0.614 | 0.222 | 0.072 | 0.033 | 1.680 | 0.617 | 0.215 | 0.130 | 0.032 | 1.616 | 0.619 | 0.222 | 0.045 |
| 169 | 0.020 | 1.680 | 0.670 | 0.219 | 0.087 | 0.034 | 2.044 | 0.649 | 0.188 | 0.116 | 0.031 | 2.044 | 0.559 | 0.214 | 0.065 |

Columns with different letters for the same nutrient are significant different among sites ($P<0.05$). JS: Jardín subsoiling, JNS: Jardín No-till, RS: Río Rayo subsoiling.

**Table 6. Quality parameters of leaf litter placed into litter bags on the three sites studied in *A. mangium* plantations in the Bajo Cauca region (Colombia)**

| Initial values (%) | | | | | | | | |
|---|---|---|---|---|---|---|---|---|
| Sites | C | N | P | C/N | N/P | Lignin | Lignin/N | Total phenols* |
| JS, JNS and RS | 50.01 | 1.35 | 0.008 | 37.14 | 169.04 | 12.40 | 9.21 | 67.14 |
| Final values (%) | | | | | | | | |
| Sites | C | N | P | C/N | N/P | Lignin | Lignin/N | Total phenols* |
| JS | 44.51 | 1.68 | 0.02 | 26.49 | 83.17 | 19.40 | 11.55 | 10.17 |
| JNS | 39.60 | 2.04 | 0.03 | 19.37 | 60.51 | 21.30 | 10.42 | 16.63 |
| RS | 47.92 | 2.04 | 0.03 | 23.44 | 65.56 | 19.9 | 9.73 | 6.26 |
| Mean values | 44.01 | 1.92 | 0.03 | 23.10 | 69.75 | 20.20 | 10.57 | 11.02 |

*Expressed as % equivalent of tannic acid. JS: Jardin subsoiling, JNS: Jardín no-till, RS: Río Rayo subsoiling.

At the end of the study, lignin concentration increased (12.4-20.2%) with higher values in JS and JNS. The majority of the quality parameters in litter were associated significantly to RDM values (Table 7). The Pearson correlation coefficients showed a positive association between RDM and C/N, N/P and total phenols. Rainfall showed an inverse correlation ($P<0.05$) with RDM in all sites. N/P ratio and total phenols decreased abruptly at the beginning of the study (day 28), with a reduction in phenols of up to 70% (Figure 6).

## Soil Parameters

Soil parameters were improved by the 11-years old plantations (Table 8). By comparing soil conditions between unplanted sites and under *A. mangium* plantations, we detected significant ($P<0.05$) increases in soil N content (0.24 vs. 0.5%), soil available P content (2.48 vs. 6.53 mg $kg^{-1}$), and soil cation exchange capacity (CEC) (6.12 vs. 11.19 $cmol_c$ $kg^{-1}$). Additionally, significant increases in soil organic matter content (OM) (6.14 vs. 18.66%) and soil aggregate stability (AS) (73.0 vs. 85.35%) were also found.

**Table 7. Pearson correlation coefficients between RDM, rainfall and quality parameters of the leaf litter in the three sites studied on the *A. mangium* plantations in the Bajo Cauca region (Colombia)**

| Parameters | JS | JNS | RS |
|---|---|---|---|
| N | -0.76 | -0.87 | -0.83 |
| P | -0.94 | -0.93 | -0.93 |
| C/N | 0.81 | 0.91 | 0.89 |
| N/P | 0.86 | 0.72 | 0.78 |
| Lignin | -0.74 | -0.89 | -0.78 |
| Lignin/N | -0.24* | -0.38* | 0.04* |
| Total phenols | 0.71 | 0.64 | 0.74 |
| Total rainfall | -0.70 | -0.75 | -0.46 |

*No significant ($P$>0.05). JS: Jardín subsoiling, JNS: Jardín no-till, RS: Río Rayo subsoiling.

**Table 8. Mean values of the soil parameters of the plots located in the stands and outside the stands (unplanted plots) of *A. mangium* plantations in the Bajo Cauca region, Colombia. (Standard deviation is shown in brackets)**

| Parameters | Sites | |
|---|---|---|
| | Plantation (n=12) | Unplanted (Control) (n=6) |
| pHa (1:1) | 4.50 (0.18) | 5.39 (0.53) |
| OMa (%) | 18.66 (7.60) | 6.14 (0.67) |
| ASa (%) | 85.35 (7.49) | 73.00 (12.78) |
| MWDb (mm) | 2.85 (0.64) | 2.12 (0.78) |
| Na (%) | 0.50 (0.17) | 0.24 (0.05) |
| Pa (mg kg-1) | 6.53 (3.57) | 2.48 (0.23) |
| CECa (cmolc kg-1) | 11.19 (2.64) | 6.12 (0.35) |

OM: Organic matter content, AS: Aggregate stability, MWD: mean weight diameter of soil aggregates, N: nitrogen, P: phosphorous, CEC: Cation Exchange Capacity. a and b denote the existence and inexistence, respectively, of significant differences (Mann-Whitney, $P$<0.05).

## Discussion

The fine litter production in the sites studied was relatively high compared with the values found in plantations of tropical lowlands (Parrotta 1999, Goma-Tchimbakala and Bernhard-Reversat 2006, Barlow *et al.* 2007), but comparable with those reported in degraded and poor soils (Lim 1988, Bernhard-Reversat 1993).

The greatest fine litterfall values occurred with the high quality treatments (SH and NH), as well as the highest basal area. Biomass values in the stands were similar to those reported by Brown and Lugo (1982), who found a positive lineal correlation between total litterfall and stand biomass. In addition, the absence of significant differences between high quality treatments (SH and NH) indicates the low influence of the subsoiling practice on fine litterfall (Table 2).

Total fine litterfall was composed mainly of leaf fraction (54%), followed by reproductive material (RM: 24%) and in minor proportion by other rests (OR: 6%) and leaves from other species (OL: 1.5%). Similar tendencies in fine litterfall composition have been reported in other plantations and tropical forests (Proctor *et al.* 1983, Weaver *et al.* 1986, Lim 1988, Proctor *et al.* 1989, Veneklaas 1991, Swamy and Proctor 1997, Saharjo and Watanabe 2000).

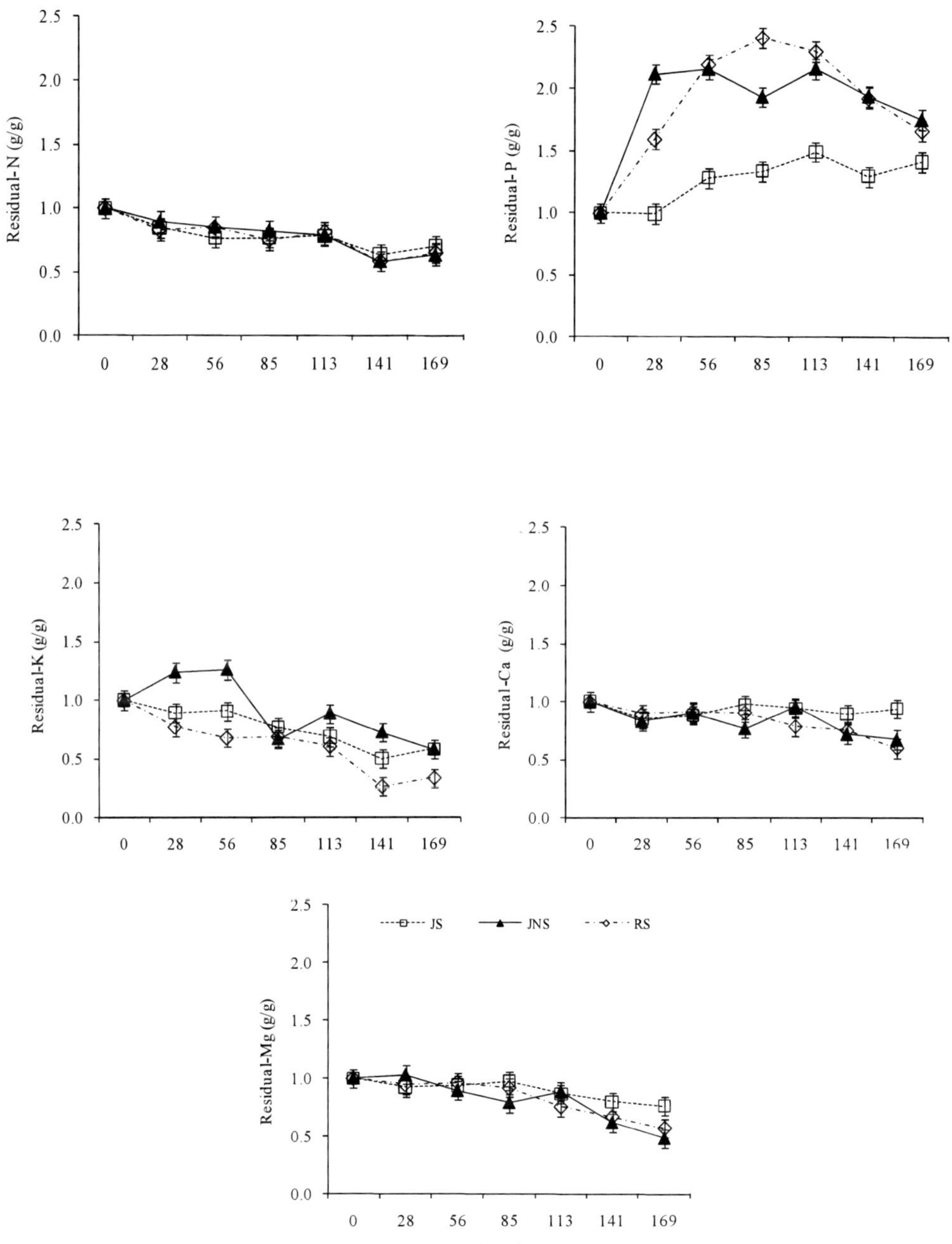


Figure 5. Residual dry matter of nutrients in the leaf litter of *A. mangium* deposited in litter bags during the period studied. Bars indicate a 95% confidence interval. JS: Jardín subsoiling, JNS: Jardín no-till, RS: Río Rayo subsoiling.

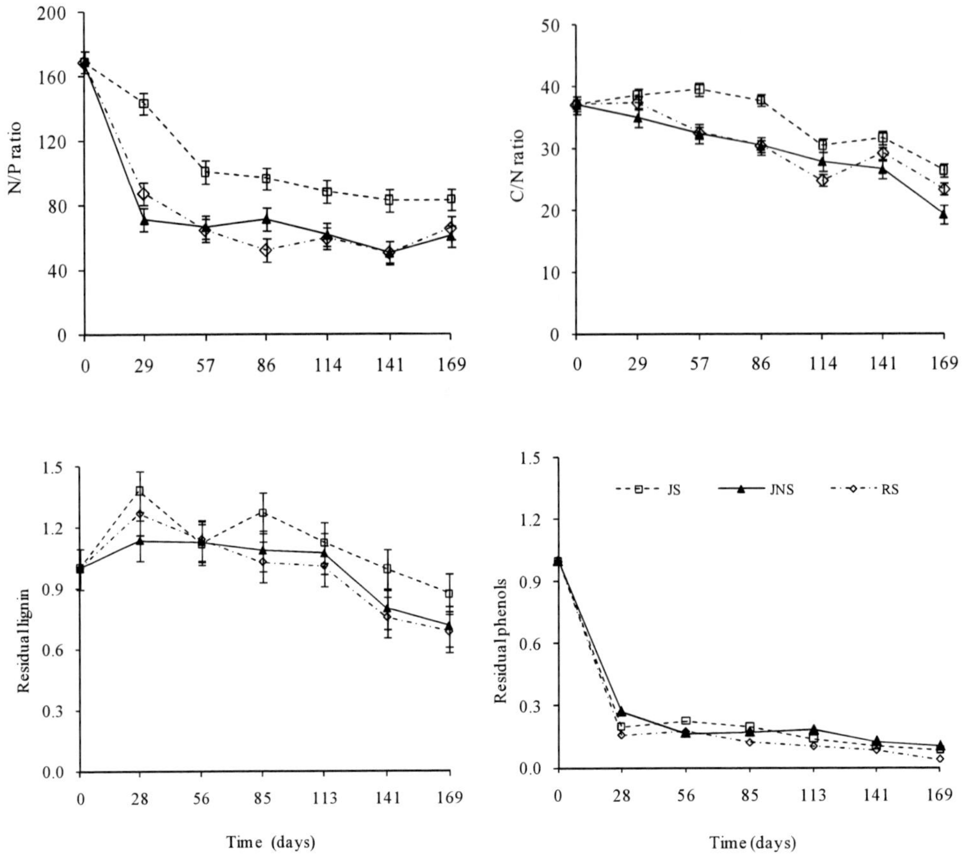


Figure 6. Evolution of RDM quality parameters in the three *A. mangium* plantation sites in the Bajo Cauca region (Colombia) during the period studied. JS: Jardín subsoiling, JNS: Jardín no-till, RS: Río Rayo subsoiling. Bars indicate a 95% confidence interval.

However, although a high fine litter production was registered during the drier periods (March-June 2007), rainfall did not significantly correlate to litterfall in any treatment. Since rain water was available along the year, even in the driest months, likely other factors determined fine litter production (soil conditions or stand structure) in these plantations (Mall *et al.* 1991).

The high leaf litter N concentration reflects the ability *A. mangium* to form symbiosis with $N_2$ fixing bacterica, particularly in a soil with low contents of N at the time of the plantation establishment. N concentrations in leaf litterfall coincided with the values reported in forests and plantations in tropical lowlands. However, our values were low compared to those reported by Seneviratne *et al.* (1998) and Ngoran *et al.* (2006) in young and mature *A. mangium* plantations in India and Africa, respectively. In general, leaf concentrations of P, K and Mg were lower than the reported values in forests and plantations in tropical lowlands. This situation could reflect a low nutrient demand by the species and/or low soil nutrient availability. In particular, low leaf concentrations of P represent a major limiting factor in nutrient cycling and plant nutrition.

According to Aerts (1997) the critical value of the N/P ratio in leaf litterfall is 11.9, in this study it was considerably higher (N/P= 124), which indicates a serious deficit in P. This confirms the experimental findings regarding the biological immobilization of this nutrient and its retranslocation in the trees (Castellanos-Barliza and León 2010). The leaf P concentration differed between the S1-N1 treatments and the S2-N2 treatments, indicating the influence of plantation quality on plant P concentration.

Significant differences were observed in N leaf litter concentration between treatments S1 and N1; however, neither the subsoiling influence nor the plantation quality influence was completely clear (Table 3). K showed higher variability among treatments, likely as a result of its high mobility this is easily removed from leaf organs by rainfall (Hagen-Thorn *et al.* 2006). N concentration had a minimum variability, which implies a higher stability and suggests that its presence in the leaves fluctuated in a narrow rank during this study. Also, the variability found in leaf litter P concentration was high. Leaf litter nutrient concentrations were generally higher during the drier months, a time period that also corresponded to the peak of fine litter production. The highest nutrient return was found with the high quality treatments (S1 and N1). With respect to N, Ca, Mg and K returns, no significant differences were found, but a correspondence between high quality treatments (S1 and N1) and high values of these nutrient returns was evident. Nutrient return had low variability in all treatments; N, Ca, and K returns corresponded to those reported in forests and plantations in low tropical lands, in contrast with P and Mg returns, which presented lower values.

RDM evolution in the three sites showed a first stage in which the loss of weight was higher until day 141. A higher weight loss in the first stages of the decomposition process was also reported by Arellano *et al.* (2004) and Goma-Tchimbakala and Bernhard-Reversat (2006). These losses were produced as a result of the splitting and use of labile compound in litterfall by decomposers. The *k* values in these degraded soils were higher than the values reported by McGrath *et al.* (2000) in *Theobroma grandiflorum* plantations in the Brazilian Amazon ($k$=0.4). The mean life ($t_{0.5}$) found here was similar to that reported by Singh *et al.* (1999) in India with *Dalbergia sissoo* leaf litter ($t_{0.5}$=0.41 years). The time needed for 99% of the leaf litter to disappear ($t_{0.99}$) was about 2.6-3.7 years, which is similar to the reports of Del Valle (2003) in the Colombian Pacific flood forests, considered to be intermediate tropical forests.

This first stage was followed by a second stage in which decomposition was slower, likeley occasioned by the dominance of recalcitrant materials. In searching for a possible association between rainfall and material weight loss in the litter bags, an inverse correlation was found being this stronger in Jardin sites with lower rainfall regime (JS and JNS) (Table 7). However, this relatioship was not found in the ANOVA with previous subsoiling treatments. This likely means that other factors such as: weather, chemical litter composition, and microorganism activity influenced litter decomposition on these sites (Swift *et al.* 1989, Aerts and Chapin 2000).

We found that litter quality indexes (lignin, phenols N, C, P content, C/N and C/P ratios) were good predictors of litter decomposition as reported by several authors (Xuluc-Tolosa *et al.* 2003, Ngoran *et al.* 2006, Martínez-Yerízar *et al.* 2007, Prause and Fernández 2007). Organic substrate mineralization generally can be predicted from the N content and the C/N ratio. For instance, when the N content are between 1.0 and 2.5% and the C/N ratio is < 25-40, organic matter decomposition and N mineralization occur rapidly (Seneviratne *et al.* 1998, Torreta and Takeda 1999, Ngoran *et al.* 2006), particularly when the lignin contents are low (Ribeiro *et al.* 2002). The values of these parameters in the leaf litter, led us to expect a low decomposition rate and scarce N release (Bubb *et al.* 1998).

However, this relative rapid decomposition rate did occur. In other complementary studies (León *et al.,* 2008) found active soil functional groups that play an important role in the biochemical cycle ($N_2$ fixing bacteria in rhizobial nodules, phosphorous solubilizing microorganisms, and mycorrhizal arbuscular fungi).

N and P concentration showed a strong inverse correlation with RDM (Table 7), which denote their individual influence on decomposition. Leaf litter P concentration was lower than in other tropical leguminous plants (Palm and Sánchez 1990), which probably indicates its low circulation in the plant-soil systems due to (i) scarce availability in these mining-degraded soils and (ii) a high mobility inside the plants.

On the other hand, lignin content in the litter bag materials was lower than that reported in other tropical and subtropical plantations of *A. mangium* (Bernhard-Reversat and Schwartz 1997, Kadir *et al.* 2001). Although lignin content may regulate the decomposition rate (Bernhard-Reversat 1993), this did not ocurr in our study. The negative correlation between lignin and RDM suggests that a higher decomposition rate leads to higher lignin content, which contrasts with the recalcitrant character of this compound; making the degradation of organic matter more difficult to achieve (Meentemeyer 1978, Arunachalam and Singh 2002, Prause and Fernández 2007). In fact, at the end of this study we detected an increase in lignine concentration respect to intial values (Table 6). This suggests that recalcitrant compounds increase over time. Despite of that, this apparently did not affect the leaf litter weight loss in the litter bags (Figure 4).

Palm and Sánchez (1990) indicated that low soluble phenol concentration in a leaf litter produced a faster decomposition than materials with high phenol values. The phenol content found in this study was higher than that reported by Kadir *et al.* (2001) and Palm and Sánchez (1990) in *A. mangium* and other tropical leguminous plants (Table 7). Decomposition rates obtained on this study according to the simple exponential model corresponded to intermediate values of the reported interval in tropical forest and plantations ($k$=0.1-4.8) (Sundarapandian and Swamy 1999, Singh *et al.* 1999, Kurzatkowski *et al.* 2004). These results could have important repercussions from the perspective of degraded land restoration, since the decomposition rates found indicate that litterfall contribute in the process with high levels of organic matter and nutrients, which in turn contributes to the reestablishment of the soil physicochemical properties as well as biological activity (León *et al.* 2010).

RDM nutrient concentration showed a decreasing sequence: N>Ca>Mg>K>P, which is coincident with previous reports regarding *A. mangium* plantations (Ngoran *et al.* 2006; Singh *et al.* 2004). Among other factors, this situation may be found to be the result of atmospheric depositions washed from the canopy on the soil surface, leaf washing, mycelial invasion, and abundant microorganism presence in the topsoil. Nutrient release followed the decreasing sequence K>N>Mg>Ca>P (Figure 5). Rapid release of K is widely reported due to its mobile condition. For example, Villela and Proctor (2002) reported that in tropical forests in Pará (Brasil), K loss of 70% was found in *Ecclinusa guianensis* leaves. Similar tendencies were reported by Ngoran *et al.* (2006) in *A. mangium*, reaching a K loss over 80%.

Three phases were distinguished in N release: an initial slow phase (until day 113), a fast phase (day 141) and a final phase dominated by immobilization (Figure 5). This N release pattern was different than that found in other studies (Bubb *et al.* 1998) and suggests that N release are regulated by site conditions, fundamentally by microbial activity and litterfall quality (Gallardo-Lancho 2000). A slow N liberation possibly resulted from low soil microbe levels.

It is outstanding the increase in the soil organic matter, N, and P contents; as well as the improvement in soil physical parameters. N concentration in the soil was quite high in the plantations (with a mean of 0.5%), more than two times the values found on the unplanted plots, likely as a result of symbiotic associations established by the species with bacteria fixing atmospheric nitrogen. Undoubtedly, these changes in soil parameters, after 11-years, support our hypothesis that *A. mangium* plantations help to remediate soil conditions in these degraded lands.

The positive effect of the N content of leguminous species on soil has been highly recognized (Pearson and Vitousek 2001). In the Acacia genus, there have been reports on *A. albida* in arid soils in the Oriental Africa and Center-Sur Africa, where Acacia has often been introduced as part of desertification control programs. There, Dunham (1991) encountered N concentrations from 0.08% to 0.15% on sites outside the plantations, with all of these values being inferior to the findings of this study. The species' capacity to establish symbiotic relationships with bacteria-fixing atmospheric nitrogen offers edaphic N the independency to grow and develop, reaching higher leaf concentrations with a mean value of 6.4% (León *et al.* 2010). P increased significantly in the soils in the plantations in front of the unplanted plots (Table 8). Dunham (1991) found statistically significant increases in soil P under *Acacia albida* plantations – as much as 60% higher than on unplanted plots. The limited availability of P in the soil was reflected in its concentration in leaves, especially in the plantation's mature leaves, which according to León *et al.* (2010) ranged between 0.036% and 0.047%.

Soil organic matter and aggregate stability were improved by the plantation's development. This was likely produced as a result of (i) a better aggregate soil structure by fine roots that occupy the top soil, and (ii) contributions of organic matter derived from fine litter decomposition.

As a consequence of the later process, higher soil acidity can be developed as found in the plantation soils, due probably to the fact that during the decomposition process organic acids are liberated and acidic compounds are also produced by the roots (Marschner 1997).

## Conclusion

The results of this study clearly show that *A. mangium* plantations have a great capacity to reestablish the biogeochemical cycles in soils degraded by mining activities and to improve soil physicochemical and biological parameters. The large amount of fine litterfall in these plantations represented a potential source of organic matter and nutrients. Litter decomposition was considered rapid and may be predicted at least partially from some litter quality indexes and rainfall. The annual decomposition constant (*k*) ranged between 1.25 and 1.80, the C return onto the soil fluctuated between 2.0 and 2.4 Mg $ha^{-1}$. These return rates measured were probably lower than the real values, due to the restrictive conditions imposed by the litter bags and because only the leaf fraction of the fine litterfall was considered in this estimate. The contributions of organic matter, carbon, and nutrient release from fine litter decomposition substantially improved some soil properties (soil organic matter, N and P) by 2-3 times respect to unplanted sites. In general, effective release of Ca, Mg and K was registered with similar patterns. P was the most restrictive element in the decomposition leaf litter and plantation productivity.

The high contributions of organic matter and nutrients via fine litterfall and decomposition indicate that even without forestry management practices, the biogeochemical cycles could be reactivated and soil improved.

## References


Aerts, R., 1997. Climate, leaf litter chemistry and leaf litter decomposition in terrestrial ecosystems: a triangular relationship. *Oikos* 7, 439-449.

Aerts, R. and Chapin, F. S., 2000. The mineral nutrition of wild plants revisited: a re-evaluation of processes and patterns. *Adv. Ecol. Res.* 30, 1-67.

Arellano, R., Paolini, J., Vásquez, L., Mora, E., 2004. Producción y descomposición de hojarasca en tres agroecosistemas de café en el estado de Trujillo, Venezuela. *Rev. Forest. Ven.* 48, 7-14. (In Spanish).

Arunachalam, A. and Singh, N. D., 2002. Leaf litter decomposition of evergreen and deciduous Dillenia species in humid tropics of north-east India. *J. Trop. Forest Sci.* 14, 105-115.

Barlow, J., Gardner, T. A., Ferreira, L. V., Peres, C. A., 2007. Litter fall and decomposition in primary, secondary and plantation forests in the Brazilian Amazon. *Forest Ecol. Manag.* 247, 91-97.

Benton, J. J., Wolf, J. B., Mills, H. A., 1991. Plant Analysis Handbook: A Practical Sampling, Preparation, Analysis and Interpretation Guide. *Micro-Macro Publishing.*

Bernhard-Reversat, F., 1993. Dynamics of litter and organic matter at the soil-litter interface in fast-growing tree plantations on sandy ferrallitic soils (Congo). *Acta Oecol.* 14(2), 179-195.

Bray, R. H. and Kurtz, L. T., 1945. Determination of total, organic, and available forms of phosphorus in soil. *Soil Sci.* 59, 39-45.

Brown, S. and Lugo, A. E., 1982. The storage and production of organic matter in tropical forests and their role in the global carbon cycle. *Biotropica.* 14, 161-187.

Bubb, K. A., Xu, Z. H., Simpson, J. A., Safigna, P. G., 1998. Some nutrient dynamics associated with litterfall and litter decomposition in hoop pine plantations of southeast Queensland, Australia. *Forest Ecol. Manag.* 110, 343-352.

Castellanos-Barliza, J. and León, J. D., 2010. Caída de hojarasca y dinámica de nutrientes en plantaciones de Acacia mangium (Mimosaceae) de Antioquia, Colombia. *Acta Biol. Colomb.* 15, 289-308. (In Spanish).

Del Valle, J. I., 2003. Cantidad, calidad y nutrientes reciclados por la hojarasca fina en bosques pantanosos del Pacífico Sur Colombiano. *Interciencia.* 28, 443-449. (In Spanish).

Dunham, K. M., 1991. Comparative effects of Acacia albida and Kigelia africana trees on soil characteristics in Zambezi Riverine woodlands. *J. Trop. Ecol.* 7, 215-220.

Duponnois, A. and Ba, A. M., 1999. Growth stimulation of Acacia mangium Willd by Pisolithus sp. in some Senegalese soils. *Forest Ecol. Manag.* 119, 209-215.

Ewel, J. J., 1976. Litter fall and leaf decomposition in a tropical forest succession in eastern Guatemala. *J. Ecol.* 64, 293-308.

Gallardo-Lancho, J. F., 2000. Biogeochemistry of Mediterranean forest ecosystems, a case study. En: J. M. Bollag and G. Stotzky (eds). *Soil Biochemistry* 10, Marcel Dekker, Nueva York. 423-60.

Garay, I., Pellens, R., Kindel, A., Barros, E., Franco, A., 2004. Evaluation of soil conditions in fast-growing plantations of Eucalyptus grandis and Acacia mangium in Brazil: a contribution to the study of sustainable land use. *Applied Soil Ecology*. 27, 177-187.

Garten, C., 2002. Soil carbon storage beneath recently established tree plantations in Tennessee and South Carolina, US. *Biomass and Bioenergy*. 23, 93-102.

Goma-Tchimbakala, J. and Bernhard-Reversat, F. 2006. Comparison of litter dynamics in three plantations of an indigenous timber-tree species (Terminalia superba) and a natural tropical forest in Mayombe, Congo. *Forest Ecol. Manag.* 229, 304–313.

Hagen-thorn, A., Varnagiryte, I., Nihlgard, B., Armolaitis, K., 2006. Autumn nutrient resorption and losses in four deciduous forest tree species. *Forest Ecol. Manag.* 228,33-39.

Higa, A. R. and Higa, R. C. V., 2000. Indicações de espécies para o reflorestamento. In: Galvão, A. P. M. (Org.), Reflorestamento de Propriedades Rurais para Fins Produtivos e Ambientais. Um Guia para Ações Municipais e Regionais. Embrapa Comunicação para Transferencia de Tecnologia e Embrapa Florestas, Brasilia, Colombo. 101–124.

Horwitz, W., 2003. Official methods of analysis of AOAC International. Association of official analytical chemists. Arlington, VA, US.

IGAC., 2007. Estudio general de suelos y zonificación de tierras del Departamento de Antioquia. Instituto Geografico Agustin Codazzi. Gobernación de Antioquia. (In Spanish).

Jordan, C. F., 1985. Nutrient cycling in tropical forest ecosystems. Principles and their application in management and conservation. John Wiley and Sons, Chichester.

Kadir, W. R., Cleemput, O. V., Zaharah, A. R., 2001. Microbial respiration and nitrogen release patters of decomposing Acacia mangium leaf litter from Kemasul forest reserve, Malaysa. *J. Trop. Forest Sci.* 13, 1-12.

Kalra, Y. and Maynard, D., 1991. *Methods manual for forest soil and plant analysis. Information Report NOR-X-319,* Northern Forestry Centre, Alberta, Canadá, 113 p.

Kumar, R. and Agrawal, M. 2001. Litterfall, litter decomposition and nutrient release in five exotic plant species planted on coal mine spoils. *Pedobiologia*. 45, 298-312.

Kurzatkowski, D., Martius, C., Höfer, H., Garcia, M., Förster, B., Beck, L., Vlek, P., 2004. Litter decomposition, microbial biomass and activity of soil organisms in three agroforestry sites in central Amazonia. *Nutr. Cycling Agroecosyst.* 69, 257-267.

Laclau, J. P., Bouillet, J. P., Ranger, J., 2000. Dynamics of biomass and nutrient accumulation in a clonal plantation of Eucalyptus in Congo. *Forest Ecol. Manag.* 128, 181-196.

Landsberg, J. J. and Gower, S. T., 1997. *Applications of physiological ecology to forest management.* New York: Academic Press.

León, J. D., Díez, M. C., Castellanos, J., Osorio, L. F., Marín, N., 2008. Grupos funcionales de microorganismos en suelos degradados por minería de aluvión plantados con Acacia mangium. *Suelos Ecuat.* 38, 75-80. (In Spanish).

León, J. D., Osorio, N. W., Castellanos, J., Osorio, L. F., 2010. Recuperación de suelos degradados por minería aluvial de oro con plantaciones de Acacia en la región del Bajo Cauca (Antioquia, Colombia). *Suelos Ecuat.* 40, 62-67. (In Spanish). (In Spanish).

Liao, J. H., Wang, H. H., Tsai, C. H., Hseu, Z. Y., 2006. Litter production, decomposition and nutrient return of uplifted Coral Reef tropical forest. *Forest Ecol. Manag.* 235, 4-185.

Lim, M. T., 1988. Studies on Acacia mangium in Kemasul forest, Malaysia. I. Biomass and productivity. *J. Trop. Ecol.* 4, 293-302.

Lugo, A., Cuevas, E., Sánchez, M., 1990. Nutrients and mass in litter and top soil of ten tropical tree plantations. *Plant Soil.* 125, 263-280.

Lugo, A., 1997. The apparent paradox of reestablishing species richness on degraded lands with tree monocultures. *Forest Ecol. Manag.* 99, 9-19.

Mall, L. P., Singh, V. P., Garge, A., 1991. Study of biomass, litter fall, litter decomposition and soil respiration in monogeneric mangrove and mixed mangrove forest of AndamanIsland. *Trop. Ecol.* 32, 144-152.

Marschner, H., 1997. *Mineral nutrition of higher plants.* Academic Press, San Diego, California. 889 p.

Martínez-Yrízar, A., Nuñez, S., Búrquez, A., 2007. Leaf litter decomposition in a southern sonoran desert ecosystem, northwestern México: Effects of habitat and litter quality. *Acta Oecol.* 32, 291-300.

McGrath, A. D., Comerford, N. B., Duryea, M. L., 2000. Litter dynamics and monthly fluctuations in soil phosphorus availability in an Amazonian agroforest. *Forest Ecol. Manag.* 131, 167-181.

Meentemeyer, V., 1978. Macroclimate the lignin control of litter decomposition rates. *Ecology.* 59, 465-472.

Montagnini, F. and Jordan, C. F., 2002. Reciclaje de nutrientes. En: M. R. Guariguata and G. Kattan (eds). *Ecología y conservación de bosques neotropicales.* Cartago, Costa Rica. 167-191. (In Spanish).

Ngoran, A., Zakra N., Ballo, K., Kouamé, C., Zapata, F., Hofman, G., Van Cleemput, O., 2006. Litter decomposition of Acacia auriculiformis Cunn. Ex Benth. and Acacia mangium Willd. under coconut trees on quaternary sandy soils in Ivory Coast. *Biol. Fertil. Soils.* 43, 102-106.

Norgrove, L. and Hauser, S., 2000. Production and nutrient content of earthworm casts in a tropical agrisilvicultural system. *Soil Biol. and Biochem.* 32, 1651-1660.

Olson, J. S., 1963. Energy storage and the balance of producers and decomposers in ecologycal systems. *Ecology.* 44, 322-331.

Osorio, N. W. and Casamitjana, M. 2011. Toma de muestras de suelos para evaluar la fertilidad del suelo. *Suelos Ecuatoriales* 41, 23-28.

Osorio, P. I., 2000. Efecto de la Acacia mangium en algunas propiedades físicas y químicas en suelos degradados por minería de aluvión en el Bajo Cauca Antioqueño. Trabajo de grado en la Maestría de Bosques y Conservación Ambiental. Facultad de Ciencias Agropecuarias, Universidad Nacional de Colombia, Medellín, 97 p. Msc Thesis. (In Spanish).

Palm, C. A. and Sánchez, P. A., 1990. Decomposition and nutrient release patterns of the leaves of three tropical legumes. *Biotropica.* 22, 330-338.

Palma, R. M., Prause, J., Fontanive, A. V., Jimenez, M. P., 1998. Litterfall and litter decomposition in a forest of the Parque Chaqueño Argentino. *Forest Ecol. Manag.* 106, 205-210.

Parrota, J., 1992. The role of plantation forests in rehabiliting degraded ecosystems. *Agric. Ecosyst. Environ.* 41, 115-133.

Parrotta, J., 1999. Productivity, nutrient cycling, and succession in single and mixed species plantations of Casuarina equisetifolia, Eucalyptus robusta, and Leucaena leucocephala in Puerto Rico. *Forest Ecol. Manag.* 124: 45 -77.

Pearson, H. and Vitousek, P. 2001. Stand dynamics, nitrogen accumulation and symbiotic nitrogen fixation in regenerating stands of Acacia koa. *Ecol. Appl.* 11, 1381-1394.

Prause, J. and Fernández, C., 2007. Litter decomposition and lignin/cellulose and lignin/total nitrogen rates of leaves in four species of the Argentine Subtropical forest. *Agrochimica.* 51, 294-300.

Proctor, J., Phillips, C., Duff, G. K., Heaney, A., Robertson, F. M., 1989. Ecological studies on Gunung Silam, a small ultrabasic mountain in Sabah, Malaysia. Some forest processes. *J. Ecol.* 77, 317-331.

Proctor, J., 1983. Tropical forest litterfall. I. Problems of data comparison. En: Sutton, S. L., Whitmore, T. C. and Chadwick, A. C., editors. *Tropical rain forest.* Oxford, Blackwell Scientific Publications. 267-273.

Ramírez, J. A., Zapata, C. M., León, J. D., González, M. I., 2007. Caída de hojarasca y retorno de nutrientes en bosques montanos andinos de Piedras Blancas, Antioquia, Colombia. *Interciencia.* 32 (5), 303-311. (In Spanish).

Ribeiro, C., Madeira, M., Araújo, M. C., 2002. Decomposition and nutrient release from leaf litter of Eucalyptus globulus grown under different water and nutrient regimes. *Forest Ecol. Manag.* 171, 31-41.

Ribet, J. and Drevon, J. J., 1996. The phosphorus requirement of N2 fixing and urea-fed Acacia mangium. *New Phytol.* 132, 383–390.

Saharjo, B. H. and Watanabe, H., 2000. Estimation of litter fall and seed production of Acacia mangium in a forest plantation in south Sumatra, Indonesia. *Forest Ecol. Manag.* 130, 265-268.

Schlesinger, W. H., 2000. *Biogeoquímica: un análisis global.* Ariel Ciencia. Barcelona. (In Spanish).

Seneviratne, G., Van Holm, L. H. J., Kulasooriya, S. A., 1998. Quality of different mulch materials and their decompositionand N release under low moisture regimes. *Biol. and Fertil. Soils.* 26, 136-140.

Singh, G., Singh, B., Kuppusamy, V., Vala, N., 2002. Variations in foliage and soil nutrient composition in Acacia tortilis plantations of different ages in North-Western Rajashtan. *Indian Forester.* 128, 514-521.

Singh, K. P., Singh, P. K., Tripathi, S. K., 1999. Litterfall, litter decomposition and nutrient release patterns in four native tree species raised on coal mine spoil at Singrauli, India. *Biol. and Fertil. Soils.* 29, 371-378.

Singh, R., Kumar, R., Agrawal, M., 2004. Litter decomposition and nutrient release in relation to atmospheric deposition of S and N in a dry tropical region. *Pedobiologia.* 48, 305-311.

Singleton, V. L., Rossi, J. A., 1965. Colorimetry of total phenolics with phosphomolybdic-phosphotungstic acid reagents. *Am. J. Enol. Vitic.* 16, 144-158.

Spain, A. V., 1984. Litterfall and the standing crop of litter in three tropical Australian rainforests. *J. Ecol.* 72, 947-961.

Sundarapandian, S. M. and Swamy, P. S., 1999. Litter production and leaf-litter decomposition of selected tree species in a tropical forest at Kodayar in the western Ghats, India. *Forest Ecol. Manag.* 123, 231-244.

Swamy, S. and Proctor, J., 1997. Fine litterfall and its nutrients in plantation of Acacia auriculiformis, Eucalyptus tereticornis and Tectona grandis in the Chikmagalur district of the western Ghats, India. *J. Trop. Forest Sci.* 10(1), 73-85.

Swift, M. J. and Anderson, J. M., 1989. Decomposition. In: H. Lieth and M. J. A. Werger (eds). *Tropical rain forest ecosystems: Biogeographical and ecological studies. Ecosystems of the world 14A*. Elsevier Science, Nueva York, Nueva Cork. 547-569.

Tejada, I., 1992. Control de calidad y análisis de alimentos para animales. Secretaría de Educación Pública. Dirección General de Derechos de Autor. Número de registro 17222. México D.F. (In Spanish)

Torres, D. A. and Del Valle, J. I., 2007. Growth and yield modelling of Acacia mangium in Colombia. *New Forest.* 34, 293-305.

Torreta, N. K. and Takeda, H., 1999. Carbon and nitrogen dynamics of decomposing leaf litter in a tropical hill evergreen forest. *Eur. J. Soil Biol.* 35, 57-63.

Veneklaas, E. J., 1991. Litterfall and nutrient fluxes in two montane tropical rain forests, Colombia. *J. Trop. Ecol.* 7, 319-336.

Villela, D. M. and Proctor, J., 2002. Leaf litter decomposition and monodominance in the Peltogyne forest of Maracá island, Brazil. *Biotropica.* 34, 334-347.

Vitousek, P. M., Sanford, R. L., 1986. Nutrient cycling in moist tropical forest. *Annu. Rev. Ecol. Syst.* 17, 137-167.

Wang, Q., Wang, S., Huang, Y., 2008. Comparisons of litterfall, litter decomposition and nutrient return in a monoculture Cunninghamia lanceolata and a mixed stand in southern China. *For. Ecol. Manage.* 255, 1210-1218.

Weaver, P. L., Medina, E., Pool, D., Dugger, K., González-Liboy, J., Cuevas, E., 1986. Ecological observations in the dwarf cloud forest of the Luquillo mountains in Puerto Rico. *Biotropica.* 18, 79-85.

Wieder, R. K. and Lang, G. E., 1982. A critique of the analytical methods used in examining decomposition data obtained from litter bags. *Ecology.* 63, 1636-1642.

Xuluc-Tolosa, F. J., Vestera, H. F. M., Ramírez-Marcial, N., Castellanos-Albores, J., Lawrence, D., 2003. Leaf litter decomposition of tree species in three successional phases of tropical dry secondary forest in Campeche, México. *Forest Ecol. Manag.* 174, 401-412.

Yoder, R., 1936. A direct method of aggregate analysis of soils and a study of the physical nature of erosion losses. *J. Amer. Soc. Agron.* 28, 337-351.

In: Plantations
Editor: Hai Ren

ISBN: 978-1-62808-090-2


*Chapter 10*

# EFFECTS OF INCORPORATING *FURCRAEA* SPECIES BIOMASS INTO ACIDIC ANDISOLS

***Adriana M. Quinchía Figueroa[1*], Juliana Uribe Castrillón[1] and Carolina Mesa Muñoz[1]***

[1] Escuela de ingeniería de Antioquia (EIA).Unidad Académica Ing. Civil, Industrial y Ambiental. Sede de Las Palmas, Antioquia, Colombia

## ABSTRACT

*Furcraea* species are native to the Andes mountain region of Colombia and Venezuela, and are known by the names hemp, sisal, stalk, cabuya, or maguey. These species grow at any altitude, but they grow best between 1000 and 2000 meters above sea level. *Furcraea spp.* are considered effective at rehabilitating poor-quality and eroded soil, at retaining moisture in their fleshy leaves, thus requiring little water, and at capturing $CO_2$. However, the process of extracting the long fibers of *Furcraea spp.* produces a high volume of biomass waste with high concentrations of saponins, sapogenins, and steroids, substances that are toxic to some organisms. Therefore, it is important to evaluate the specific effects of incorporating biomass waste into soils according to the region where it is grown and processed. This chapter evaluates the effects of incorporating *Furcraea spp.* biomass generated during the artisanal extraction of long sisal fibers on the physical and microbiological properties and chemical conditions of andisols. The biomass was analyzed for its composition, and significant amounts of lignin and cellulose, both of which are important in the generation of labile organic matter, were found. The experiment was conducted in separate reactors with biomass mixtures and sisal soils in proportions ranging between 10% and 40% (w/w). The tests showed that biomass generates a positive change in soil porosity, organic matter levels, and cation exchange capacity (CEC), with mean increases of 40%, 6% and 10 CEC Cmol $kg^{-1}$, respectively.

**Keywords**: *Furcraea spp.*, Biomass, Andisol, Soil properties, Native forest

* Corresponding Author. Escuela de Ingeniería de Antioquia (EIA).Calle 25 sur # 42 -73 Envigado – Antioquia – Colombia. Email: aquinchia@eia.edu.co.

# INTRODUCTION

Plants of the genus *Furcraea* originate from tropical regions of South and Central America, with the *Furcraea andina* species originating in the Andean region of Colombia and Venezuela. The plants have long been used by indigenous inhabitants as living fences, for the production of alcoholic beverages, and as medicine (Aristizabal & Zapata, 1990). The genus *Furcraea* includes nearly 20 species and belongs to the family *Agavaceae*, the calss Monocotyledonae and the Angiosperm (Montoya & Tobón, 1979). These rugged plants can adapt to many climates. They are found from 0 to 3000 meters above sea level and grow in soils derived from volcanic ash, as well as gravelly and loose soils, although they prefer porous and moderately acidic soils. *Furcraea* species have deep perennial root systems that bind loose soils together, allowing them to fix atmospheric nitrogen and to assist in the regeneration of eroded soils with low fertility (Bonilla et al., 2000, Jimenez, 1968). *Furcraea* leaves are persistent, lanceolate, and generally ten times longer than they are wide, often with entire margins serrated or festooned. When the leaves are developed, they are fleshy, with parallel veins, and in their mature stage some species exhibit leaves that can measure up to three meters in length and produce long, commercially attractive fibers (SDAFE, 2002) (Figure 1).

Figure 1. Sisal cultivation in the Andean region of Colombia.

From the sisal leaf, known colloquially in Colombia as "fique", a long, natural and biodegradable fiber known locally as "cabuya" is extracted and used in the manufacture of nets, ropes, bags, and facial tissues. Besides having a positive effect on mitigating soil erosion, the plant retains soil moisture and fixes atmospheric $CO_2$. In the plant biomass, the stored carbon is transformed into substances such as lignin, cellulose, hemicellulose, protein, and sugar. During the process of manufacturing products, a significant amount of biomass waste is generated, including the approximately 96% of leaf biomass that is discarded after defibration. This can generate environmental hazards (particularly for aquatic organisms) due to the waste's high content of sapogenic substances, proteins, steroids, and toxic minerals (MAVDT, 2006).

Biomass production and decomposition are two important processes that aid soils by providing organic matter, regulating nutrient cycling, improving physical properties and providing protection against the effects of climatic changes and land degradation (Singh et al., 1999; Weltzin et al., 2005). Biomass quality and composition vary among species and locations depending on the soil's physio-chemical properties, climatic conditions, and microbial activity in the soil (Polyakova and Billor, 2007; Wang et al., 2008). Differences in these factors influence biomass pools. Forest ecosystems have been widely recognized as a key factor in nutrient cycling in global terrestrial ecosystems (Meentemeyer, 1978; Vitousek, 1982; Van Vuuren et al., 1993; Vitousek et al., 1994; Aerts and De Caluwe, 1997; Aerts and Chapin, 2000). In Colombia, large expanses of the Andean forest highlands have been removed for agricultural land use (Etter et al., 2006), and forestry plantations have altered both biodiversity and endemism in the region (Gentry & Dodson, 1987; Henderson et al., 1991).

Some of the current problems of applied ecology include a lack of effectiveness and environmental safety in the process of waste reduction and disposal, as well as the environmental impact of the waste. In this regard, the "fique" crop's plantroot provides physical environmental benefits that increase the soil's ability to retain the low-mobility carbon present in the lignocellulosic material originating from the leaf biomass. Additionally, the plantroot can improve the physical and chemical properties of these soils.

This study examines the effect of applying *Furcraea* Andina Trel biomass produced during the manual extraction of natural fibers to acidic andisols. The biomass, as an important source of fixed carbon originating from atmospheric $CO_2$, was analyzed to determine its chemical composition and cellulose, hemicellulose and lignin contents. Changes in physical, chemical and biological properties in an andisol (source of the test) as a result of biomass application were studied for a period of four months under controlled conditions.

## METHODS

This study used an andisol located in the northern region of the Central Cordillera of the Colombian Andes at 2230 meters above sea level (6° 22' 2019,75''N – 75° 21' 38,91''W). The study took place in the municipalities of San Vicente and Barbosa in the Antioquia department. The annual average temperature is 25°C and the annual average rainfall is 2800 mm (IGAC, 2007) (Figure 2). The soil of this region is characterized as acidic, and was physically, chemically and biologically analyzed to determine its initial conditions. The residual biomass resulting from fiber extraction was supplied by farmers from the same region, and its chemical composition was analyzed as well. Physical, chemical, microbiological, bromatolological and infrared analyses, as well as scanning electron microscopy, were carried out in the laboratories of the National University of Colombia-Medellín and the Engineering School of Antioquia (UN-EIA).

The experiment consisted of five tests that used superficial soil mixtures and fique biomass in different proportions. The tests were based on a univariate experimental design (analysis of variance). Four samples consisting of fique biomass proportions of 10, 20, 30 and 40%, were analyzed. The tests, which were carried out over the course of four months, assessed changes in the soil properties. The average temperature throughout the test was 27 ± 2°C. For each sample, a control corresponding to the soil was used with no addition of fique bagasse. All tests were performed in triplicate.

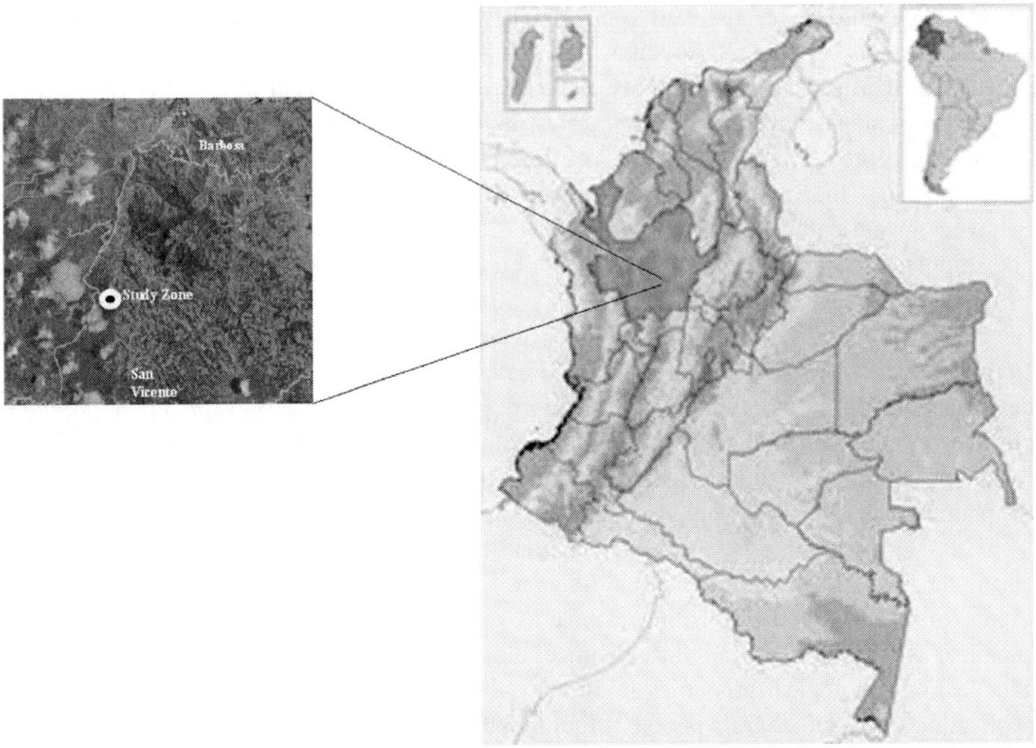


Figure 2. Study area in west central Antioquia, Colombia.

As a complement to the laboratory analysis of the biomass, an analysis was carried out using Fourier Transform Infrared Spectroscopy (FTIR) to determine the main functional groups of the fibers analyzed. Additional images of the surface material were generated using Scanning Electron Microscopy (SEM).

## RESULTS

Due to the fact that biomass waste consists not only of short fibers but also of remaining parts of the leaf parenchyma and its juice (which explains why the lignocellulosic content percentages are not very high; 51.1% in total), organic solvent extractibles, proteins, sugars and starches, as well as other organic and inorganic products of low molecular weight, can constitute a significant proportion of the raw material (Barba, 2002).

The biomass content consisted of 8% lignin, 41.5% cellulose and 1.6% hemicellulose, values comparable to those of soft wood or plant fibers such as those of coconut (David & Hon, 2003). The fixed carbon content found in lignocellulosic sisal compounds can be disposed of in the soil, becoming part of labile carbon. The lignocellulosic characteristics are verified through the infrared biomass spectrum (FTIR), as shown in Figure 3, where the band located at 3264 $cm^{-1}$ corresponds to alcohol groups, the band at 2916 $cm^{-1}$ to vibration characteristic C-H methyl and methylene groups, the band at 1609 $cm^{-1}$ to carbonyl groups

C=O and groups C=C in aromatic rings which can also generate vibrations in bands near 1412 $cm^{-1}$. The bands 1412 and 1316 $cm^{-1}$ are assigned to -CH3, CH2, double bonds, aldehydes, ketones, organic acids, esters and amides, as well as secondary alcohols and CO carboxylate groups. The 1025 $cm^{-1}$ band can be assigned to groups of alcohols (R-OH). These functional groups are typical of the structures of cellulose (CH, R-OH, CH2), hemicellulose (CH, R-OH, CH2) and lignin (C=C, C=O, CO), which are important components of sisal fiber.

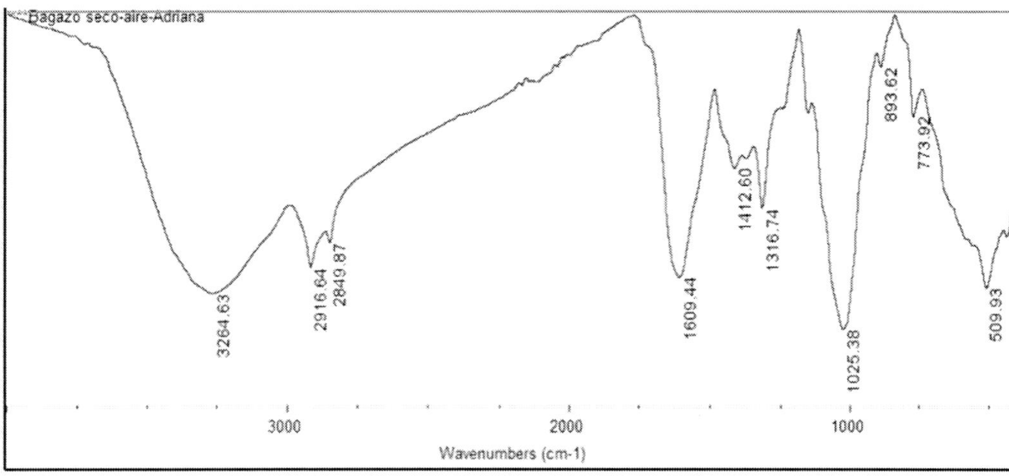


Figure 3. Fourier Transform Infrared (FTIR) Analysisof Biomass.

Analysis of the functional groups using Boehm titration (Table 1) confirmed the existence of basic structures that allow fixing of atmospheric carbon, as well as an increase in CEC values, as a result of the application of the biomass.

**Table 1. Content of functional groups in the biomass of sisal**

| Basic (Meq $H^{+}g^{-1}$) | Acid (Meq $H^{+}g^{-1}$) | | |
|---|---|---|---|
| | Carboxylic | Lactonic | Phenolic |
| 1.119 ± 0.047 | 0.574 ± 0.075 | 0.235 ± 0.0240 | 0.910 ± 0.198 |

Scanning electron microscopy analysis (SEM) (Figure 4) shows the surface structure and porosity of the biomass fiber, which provided the soil mixtures with enough empty spaces to improve porosity and dry density, and also generated stable and enduring structures during the tests. The chemical composition of the biomass is shown in Table 2, the elemental contents supplied by the material when it is added to the surface of a degraded soil can be observed. The effect of this contribution was demonstrated through subsequent evaluation of the mixtures.

The initial soil (shown in Table 3) displayed moderate acidity, low CEC, high bulk density (BD) and low porosity, the last of which is characteristic of andisols. Changes that occurred during the disposal of biomass were observed to be positive in all cases.

Figure 4. Biomass sisal, scanning electron microscopy (SEM) with approach of the fiber.

**Table 2. Results of residual foliar fique biomass analysis**

| N | P | S | Ca | Mg | K | Fe | Mn | Cu | Zn | B |
|---|---|---|---|---|---|---|---|---|---|---|
| % | | | | | | g g-1 | | | | |
| 1.92 | 0.21 | 0.10 | 4.27 | 0.30 | 2.91 | 394 | 18 | 5 | 17 | 23.31 |

**Table 3. Physical and chemical analysis of soil**

| DD | Porosity | Texture | | | | pH | O.M | Al | Ca | Mg | K | CECE | P |
|---|---|---|---|---|---|---|---|---|---|---|---|---|---|
| g $cm^3$ | % | A % | L % | Ar % | Clase | | % | Cmolc $kg^{-1}$ | | | | | Mg $kg^{-1}$ |
| 2,32 | 11,54 | 84 | 8 | 8 | AF | 4,8 | 14 | 1.1 | 3,4 | 0,2 | 0,18 | 4,9 | 7 |

Soil without biomass showed a coloration of 10YR2/1 (black). Soils with 10 and 20% proportions of bagasse, on the other hand, displayed a lightening in color, to 2.5Y3/1 (dark gray). This tone remained constant for the duration of the study. The same situation occured in the soil samples with 30 and 40% proportions of bagasse, which changed color to 2.5Y3/2 (very dark grayish-brown). It is important to note that the color was measured with air dried samples. The variations in color may be explained by the oxidation process of organic acids present in the biomass.

Decreases in the soil's dry density corresponded to the proportion of biomass in the soil. This change occurred in all the mixtures, with the 30% biomass sample presenting the most significant reduction. In general, a reduction in dry density is related to porosity and soil aeration and is observed with an increase in pore spaces required for aerobic organisms. According to Jaramillo (2002), the reduction in dry density correlates with the content of organic matter, and it can improve soil structure by increasing particle aggregation. This characteristic has been shown in physical degradation and resource degradation studies. The biomass/soil mixtures showed dry density reductions of approximately 40%, as can be seen in Figure 5.

Changes in density were indirectly correlated to the percentage of porosity, indicating that a direct supply of sisal biomass may increase an andisol´s porosity from 11.54% to 56.82%. Good porosity development of soil micro- and macrobiota requires values close to 50%, as suggested by Jaramillo (2002). The test results are shown in Figure 5.

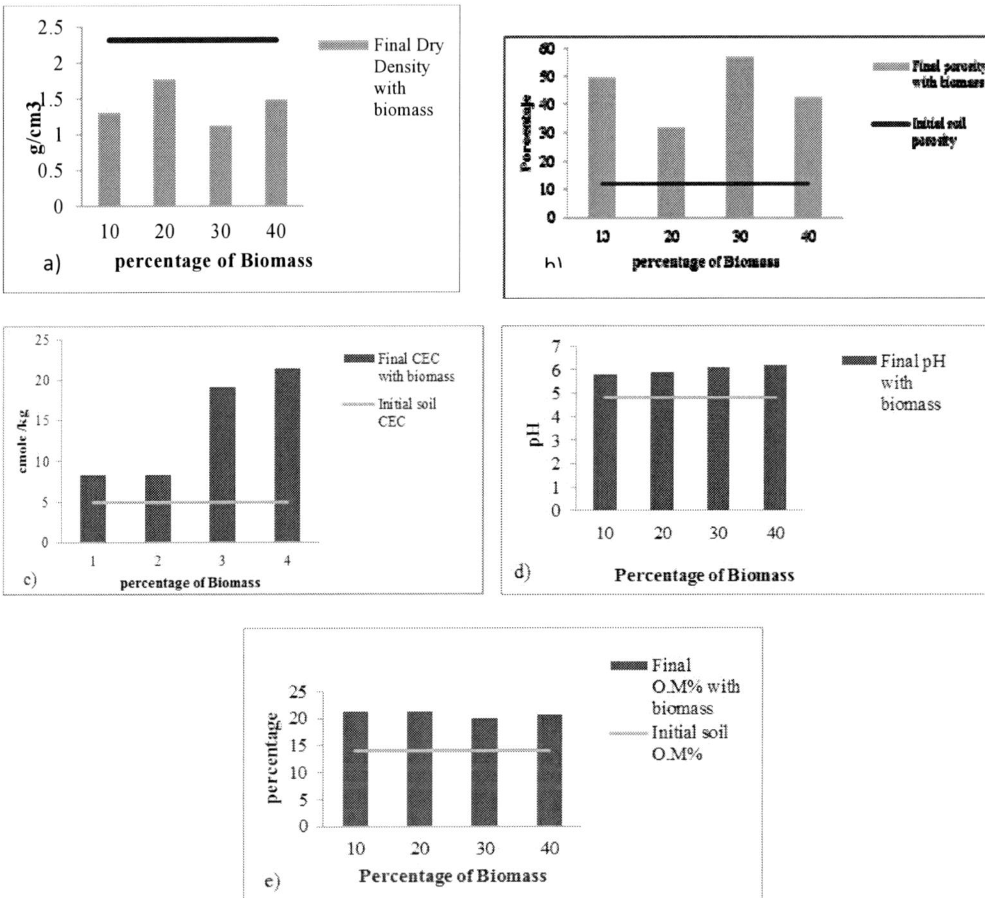


Figure 5. Changes in a) dry density of soil (g/cm3), b) porosity of soil (%), c) CEC of soil (cmol/kg), d) pH of soil and e) O.M. of soil (%) applying different percentages of biomass.

The tests also showed an increase in the Cationic Exchange Capacity (CEC) of the mixtures when sisal biomass was applied. Using the initial soil CEC as a reference, Figure 6 shows the level of change with respect to the percentages of biomass applied. The soil with 10 and 20% biomass proportions maintained CEC values between 8 and 10 cmol $kg^{-1}$ over time, while the values for soil with 30 to 40% proportions varied between 16 and 26 cmol $kg^{-1}$, a figure consistent with the greater amount of organic matter added. Carboxylic groups and amino and OH radicals present in biomass (Figure 3) provide a significant quantity of pH-dependent negative charge, which can increase a soil´s total CEC. This change enhances the absorption of elements by biota and the retention of nutrients needed to maintain the environmental conditions of the soil as a natural resource.

One of the most important results of this study was the change in soil pH following the addition of sisal biomass. On average, pH increased from an initial value of 4.8 to a post-biomass addition value of 6.5 units. In the mixtures with 30 and 40% residual biomass, soil pH appeared to increase by almost 30% compared to its initial value. As in the case of the CEC, the pH of the mixtures tended to stabilize over time during the tests, depending on the buffering capacity of the soil (SCCS 2000). Figure 4 shows the magnitude of pH change.

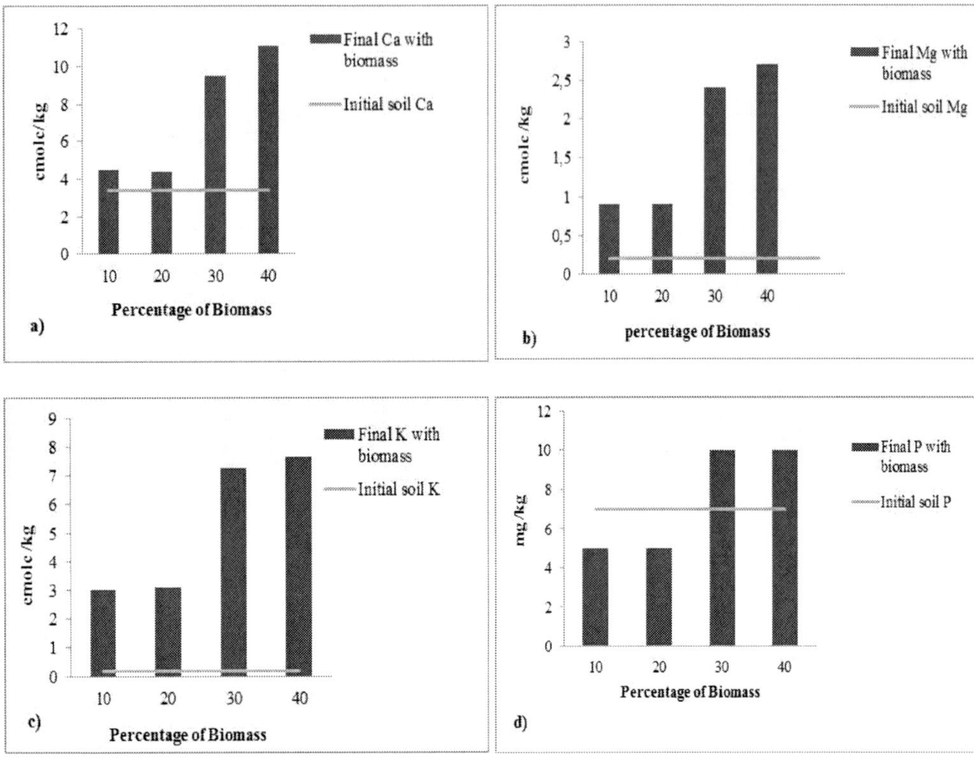


Figure 6. Changes in the levels of component nutrients depending on biomass percentage: a) changes in Ca; b) Changes in Mg; c) Changes in K, d) Changes in P.

Soil changed from acidic to slightly acidic or neutral as a result of biomass incorporation. This change is considered agronomically beneficial, because most elements can easily be absorbed by plants at a pH of 5.5 to 6.7 (Jaramillo, 2000). Additionally, in conditions close to neutral pH, the decomposition process of organic matter helps to increase the availability of nutrients, positively influencing biota development. The percentage of organic matter in the samples increased, as anticipated, following the application of biomass, from 14% in the initial samples to an average of 20% in the biomass samples (Figure 5).

The levels of Ca, Mg, K and P increased with biomass incorporation, compared to the initial values found in the soil. This change was most evident in the sample with the highest proportion of biomass, which is consistent with that sample's higher level of organic matter (Figure 6).

With respect to the toxin content of the biomass and the environmental hazards posed by it, an initial microbiological analysis was performed. The fungi in the soil were identified as gram-negative *Mucor sp.*, *Fusarium sp.* and *Rhizopus sp.* At the end of the test, results showed that no inhibition of microbial biomass distribution had occurred and that the microbial composition of the soil was similar to that of the original soil. The results also showed an increase in gram-positive bacterias, filamentous fungi and *Candida sp.* yeasts.

Through an analysis of variance in the physical and chemical soil conditions, a statistically significant difference was found between the initial soil and soil mixed with the residual biomass of manually extracted sisal fiber. In the case of dry density, a value of $p = 0.58$ was established, with confidence levels of 95%. This is greater than the predetermined significance level a = 0.05, meaning that no statistically significant difference was found between mean values of dry density based on biomass proportion (Figure 7).

CEC analysis showed a p valued of 0.000, which is lower than the predetermined significance level, a = 0.05; therefore, at least two of the sisal biomass samples analyzed exhibited a statistically significant difference between CEC mean values for the samples analyzed. Based on the analysis of two homogeneous groups containing 10-20% and 30-40% biomass, it can be assumed that the effect of biomass application on CEC will be similar between the 10 and 20% and the 30 and 40% biomass samples (Figure 7).

In the pH analysis, a p-value of 0.000 was obtained. This leads to the conclusion that homogeneous groups do not exist. Additionally, the confidence intervals of the proportions do not overlap, evidence that there are statistically significant differences between all biomass samples.

Similarly, the analysis of organic matter showed a p-value of 0.0022, indicating that the proportions of 10% and 20% biomass present a statistically significant difference between their mean values (Figure 7).

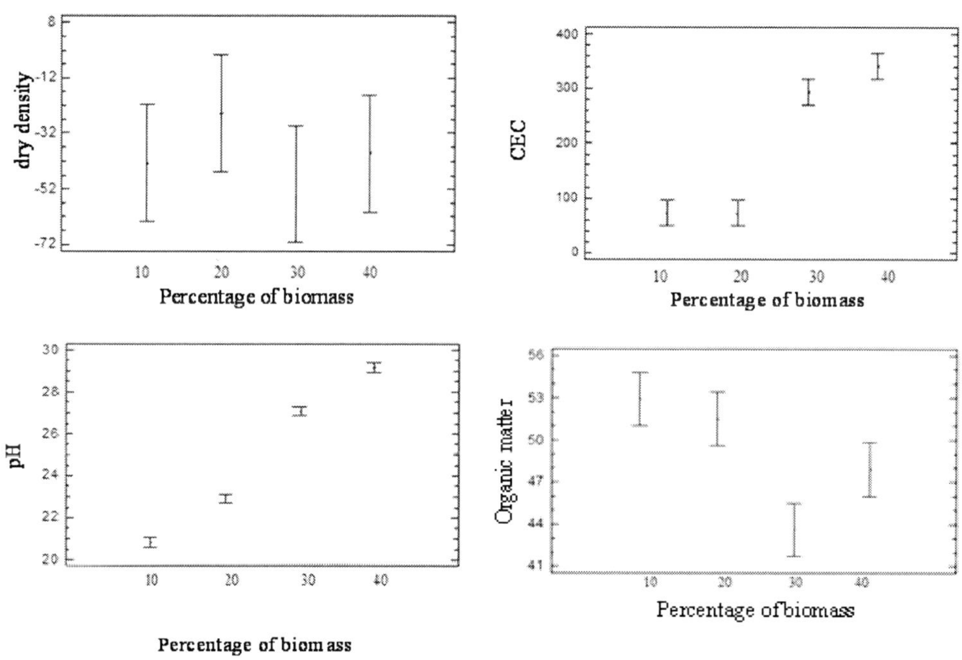


Figure 7. Graphs of statistical significance with LSD method.

## Conclusion

The commercial cultivation of *Furcraea spp.* native to the Colombian Andes generates a solid waste, or biomass, that, when incorporated into soils, improves the physical and chemical conditions of poor-quality soils. Based on the present study, the best results appear to occur when 30% biomass is mixed into the soil. Due to the characteristics of the biomass composition, incorporating it into soil has the potential not only to contribute to an increase in organic matter content, but also to provide a carbon retention mechanism present in the lignocellulosic structure of this residue. Based on the results obtained, the use of 30% and 40% proportions of biomass appear to cause the most significant positive changes in the properties analyzed, creating benefits such as green manure, soil recovery and the fixing of atmospheric carbon.

## ACKNOWLEDGEMENTS

The authors gratefully acknowledge the financial support of the EIA; Dr. Prof. Juan Carlos Loaiza, National University of Colombia and Ing. Prof. Santiago Ortega EIA.

## REFERENCES

Aerts, R., Chapin, F.S. 2000. The mineral nutrition of wild plants revisited: are-evaluation of processes and patterns. *AdvEcolRes* 30: 1-67

Aerts, R., De Caluwe, H. 1997. Nutritional and plant-mediated controls on leaf litter decomposition of Care species. *Ecology* 78: 244-260

Aristizabal, A & Zapata, A. 1990. Diagnóstico agrotecnológico del cultivo de fique (Furcraea spp.) en dos núcleos veredales del departamento de Antioquia. Universidad Nacional de Colombia. Facultad de Ciencias Agropecuarias. MscThesis. 144p. (In Spanish)

Bakouri Hicham, et al. Potential use of organic waste substances as an ecological technique to reduce pesticide ground water contamination. *Journal of Hydrology*. 2008. vol. 353, p. 335 – 342.

Barba, C. 2002. Síntesis de carboximetilcelulosa (CMC) a partir de pastas de pantas anuales. UniversitatRivira I Virgili. Departament, dÉnginyeria. Tarragona.(In Catalan).

Bonilla, J., Peinado, J; Urdaneta, M., Carrascal E. 2000. Informe nacional sobre uso y manejo de plaguicidas en Colombia Tendiente a identificar y proponer alternativas para reducir el escurrimiento de plaguicidas al mar Caribe. Ministerio de Medio Ambiente de Colombia. Bogota. 108p. (In Spanish).

David, N &S.Hon. 2003. Chemical Modification of Lignocellulosic Materials. Clemson University. United States of America. ISBN 0-8247-9472-9. 370 p

Etter, A., McAlpine, C., Wilson, K., Phinn, S., Possingham, H. 2006. Regional patterns of agricultural land use and deforestation in Colombia. *Agriculture, Ecosystems and Environment* 114: 369–386.

Gentry, A., Dodson, D. 1987. Contribution of nontrees to species richness of a tropical rainforest. *Biotropica* 19:149-156

Henderson, A., Churchill, S.P., Luteyn, J.L. 1991. Neotropical plant diversity. Nature 351:21-22

IGAC., 2007. Estudio general de suelos y zonificación de tierras del Departamento de Antioquia. Instituto Geografico Agustin Codazzi. Gobernación de Antioquia. (In Spanish).

Jaramillo, D. 2002. Introducción a la ciencia del suelo. Universidad Nacional de Colombia. Leo digital. 53 p. (In Spanish).

Jiménez, G. 1968. Estudio general del fique (Furcraea spp.). Universidad Nacional de Colombia. Facultad de Ciencias Agrícolas. Medellín. Msc Thesis. 63p. (In Spanish).

MAVDT. 2006. Guia ambiental para el subsector fiquero. Ministerio de Ambiente, Vivienda y Desarrollo Territorial; Ministerio de Agricultura y Desarrollo Rural de Colombia. 2006. 121 p. ISBN 958-97785-3-4. (In Spanish).

Meentemeyer, V. 1978. Macroclimate and lignin control of litter decomposition rates. *Ecology* 59: 465-472

Montgomery, Douglas C. Design and analysis of experiments. Arizona State University. 5th ed. 2001. ISBN 0-471-31649-0.

Montoya, R.Á., Tobón-D, H. 1979. Algunos aspectos sobre el cultivo de fique (*Furcraea spp.*) y control de la calidad de la fibra. Universidad Nacional de Colombia. Facultad de Agronomía. MSc Thesis. 63p. (In Spanish).

Polyakova, O., Billor, N. 2007. Impact of deciduous tree species on litterfall quality, decomposition rates and nutrient circulation in pine stands. For Ecol Manage 253: 11-18

Said, A.El-Aziz; Ludwick, A. G. and Aglan, H. A. Usefulness of raw bagasse for oil absorption: a comparison of raw and acylated bagasse and their components. *Bioresource Technology*. 2009. vol. 100, p. 2219-2222

SCCS.2000.Fundamentos para la interpretación de análisis de suelos, plantas y aguas para riego. Sociedad Colombiana de la Ciencia del Suelo. Colombia. 324 p. ISBN 958-95299-1-7.(In Spanish).

SDAFE. 2002. Comité Cadena Productiva del Fique. Cadena productiva del fique. Secretaría de Desarrollo Agropecuario y Fomento Économico. Caldas – Colombia. [Cited 19 abril 2008]. http://www.ces.unicauca.edu.co/archivos/aspectos_del_cultivo_del_fique.doc. (In Spanish).

Shen, W; LI, Z. and LUI, Y. Surface Chemical Functional Groups Modification of Porous Carbon. *Recent patents on Chemical* Engineering 2008. vol.1, p.27-40

Silverstein R. and Webster F., Spectrometric Identification of Organic Compounds, John Wiley and Sons, New York, Sixth edition, 1998, ISBN 0-471-13457-0, p. 482

Singh, K.P., Singh, P.K., Tripathi, S.K. 1999. Litterfall, litter decomposition and nutrient release patterns in four native tree species raised on coal mine spoil at Singrauli, India. *Biol Fert Soils* 29: 371-378

Van Vuuren, M.M., Berendse, F., De Visser, W. 1993. Species and site differences in the decomposition of litter and roots from wet heatlands. *Can J Bot* 71: 167-173.

Vitousek, P.M. 1982. Nutrient cycling and nutrient use efficiency. Amer Nat 119: 553-572

Vitousek, P.M., Turner, D.R., Parton, W.J., Sanford, R.L. 1994. Litter decomposition on the Mauna Loa environmental matrix, Hawaii: patterns, mechanisms and models. *Ecology* 75: 418-429

Wang, Q., Wang, S., Huang, Y. 2008. Comparisons of litterfall, litter decomposition and nutrient return in a monoculture Cunninghamia lanceolata and a mixed stand in southern China. *For Ecol Manage* 255: 1210-1218

Weltzin, J.F., Keller, J.K., Bridgham, S.D., Paster, J., Allen, B.P., Chen, J. 2005. Litter controls plant community composition in a northern fen. *Oikos* 110: 537–546.

In: Plantations
Editor: Hai Ren

ISBN: 978-1-62808-090-2


*Chapter 11*

# EFFECTS OF FOREST PRACTICES ON TORTOISES WILD POPULATION IN A FOREST AREA

***Maria Casamitjana-Causa,[1*] Juan C. Loaiza[2] and Pere Frigola Vidal[3]***

[1]Universitat de Girona, LEQUIA Laboratory of Chemical and Environmental Engineering. Girona, Spain
[2]Biosciences School, National University of Colombia, Medellin, Colombia. Forestry and Technology Centre of Catalonia, Solsona, Spain
[3]Departament d'Agricultura, Ramaderia, Pesca, Alimentació i Medi Natural, Catalan Governement, Spain

## ABSTRACT

Forestry management effects has been studied by means of radiotracking 21 wild adult tortoises *Testudo hermanni hermanni* on three different forestry treatments during 6 months (July 2009 - January 2010). During the study, the main characteristics of the home range, as well as the interactions between individuals, vegetation, and climate were studied. Results show that tortoises were present in open areas (pastures, sparse corks) for 30 % of locations, whereas mostly (70%) in dense forest and heath scrubland. Tortoises mortality for manual forestry methods was found to be lower than mechanized forestry practices, both of which are lower than mortalities associated with forest fires.

**Keywords:** Forest Management, Mountain tortoises, Biodiversity, Conservation, Ecological Interaction

* Corresponding autor: Maria Casamitjana Causa. Universitat de Girona LEQUIA, Spain. Carrer Puig Salner 14, 17486. Castello D'Empuries, Spain. E-mail: casmi242@hotmail.com.

## INTRODUCTION

Hermann's tortoise (*Testudo hermanni hermanni*) has an isolated distribution in several Mediterranean countries: Spain, France, Italy, Balearic Islands, Corsica, Sardinia and Sicily (Sos et al., 2008). Several studies have determined population distributions, densities and sex-ratios of this species (Fèlix et al., 1990, Franch and Oller, 1998, Budo et al., 2003; Bertolero, 2010), as well as the size of its home range (Hailey, 1989; Calzolai and Chelazzi, 1991; Bertolero, 2010). Complete population estimates are difficult to compile, particularly in some areas such as the Albera Natural Park, due to their crypsis and that a high percentage of the year they are neither active nor visible, as they dig themselves in for hibernation (winter), or for a period of aestivation when hot in midsummer (Franch and Oller, 1998, Budo et al., 2003, Bertolero, 2010). Key to the conservation of some species is a requirement to retain an ability to detect a trend (usually a decline) in a population over time (Anon, 2002). The Hermann's tortoise has survived probably due to the fact that some areas were an effective refuge for Testudo during periods of climatic shifts (Morales and Sanchis, 2009). The only remaining native populations of the tortoise in the Iberian Peninsula are found in the Albera massif, a protected area in the Eastern Pyrenees, Catalonia, Spain (Llorente et al., 2002; Vilardell et al., 2008). At present, in the Albera mountain range, it is estimated that the population of Hermann's tortoise is between 6000 and 7000 specimens (Martinez and Soler, 2005). As a result of human pressure (land use changes, wildfires, illegal harvesting, and forest management) and predation, the subspecies T. h. hermanni has disappeared from most regions, and is currently considered to be globally endangered (Madec, 1996; Guyot and Clobert. 1997; Budo et al., 2003). Egg predation by junior tortoises or adults can occur, or by several mammals and bird species (Budo et al., 2003; Vilardell et al., 2008; Bertolero, 2010).

One factor affecting the recovery of this species is of the effect of forest fires (Cheylan, 1981), as well as the development of dense scrub in forestry zones (Bertolero, 2010). Some studies in the south of France report up to 85% of tortoise mortality associated with forest fires (Cheylan, 1984). Studies in a northern Greek zone (Alyki Lake) report fire mortalities of 46 % and 70 %, respectively (Stubbs et al., 1981; Hailey, 2000). During the forest fires in the summers of 1986 and 1994 in the Albera Natural Park and in the Garraf massif in Spain, a mortality of 30.4% and 76 % were estimated respectively (Fèlix et al. 1990; Martínez and Soler, 1998). The high mortality of Testudo species after a fire is because of difficulties of thermoregulation, land aridification, deforestation, burrow absence, food scarcity, inhalation of toxic fumes and some with burns (Merchán and Martínez, 1999; Martinez and Soler, 2005).

On one hand, Mediterranean forests need appropriate silvicultural and management strategies for effective forest conservation in the region; otherwise it is impossible to control forest fires and landscape degradation (Scarascia et al., 2000).

On the other hand, studies particularly in mountainous areas showed that forestry management practices have negative impacts on amphibians and reptiles (Ash 1997; deMaynadier and Hunter 1998; Yahner et al., 2001). The direct influence of forestry practices on Hermann's tortoise and its influence on his home range have not been studied.

The primary aim of this chapter was to determine the effect of the change of vegetation on tortoises home range, (resulting three different vegetation treatment plots: bushes totally cut plot, scrub cut only 3m around the trees and no actuation plot). To achieve this radio tracking was used to know the tortoises' weekly movements, and to evaluate the impact of forestry practices on the tortoises. The secondary aims were to compare the effect of forestry practices; mechanical (tractor) and hand cutting treatments on Hermann's tortoises, studying the response of tortoises to the different machines used, as well as the tortoise mortality associated directly with the respective forestry practices.

# Materials and Methods

## Study Site

Hermann's tortoise was studied in La Balmeta valley, Spain (grid ref). This was on land administered by the Spanish government named Mas Guanter (GI-3002), belonging to the municipality of Llançà in the Oriental Pyrenean mountains area in the Northeast of Spain, (Figure 1). Mas Guanter is 57857 ha, with 35907 ha wooded forest and 2195 ha scrubland and pastures. Study plots were established on the mountainside of the highest part of the Balmeta mountain stream area, with a total area of 1113 ha. The area has a Mediterranean-type climate with the last 17 years having a mean annual rainfall of 563 mm. Mean temperature during summer is 22.5°C and 6.4 °C during winter.

The relief is has a V-shaped valley with pronounced slopes. 85% of the studied zone (agronomical terraces and fields) have slopes often higher than 30% with an altitude ranging from 104 to 605 m. The substrate consists of graywacke, ocher sandstorm, limonites, and some pelites and psammites, with a low metamorphism degree.

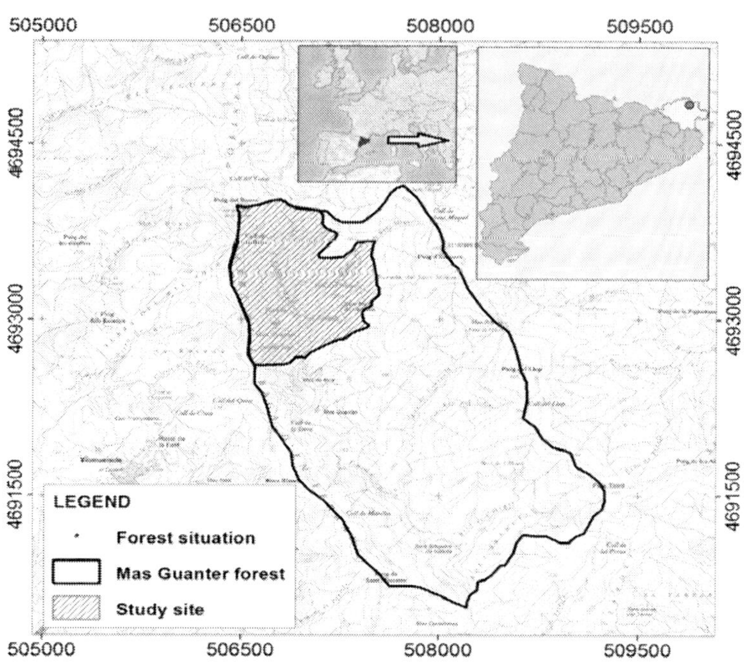


Figure 1. Location of the study zone into the public forest Mas Guanter, the underlined zone indicates the study site.

The soils at the experimental sites are classified as Lithic and Typic Dystroxerepts (SSS, 2006). The forestry practices to avoid forest fires included firebreaks, elimination and crush of scrublands in the roadsides (25 m close to the way), forestations with cork (Quercus suber) and reforestations with pine.

In the wooded forests (cork) pruning was undertaken to improve tree form. To analyse the effect of different management practices, the study area (111,3 ha) was divided into different zones with different forestry treatments, controlling and analysing the effect of these treatments on the tortoises populations. One zone (30 ha) had total elimination of bush below trees (manual), and pruning of trees.

The second witness plot "untreated" (20 ha), and the third had micro plots (30 ha), where bushes 3 m around cork trees were removed to create open zones for tortoises to sunbathe. 30 ha of firewalls (total elimination of bushes but mechanized) and river forestry clearance practice (1.3 ha). Distribution of the treatment plots preserved the witness plot in the center of the study zone as a possible refuge for tortoises. Percentages of the different forestry treatments in the study zone are shown in Figure 2.

Before the start of the forestry work began, forest workers received an induction about how to cut the bushes reducing the risk of damage to a tortoise, by first cutting horizontal 15 cm above the soil to check for tortoise presence. To ensure food zones and refuge zones for tortoises in the valley, some protection zones were preserved mostly near elm leaf blackberry (Rubus ulmifolius). These zones were identified as tortoise refuges (Observed data).

## Capture Techniques and Radio Tracking

Radio tracking techniques were used to monitor tortoises during the implementation of the different forestry practices with fire prevention an objective, from July 2009 to January 2010. Twenty one wild tortoises were located (June 2009) and captured by hand by walking the forest area along mammal pathways. Tortoises were usually encountered in open grassy areas and old mix forest (cork and shrubs).

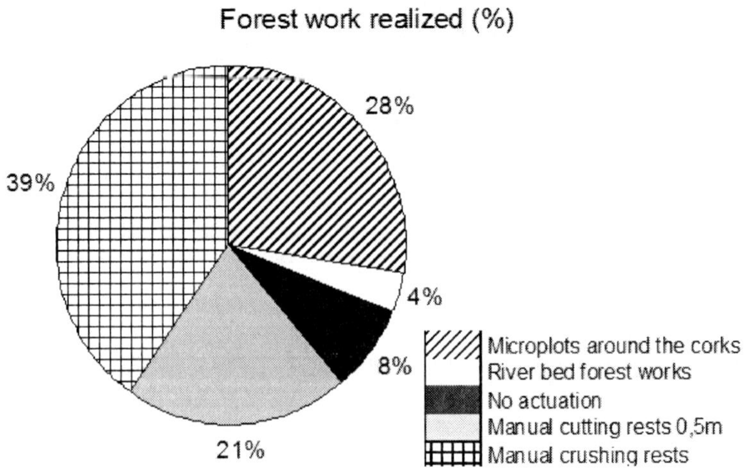


Figure 2. Percentage distribution of the different forest treatments realized on the study site (111,3 ha).

Most individuals were captured mainly during the first hours of the day and from half afternoon till it get dark (maximum activity moments during the summer period), in humid zones and under blackberry bushes. 52% of individuals were males and the rest females. Each tortoise was distinctly marked with combinations of notches on the marginal scales (Cagle, 1939).

All animals were sexed, and measurements including length, width and weight taken and age estimated from the growth annuli on the scutes of the carapace, being one growth annuli per year. We drew and photograph the marks and anomalies of the carapace, mainly due to forest fires and predators. After tortoises were marked, a transmitter was attached on each, located subsequently with a 216 MHz receptor (Panasonic, SL, PLACE). Each transmitter (Ayama, PLACE) had a durability of 6 - 8 months, 20 mS having 40 pulsations/minute and 216 MHz frequency. Transmitters were attached with resin used for the treatment of hooves (Demotec 90, PLACE). Location on the carapace was selected so as not to disturb the different biologic activities of the wild tortoises (survival, reproduction, foraging, hibernation), being between the second and third costal scute in females and on the last costal scute in males. On 31 July 2009, the 21 studied tortoises were released with seven tortoises per treatment (microplots, no actuation and total elimination of bushes), closest to their initial location. Radio tracking consisted of one or two locations weekly, according to the proximity of the forestry practice, to avoid possible damage and to evaluate the reaction of the tortoises. Per each location were recorded type of vegetation, habitat, activity of the tortoise and the geographical location (coordinates UTM) using a global positioning system (GPS, type, PLACE).

## Home Range

To estimate tortoises use of habitat (before the change of vegetation due to the respective forestry practices) the fix Kernel methodology was used because approaches more than other methodologies the reality (Powell, 2000; Kernohan et al., 2001). Location data were weighted having more importance the values close certain value and less weight to the remote values. On the polygons drawn on a map that resulted from this GIS technique application, is where there is more probability to found the studied animal. The technique is based on the locations approximation to the Kernel Gaussiana function, based on the h parameter, which determine the data dispersion. The polygons where found using a probability function of normal bivariant standart (equation 1), according the coordinates X, Y, founded:

$$h_{mf} = n^{-1/6}\sqrt{\frac{\mathrm{var}_x + \mathrm{var}_y}{2}} \qquad (1)$$

Between the informatics tools that could calculate Kernel polygons, stands out Animal Movement Extension, Biotas, Hawth's Tools, Home Range Extension (HRE, Supplier, PLACE), and Kernel Home Range (Kernohan et al. 2001). Between these options the chosen tool was HRE, due to its efficiency determining contours of home range probabilities, due to project adjusted home ranges according to a determined h (Mitchell, 2006).

To execute HRE program Arc Map3 (Supplier, Place) was used. In the studied zone different Kernel polygons were done per monitored individual as well as per groups of tortoises that shared partially the same zone and home range, studying the nucleous area (kernel 50% and 70%) and the representative zones (kernel 95%). The kernel 95, 70 i 50% determinate the area where the animal spends 95, 70 and 50% of the time, respectively, in other words is the probability to found the animal on this surface. According to this polygons obtained (Figure 3) the home range of the Hermann's tortoise was analysed to establish the parameters needed to their conservation (vegetation, slopes, state of the vegetation) (Longepierre et al., 2001).

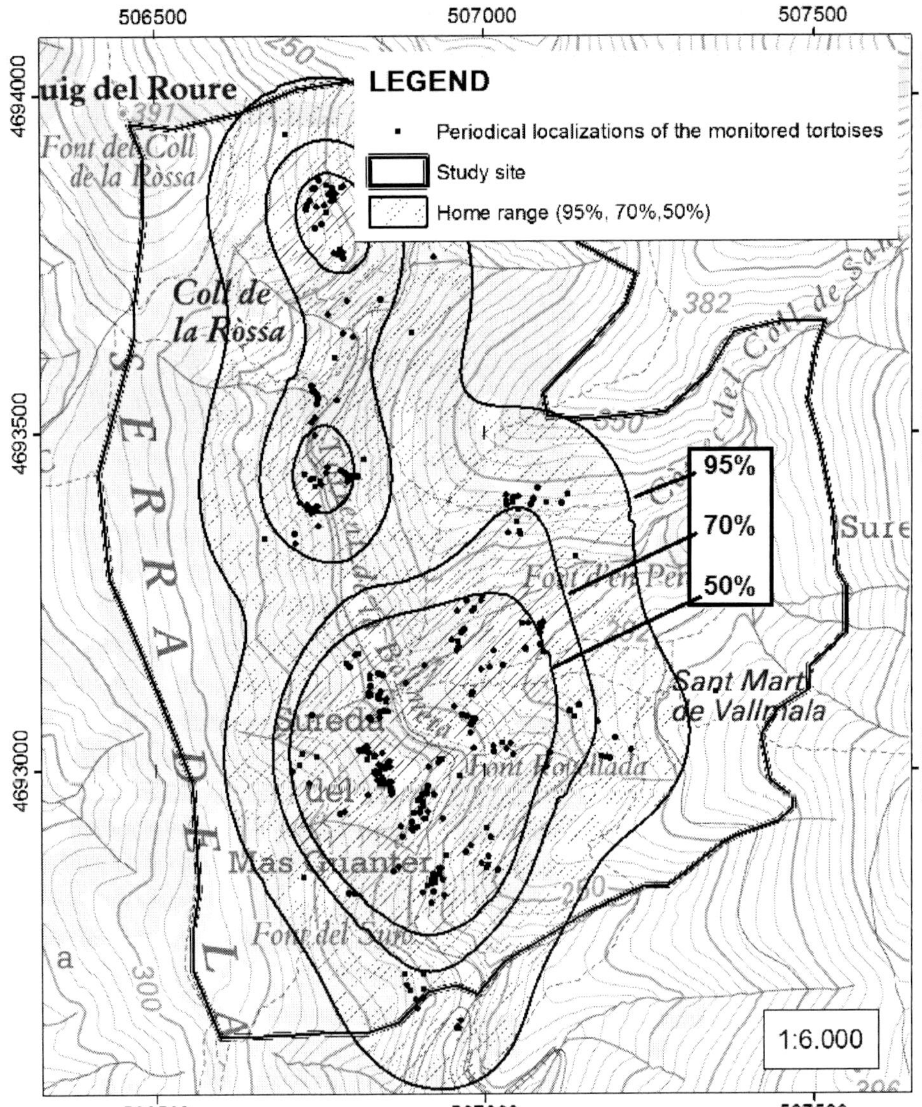


Figure 3. Home Range distribution of the 21 monitored tortoises by means of radiotracking in Mas Guanter Forest on the Albera Natural Park. The 95% and 50% of the localizations were in 70.89 ha and 17. 40 ha, respectively.

## Evaluation of Damage to Tortoises by Forestry Practices

To determine the effects of the different machines used for the respective forestry practices and their direct impact on tortoises, 60 plaster tortoises were made using mould with similar dimensions and approximate weight to a real tortoise adult. The durability of moulds allowed marking and damage to be registered. A statistical design was realized to evaluate the effect of the different forestry practices (manual with brush cutters and mechanized by tractor), as well as the influence of the vegetation and the slope as a possible parameters correlated with the impact degree. Each sampling plot had an area of 200 $m^2$. It was considered three different treatments as manual work crushing the remains into small parts, manual work cutting up into 0.5 m pieces, or mechanized work with chain tractors, considering 3 plots per treatment. On the manual treatment the studied parameter was according the vegetation cut, considering one plot of elm leaf blackberry, brierroot (Erica arborea L.) and remains of mediterranean falsebrome (Brachypodium retusum), having 9 repetitions per plot. In the mechanized forestry practice with chain tractor, relief was evaluated with one plot with gentle slope and another with moderate slope, having 9 repetitions in each. In total 24 plots with 9 repetitions each being 216 repetitions was used with tortoise models.

In each plot the 9 model tortoises were hidden under the vegetation as a wild tortoise does. After the workers cleared the undergrowth from the plot, according the equipments used, an analysis of the damaged model tortoises was done, considering different levels of damage. Data were subjected to analysis of variance (ANOVA) and Duncan test to compare effect of the different forestry practices as well as the influence of the vegetation and the slope as possible parameters correlated with the impact degree using a significance of 0.05. Statistical analyses were conducted with the software Statgraphics Centurion XV (StatPoint. Inc).

# Results and Discussion

## Home Range

Hermann's tortoise had a mean home range area of 4.35 ha (SD 3,07), being higher in males than in females. The values to Swihart Slade (2,25) and Schoener (0,22) indexes were related to home ranges overlapped, uniform distribution and coincident localization (Figure 2). The home range area of the transmitter individuals according to the sex are showed in Table 1. In most of cases the home ranges overlapped, resulting in 8 groups of tortoises (Table 2). Density from home range was 0.18 (95% probability) and 0.80 tortoise per hectare (50% probability).

The obtained data were compared by seasons, determining a seasonal effect in home range and habitat use. Home range according to the season (summer and autumn) was 36.83 ha and 33.00 (95% probability), and 9.82 and 7.80 (50% probability), respectively. In summer there was more movement and consequently a greater home range. In summer tortoises concentrated in the lowest parts of the valley, whereas in autumn individuals moved to the slopes of the valley.

**Table 1. Home range surface, according to sex**

| | Home range representative | | Nucleus area | |
|---|---|---|---|---|
| | mean | SD | mean | SD |
| Males | 5,70 | 3,42 | 1,24 | 0,83 |
| Females | 3,24 | 2,36 | 0,66 | 0,51 |
| Total invidious | 4,35 | 3,07 | 0,92 | 0,71 |

Representative zone 95% Kernel (ha); Nucleus area 50% Kernel (ha); SD: standard desviation.

**Table 2. Tortoises group having coincident home range**

| Group | Kernel 95% (ha) | Density (tortoises/ha) | Totoises (%) | |
|---|---|---|---|---|
| | | | Male | Female |
| 1 | 5,70 | 0,18 | 0 | 53 |
| 2 | 9,24 | 0,32 | 105 | 53 |
| 3 | 3,50 | 0,57 | 0 | 105 |
| 4 | 6,22 | 0,80 | 53 | 21 |
| 5 | 9,53 | 0,31 | 105 | 53 |
| 6 | 10,87 | 0,18 | 53 | 53 |
| 7 | 1,25 | 0,80 | 0 | 52 |
| 8 | 2,90 | 0,69 | 105 | 0 |

## Habitat Use

According vegetation home range of Hermann's tortoises was mainly (33.7%) heather (Erica arborea), and (33.7 %) sparse cork forests while 20.2 % was open areas under B. retusum and pasture and the remainder (8.5%) under elm leaf blackberry and river vegetation. During the active period, vegetation used to hide was 47% under elm leaf blackberry, 28% under heather and 25% under B. retusum and dead vegetation. During summer, when temperatures were high (26.7°C mean temperature) and the rainfall low (70 mm per 3 months), tortoises chose the river zones (without water but cooler) and the blackberry to hide in 53.8% of cases, under heather 30.3 % of cases and 15.9% of cases were in open areas with B. retusum or herbaceous, where the insulation was higher. In comparison during autumn tortoises sought shelter in herbaceous and more open spaces with greater sunlight, preparing the space to hibernate. The deep hibernation (from November) was classified depending on vegetation used: 41.7 % under B. retusum and herbs, 19.4% under E. arborea, 25% under R. ulmifolius, and 13.9% hibernated in cleared areas as a result of forestry clearing practice. According to sex, there was no difference between the selected vegetation in summer. In autumn males had higher occurrence in open areas out of the river-bed, whereas most females hibernated under R. ulmifolius, (Figure 4). Higher and lowest temperatures affect tortoises displacements and precipitation favoured it, Figure 5. The individuals developed their activity between 148 and 299 masl and 30 to 50% slope correspond to 77% of the area (optimal rank of slopes), this confirming the results found by Llorente et al. (1995). However, the females have a slight trend to go up during autumn resting there till the hibernation (lower slope sites 2%). During summer, the tortoises move on zones with a mean slope of 33,9% (SD 16,3) whereas in autumn and winter the rank oscillates between 41,3% and 39,3% (SD=13,2 and 14,5 respectively), it doesn't exist significant differences between sexes.

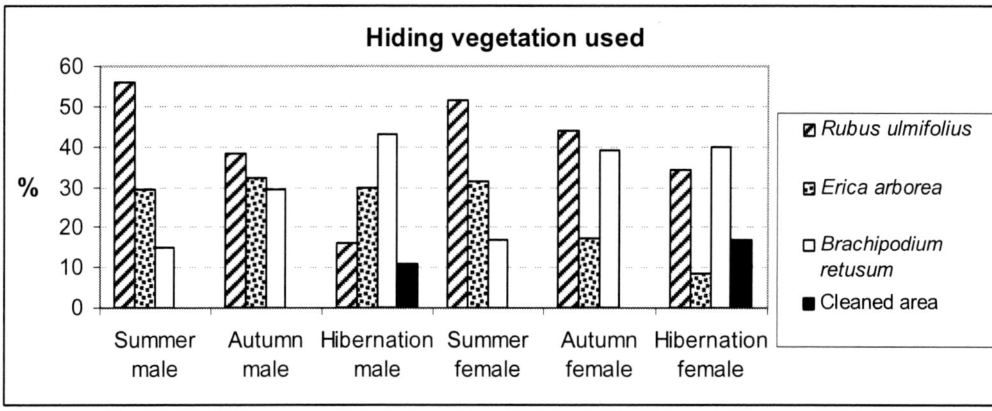


Figure 4. Hiding vegetation used according to the sex and seasons along the study zone.

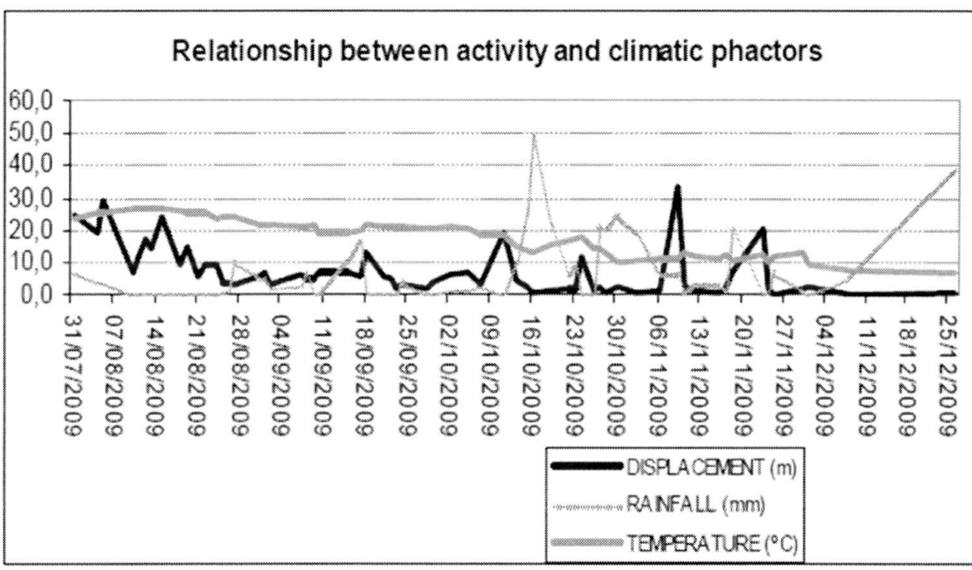


Figure 5. Temperature, rainfall and tortoises displacements relationship, during the studied period. The most influent climatic parameters into the tortoise ethology and activity had been temperature and rainfall.

## Movements of Tortoises

Analysing the activity of the tortoises it has taken into account the accumulated displacement and the mean daily movement (Table 3). Mean total distance covered during the study time, was 752.6 m having a daily mean distance covered of 7.7 m. Due to excessive hot days in summer, tortoises started an inactivity period reducing their activity and movement, resting in a semi lethargic state during mid-morning and early afternoon or that could be extended for some days (Observed data), but even so in summer the activity was slightly higher due to a second mating period (males increased the activity searching females) and some late lays, because the females have to search an adequate oviposition zone. Total movement and size of home range of 80% of tortoises was correlated (R2 = 0.92).

**Table 3. Accumulated displacement values and average daily displacement, according to the sex and different seasons studied**

| | | Accumulate displacement (m) | | Average daily displacement (m/day) | |
|---|---|---|---|---|---|
| | | Males | Females | Males | Females |
| Summer | mean | 642,0 | 477,8 | 14,1 | 10,1 |
| | SD | 334,6 | 354,7 | 7,4 | 7,7 |
| Autumn | mean | 142,9 | 132,8 | 3,7 | 3,6 |
| | SD | 61,3 | 67,2 | 1,9 | 3,0 |
| Hibernation | mean | 73,0 | 31,9 | 2,3 | 1,0 |
| | SD | 97,4 | 24,4 | 3,7 | 0,7 |
| Total displacement | mean | 857,9 | 642,4 | 8,3 | 7,0 |
| | SD | 324,9 | 408,1 | 12,4 | 14,0 |
| Global displacement | mean | 752,6 | | 7,7 | |
| | SD | 370,4 | | 13,2 | |

Displacement global: Total displacement mean (males and females), SD: standard deviation.

**Table 4. Accumulated turn angle between displacements according to the different studied periods and sex**

| | | Males | Females | Total tortoises |
|---|---|---|---|---|
| Summer | mean | 660,4 | 602,3 | 631,3 |
| | SD | 208,1 | 273,6 | 238,5 |
| Autumn | mean | 642,1 | 681,4 | 661,7 |
| | SD | 249,6 | 154,2 | 202,9 |
| Hibernation | mean | 239,1 | 246,7 | 242,9 |
| | SD | 160,1 | 163,9 | 157,7 |

There was a direct relation between the length of the movements and a higher explored area. The 20 % remaining tortoises (mostly females), showed a high activity in a reduced area. Certain movement and change in movement direction of tortoises was associated with forestry practice (Table 4).

## Forestry Practice Effects

Considering the parameters to compare different tortoise habitat established by Bertolero (2002), the effect of the forestry practices on change of habitat was low (Table 5). The 55% of radio monitored individuals showed no effect to forestry practice in the proximity, whereas the other 45% did small displacements in the opposite direction of the workers. In most of cases the rhythm of work moved forward faster than the the tortoises could. During forestry practices, 15 tortoises were found killed with one injured that survived after healing.

Also 8 old carapaces were found without any impact, possibly due to predation. Simulation with mould tortoises showed slight damage 5% - 10% in manual crushing remains and serious damages 4% - 11% on manual cutting remains under B. Retusum, E. arborea and Rubus ulmifolius, respectively. It is important to emphasize that in manual work any mold was crushed into parts (which would equal to the death of a real individual).

**Table 5. Habitat evaluation of the Hermann tortoise according to Bertolero (2002) methodology**

| Analyzed factors | Valuation | |
|---|---|---|
| | Before Forest work | Immediately after forest work |
| 1 Food possibilities | Very good | Bad |
| 2 Sunbath possibilities | Good | Very good |
| 3 Nest zones | Bad | Very good |
| 4 Refuge sites | Very good | Bad |
| 5 Habitat area | Good | Good |
| 6 Barriers between favorable habitats | Bad | Very bad |
| 7 Number of potential depredators | Very bad | Very bad |
| 8 Density of potential depredators | Bad | Bad |
| 9 Depredation probability | Bad | Very bad |
| 10 Associated human activity risk | Good | Bad |
| 11 Natural catastrophe risk (forest fires) | Very bad | Very good |
| 12 Habitat legal protection | Very bad | Very bad |
| Total | 29 | 27 |

Very good : 4; Good: 3; Bad: 2; Very bad: 1.

## DISCUSSION

According to the advance of the active phase of the tortoise, the movements decreased, linked to a decrease in temperature. The optimal rank of slopes of the home range fluctuated between 30 and 50%, corresponding to 46.9% of the study area. Optimal height rank fluctuated between 150 and 300 masl, not affected by sex of tortoise, but slightly linked to the season, being lower altitudes in the summer. In terms of habitat, T. Hermanni was founded 30 % in open areas (pastures, sparse corks), whereas the remainder (70%) was in dense forest and heath scrubland. The vegetation used as refuge depended on the year season, being under R. ulmifolius on summer, and B. retusum and forestry litter. During the execution of the forestry practices on the hibernation period, 55% of tortoises showed no reaction.

Tortoises moulds showed that manual work had an impact of 4% (Slight damages), whereas the mechanical work caused 22% of mortality and 6% of serious damages. Tortoises models showed that mechanized work by chain tractor had a mortality of 37% on flat zones, with less damage (19 %) on irregular reliefs with slow slopes (Table 6). The impact of mortality of manual work is much less than the mortality associated with forest fires on the Albera massif (30.4%) (Fèlix et al., 1990). It was verified that the mechanical work with chain tractors in flat zones had a higher effect than on the irregular slopes. The forestry practices had negative impacts at the time for T. hermanni, yet a positive impact several months later due to renovation of vegetation, open areas, and nest sites. Considering that the studied tortoise population has a home range under 300 masl, it is recommended to avoid mechanized forestry work with chain tractors under this range, and to exclude mechanized work on the river-bed. The implementation of manual forestry work (by experienced and trained workers) are recommended, due to an increase in favourable habitat and decrease in risk of forest fires. To minimize the impact of the forestry intervention it is recommended to implement vegetation strips between forest work areas.

**Table 6. Affection degree on tortoises according to the different forest work**

| Forest work | | Affection degree (%) | | | |
|---|---|---|---|---|---|
| | | No touch | Sd | Ss | Died |
| Manual work (cutting remains 0,5 m, with brush cutters) | Brachipodium retusum | 85 | 11 | 4 | 0 |
| | *Erica arborea* | 95 | 5 | 0 | 0 |
| | *Rubus ulmifolius* | 90 | 10 | 0 | 0 |
| | mean | 90 | 9 | 1 | 0 |
| Manual work (crushing remains with brush cutters) | Brachipodium retusum | 89 | 7 | 4 | 0 |
| | *Erica arborea* | 96 | 0 | 4 | 0 |
| | mean | 92 | 4 | 4 | 0 |
| Mechanized work (chain tractor) | Slope | 67 | 15 | 0 | 19 |
| | Flat zones | 56 | 7 | 11 | 26 |
| | mean | 61 | 11 | 6 | 22 |

Sd: Slight damage, Ss: Serious damage.

## Acknowledgments

The research was funded by the Spanish Government (Dirección General de Medio Ambiente y Política Forestal). We would like to thank the collaboration of TRAGSA sl. and Miguel Galan, for facilitating the research realization. The lead author also wishes to express his gratitude to the Centre de Reproducció de Tortugues de l'Albera for his support and advice and Leigh Sanders for the linguistic proofreading.

## References

Anon (2002) Report to the Range States on the development of Hawksbill Eretmochelys imbricata) Population Monitoring Protocols for the Wider Caribbean. Second CITES wider Caribbean Hawksbill Turtle Dialogue Meeting Grand Cayman (Cayman Islands). Document HT2 Doc-8.

Ash, A. N. (1997) Disappearance and return of salamanders to clearcut plots in the southern Blue Ridge Mountains. *Conserv. Biol.* 11: 983 – 989.

Bertolero, A. (2002) Biologia de la tortuga mediterranea aplicada a su conservación. Universitat de Barcelona. Departament de biologia animal. *Doctoral thesis* Pp. 226.

Bertolero, A. (2010) Tortuga mediterránea – *Testudo hermanni* In: *Enciclopedia Virtual de los Vertebrados Españoles Salvador,* A, Marco, A. (Eds) Museo Nacional de Ciencias Naturales, Madrid.

Budó, J. (2001) Regressió de la població de tortuga mediterrània (*Testudo hermanni hermanni*) i tortuga de rierol (*Mauremys leprosa*) a la Reserva Natural de Sant Quirze, dins el Parc Natural de l'Albera (Pirineu oriental, Catalunya) *Butll. Soc. Cat. Herp.*15: 45-50.

Budó, J., Capalleras, X., Mascort, R., Fèlix, J. (2003) Estudi de la depredació de postes de tortuga mediterrània (*Testudo hermanni hermanni*) a la serra de l'albera (Pirineu oriental, Catalunya) *Butll. Soc. Cat. Herp.* 16 p.

Cagle, F. R. (1939) *A system of marking turtles for future identification Copeia volume*:170 – 172.

Calzolai, R., Chelazzi, G. (1991) Habitat use in a central Italy population of *Testudo hermanni Gmelin (Reptilia Testudinidae*) *Ethol. Ecol. Evol.* 3:153-166.

Cheylan, M. (1981) Biologie et Ecologies de la Tortue d'Hermann, Testudo hermanni Gmelin. *Contribution de l'espèce à la cannaissance des climats quaternaires de la France Escole Pratique des Hautes Études Memòria i Treball de l'Institut de Montpeller* 13p.

Cheylan, M. (1984) The true Status and Future of Hermann's Tortoise Testudo hermanni robertmertensi WERMUTH 1952 in Western Europe. *Amphibia-Reptilia* 5: 17-26.

DeMaynadier, P. G., Hunter, M. L. (1998) Effects of silvicultural edges on the distribution and abundance of amphibians in Maine. *Conserv. Biol.* 12: 340 – 352.

Fèlix, J., Budó, J., Capalleres, X., Farré, M. (1990) Conseqüències dels incendis forestals en una població de tortuga mediterrània (*Testudo hermanni hermanni* GMELIN, 1789) de l'Albera *Ann. Inst. Estad. Empord.* 23: 13-36.

Franch, M., Oller, M. (1998) Estudi d'una població de tortuga mediterrània (Testudo hermanni hermanni) a la serra Balmeta (L'Albera) Document CRT de l'Albera Garriguella, Catalunya. Unpublished data.

Guyot, G., Clobert, J. (1997) Conservation measures for a population of Hermann's tortoise *Testudo hermanni* in southern France bisected by a major highway. *Biol. Conserv.* 79: 251- 256.

Hailey, A. (1989) How far do animals move? Routine movements in a tortoise. *Can. J. Zool.* 67: 208-215.

Hailey, A. (2000) The effects of fire and mechanical habitat destruction on survival of the tortoise *Testudo hermanni* in Northern Greece. Ed Elsevier. *Biol. Conserv.* 92: 321-333.

Kernohan, B. J., Gitzen, R. A., Millspaugh, J. J. (2001) Analysis of animal space use and movements. *Radio Tracking Animal Populations.* Millspaugh, J. J. and Marzluff, J. M. (Eds) Academic Press, PLACE.

Llorente, G. A., Montori, A., Santos, X., Carretero, M. A. (1995) *Atlas dels amfibis i rèptils de Catalunya i Andorra.* Ed. El Brau, Figueres, Spain.

Llorente, G. A., Montori, A., Carretero, M. A., Santos, X. (2002) Testudo hermanni (Gmelin, 1789) Tortuga mediterránea. Atlas y libro rojo de los Anfibios y Reptiles de España Pleguezuelos. *Dirección general de Conservación de la Naturaleza – AHE.* volume: 151-153.

Longepierre, S., Hailey, A., Gernot, C. (2001) Home range area in the tortoise Testudo hermanni in relation to habitat complexity: implications for conservation of biodiversity. *Biodivers. Conserv.* volume: 1131-1140.

Madec, D. (1996) La predation dans le processus de conservation de la Tortue d'Hermann, Testudo hermanni In: B. Devaux (ed.) *Proceedings International Congress of Chelonian Conservation OPTOM Gonfaron France*: 181-183.

Martínez, A., Soler, J. (1998) Criteris de selecció de la tortuga mediterrània (Testudo hermanni hermanni) reintroduïda al Parc Natural del Garraf. *III Trobada d'estudiosos del Garraf* volume: 3-29.

Martínez, A., Soler, J. (2005) *La tortuga mediterrània a Catalunya.* Ed. Col·lecció Natura núm 3: 130.

Merchán, M., Martínez-Silvestre, A. (1999) *Tortugas de España*. Ed. Antiquaria. Madrid

Mitchell, B. (2006) *Comparison of Programs for Fixed Kernel Home Range Analysis*. University of Vermont Wildlife newsletter.

Morales, J. V., Sanchis, A. (2009) The Quaternary fossil record of the genus Testudo in the Iberian Peninsula Archaeological implications and diachronic distribution in the western Mediterranean. *J. Arch. Sci.* 36: 1152–1162.

Powell, R. A. (2000) Animal home ranges and territories and home range estimators. *Research Techniques in Animal Ecology Boitani*. Columbia University Press PLACE pp: 65 – 110.

Scarascia, G., Oswald, H., Piussi, P., Radoglou, K. (2000) Forests of the Mediterranean region: gaps in knowledge and research needs. *Forest Ecol. Manag*. 132: 97 – 109.

Sos, T., Daróczi, S., Zeitz, R., Pârâu, L. (2008) Notes on morphological anomalies observed in specimens of Testudo hermanni boettgeri Gmelin, 1789 (Reptilia: Chelonia: Testudinidae) from Southern Dobrudja. Romania North-Western *J. Zool. V* 4:154-160.

SSS (2006) *Keys to Soil taxonomy* Ninth edition SSS Soil Conservation Service Agric US Gov. printing office Washington, DC.

Stubbs, D., Hailey, A., Tyler, W., Pulford, E. (1981) *Expedition to Greece 1908. A Report London Univ. of London Union Natural History Society.*

Vilardell, A., Capalleras, X., Budó, J., Molist, F., Pons, P. (2008) Test of the efficacy of two chemical repellents in the control of Hermann's tortoise nest predation. *Eur. J. Wildl. Res.*: 745–748.

Yahner, R. H., Bramble, W. C., Byrnes, W. R. (2001) Effect of vegetation maintenance of an electric transmission right-of-way on reptile and amphibian populations. *J. Arb*. 27(1).

In: Plantations
Editor: Hai Ren

ISBN: 978-1-62808-090-2


*Chapter 12*

# ECOHYDROLOGY OF AMAZONIAN RAINFOREST ECOSYSTEMS

***Conrado Tobón[1,*] and Jan Sevink[2]***

[1]Forestry and Conservation Department,
Universidad Nacional de Colombia, Medellin, Colombia
[2]Institute For Biodiversity and Ecosystem Dynamics,
University of Amsterdam, Amsterdam, The Netherlands

## ABSTRACT

The Amazonia rainforest is the most complex biome on earth, both in its structure and species diversity, besides it plays an important role in the regulation of regional and global climate. Ecohydrological studies of Amazon rainforests may serve different purposes: to understand the transfer processes between soils, vegetation and atmosphere, to assess the effects of deforestation on the local, regional and global climate and biodiversity, and to evaluate the effects of deforestation. This study represents the first long-term hydrological study for the western part of Amazonia, thus the main objective was the assessment of the ecohydrological functioning of four forest types, representative for this part of the Amazonia. This is achieving by describing and quantifying the temporal and spatial dynamics of the hydrological fluxes inside forest, through a monitoring program carried out for five years (1995-1997, 2003 - 2005) on climate and water fluxes through the forest. Collected data showed that average annual rainfall for the western part of the Amazonia was3400 mm, the average annual temperature 26 °C and mean relative humidity 87%. Net precipitation appeared to be dependent on both gross rainfall and forest structure. The presence of thick litter layer and the concentration of fine roots determine the net rainfall partitioning into water uptake and total drainage to mineral soil. Results pointed to differences between ecosystems in the forest floor water storage capacity, water content and water uptake dynamics and amounts. Results indicated that water availability in studied soils is low; however, soil water storage is high and almost constant over the period of study. The rainfall distribution and the high water storage of the mineral soil appeared to be the most important parameters in maintaining water uptake or actual transpiration very similar to reference transpiration. The annual water

* E-mail: ctobonm@unal.edu.co.

balances during the four year period showed that the fraction of evaporation of intercepted gross rainfall was about constant over this period and that there are differences between ecosystems in their net evaporation. It is the compartment approach which enables to identify the ecosystem compartments and processes involved in such changes in the overall water balance.

**Keywords:** Ecohydrology, Amazonia forest, forest floor, soil moisture, droughts

## INTRODUCTION

Amazonia rainforests are estimated to host around half of the world's undisturbed tropical forest with a high level of biodiversity (Leopoldo*et al.,* 1987), and because of its magnitude the amount of fixed $CO_2$ is very high. As such, conversion of Amazonia vegetation to pastures or crops may exert a large influence on hydrology, global climate, and biogeochemical cycles (Gash et al., 1996; Shuttleworth, 1988). The special ecohydrological significance of this forest is due to its high total evapotranspiration, resulting largely from the large proportion of incident rainfall that is intercepted and evaporated after interception by the canopy (Tobón, 1999; Tobón et al., 2000a).

It is expected that the disappearance of the Amazonia forest may have important effects on the hydrology and climate in those ecosystems lying in the eastern slope of the Andes (Tobón et al., 2000a). Andean mountains act as a barrier blocking most of the humidity coming from the Amazonia, which produces high precipitation on parts the Andes,which returns water to the Amazon via the Andean river corridors (Tobón, 1999). This is a feedback mechanism: the Andes benefits from high precipitation and partly depends on the existence of the Amazonian forest, and at the same time acts as a barrier to prevent evaporated water leaving the basin (Tobón et al., 2000a).

Ecohydrological studies of the Amazonia rainforests are of broad relevance: to evaluate effects of deforestation and incorporate the hydrological forest characteristics in global models (Salati and Vose, 1984; Sukla et al., 1990, Lesack, 1993), to understand biogeochemical fluxes through the forest (Tobón, 1999; Tobón et al., 2000a, Vorosmarty et al., 1989), and to analyze appropriate land uses and management. Although the Amazonia has been the focus of global circulation model research (Nobre et al., 1991), the lack of field data has largely restricted their application to predict local, regional, and global consequences of deforestation (Lesack, 1993; Shuklaet al., 1990; Salati and Vose, 1984). Consequently, field data collection research is needed to characterize important parameters and fluxes that play a role in the regional and global climate (e.g. Gash *et al.,* 1996).

Most of the water balance studies in the Amazonia basin have been concentrated in central Amazonia (Ubarana, 1996; Leopoldoet al., 1995; Lesack, 1993; Shuttleworth, 1988), although some have been carried out in eastern Amazonia (Hölscher et al., 1997; Jetten, 1996; Wright et al., 1992). However, little attention has been paid to the forest hydrology of ecosystem types in northwest Amazonia and the effects of forest structure on water dynamics. This chapter describes a study designed to address this lack of knowledge by measuring precipitation, its spatial variation and the partitioning after entering the forest canopy in four undisturbed rainforest ecosystems in the Middle Caquetá, Colombian Amazonia. The focus is on the analysis of long term hydrological measurements of rainfall, throughfall, stemflow, and water dynamics in the forest floor and in the mineral soils in four forest ecosystems. Understanding water fluxes, as the key factor in physical, chemical, and biotic processes, is vital to understand nutrients dynamics in these ecosystems. Thus, this hydrological study at compartment level allows for a detailed study of nutrient fluxes, contributing to the overall understanding of the ecohydrological functioning of ecosystems.

In the northwest Amazonia (e.g. Colombian Amazonia), forest structure and tree species composition vary considerably between ecosystems in different landscape units (Milliken et al., 2010; Sala, 2002; Vasconcelos et al., 2000; Duivenvoorden and Lips, 1995). The structure of the tree canopy and patterns of lower layers play a decisive role in the partitioning of gross rainfall into throughfall, stemflow, and evaporated water (Longman and Jenik, 1990). Amazonian soils are highly weathered and leached, with a very low cation exchange capacity (Chauvel et al., 1991), most of which is located in the litter layer covering the mineral soil (Lesack and Melack, 1991; Pitman, 1989). In these impoverished ecosystems, the litter layer, here referred to as the forest floor, is the main compartment with respect to nutrient stocks and nutrient cycling (Tobón et al., 2000b; Tiessen et al., 1994; Salati et al., 1979). Apparently related to the concentration of available nutrients in the forest floor, fine roots of trees often concentrate in this compartment to form a "root mat" (Jordan, 1989; Cuevas and Medina, 1988; Golley, 1983). Accordingly, the main objective of the chapter is the assessment of the ecohydrological functioning of pristine forests in Colombian Amazonia. This was undertaken by describing and quantifying the temporal and spatial dynamics of hydrological fluxes through the forest compartments, identifying the parameters controlling these fluxes and storage, and the relationship with the vegetation.

## General Methodology

The study area:The study area is part of the undisturbed forest located in the Middle Caquetá Colombia, (0° 37' 23´´ N and 70° 43' 18´´ W). The research sites lay approximately 200m.a.s.l. to 250m.a.s.l. The research plots are located in the four main landscapes units in the area: the Tertiary sedimentary plain, the upland terraces of the River Caquetá (high and low terraces), and the flood plain (Duivenvoorden and Lips, 1993). The climate of the Middle Caquetá is wet tropical, classified as Afi(SelvaAmazónica) according to Köppen (1936). The average annual rainfall is approximately 3400 mm; the daily average temperature is 26°C and the mean relative humidity is 87%. Two main seasons can be distinguished: a relatively dry period between December and February, and the rainy season lasts from March until December.

Soils in this part of the Amazonia have a very low chemical fertility, low pH, and very low base saturation, except for soils on the flood plain of Andean rivers, which have a somewhat higher base saturation and are higher in weatherable minerals (Lips and Duivenvoorden J. F. 1990). Main soil types in the research plots (according to Keys to soil taxonomy) are: typicPaleudults, typicHapludults, typicKandiudults, and typic and aquicfluventic Dystropepts (USDA, 2010).

The entire area is covered by mature rainforests classified by the FAO as belonging to the group of ombrophilous tropical forest (Milliken et al., 2010; Sala, 2002; Vasconcelos et al., 2000; Duivenvoorden, 1995). The canopy reaches to about 25 to 30 m above the forest floor with some emergent trees reaching up to 45 m in the rarely inundated flood plain. There are differences in the total standing biomass, species diversity and tree density between the landscape units (Duivenvoorden and Lips, 1995). Species richness can be referred to Leguminosae, Lauraceae, Sapotaceae, Melastomataceae and Rubiaceae, Annonaceae, Moraceae and Araceae (Londoño, 1993, Alvarez, 1993). A more detailed description and vegetation classification of the research sites is given by Duivenvoorden and Lips (1995), Alvarez (1993), and Londoño (1993). The main Indigenous land use in this part of the Amazonia was and still is "shifting cultivation", where only small areas of native forest, mostly less than one hectare, are cut and used for crop plantations during two or three years, depending on the aggregated soil fertility (ashes from the burning litter). Several other types of land use can be distinguished, part of which are connected with colonization and new techniques: grassland and some crop plantation (Botero, 1984).

*Methods:* To measure the meteorological variables, an automatic weather station (Campbell Scientific Inc.) was installed in an open area of about 20 ha, on the bank of the River Caquetá. Parameters measured in the open were gross rainfall (mm), air temperature (°C), air humidity (%), global solar radiation (W/m2), wind speed (m/s), wind direction and Class A pan evaporation (mm). A CR10 datalogger (Campbell Scientific Inc.) was programmed to measure the instruments each 30 seconds and to register mean and total values each 20 minutes. Gross rainfall in the open area and at each forest site was measured by means of a tipping bucket rain gauge with a resolution of 0.2 mm, giving information on the number and duration of showers and the total precipitation.

For the ecohydrological research we selected representative areas with natural vegetation in the main physiographical units in this part of the Amazonia basin, according to Duivenvorrden and Lips (1993). Three subplots were selected in the Tertiary sedimentary plain (SP) and two subplots in the high terrace (HT), the low terrace (LT) and flood plain (FP), to measure gross rainfall above the forest canopy, throughfall and Stemflow, litter flow and soil moisture. Throughfall was measured by means of 20 rain gauges randomly located in each subplot of 50 by 20 m (1000 $m^2$). Because of the large variability in throughfall due to the forest structure, rain gauges were randomly relocated each month over the entire period. Stemflow was measured on 15 randomly selected trees in each subplot. Collars, constructed from 8 mm thick black polyethylene plastic, were sealed to the stems in an upward spiral pattern and the water diverted into rain gauges on the forest floor.

Water fluxes through the forest floor (FF) were studied in the four forest types. To determine the water storage capacity of the FF, three plots of one $m^2$ were randomly selected within each forest type. Plots were in-situ saturated during three consecutive days and covered with a plastic sheet (to avoid evaporation). Subsequently, after 24 hours total FF samples were collected and immediately weighed for wet weight. Samples were air dried until constant weight and finally weighed, for dry weight Tobón et al., 2000b). In total 78 samples were collected during four samples periods. The thickness of the layer was measured at different points within the plots and the mean value was used as the best estimate. To evaluate the FF drainage or litterflow, flux plates were used to measure the proportion of net precipitation that passes through the FF and enters the mineral soil. Plates with an open area of 683.5 $cm^2$ and 5 cm depth were installed horizontally in the contact zone between the FF and mineral soil (Tobón et al., 2000b). In the SP plot, 45 plates were installed in three subplots (i.e. 15 per subplot) and in the HT plot, 30 plates in two subplots. Plates were measured manually on daily basis, as were gross rainfall, throughfall, and stemflow.

For the characterization of the FF water content and the spatial and temporal FF storage dynamics, Time Domain Reflectometry, TDR technique (Campbell Scientific Inc) was used. Measurements were daily performed at three different depths, depending on the thickness of the FF. In total, 31 three wire TDR sensors (50 cm length) were installed horizontally in the FF of soil profiles where soil water content was also monitored. FF water content in the SP was measured at six locations. In the other landscapes units, it was measured at four locations.

To measure soil moisture TDR Probes were horizontally installed in the upsweep of pits (2.0x1.5x1.5 m). Three plots (pits) were excavated at each landscape unit. In total eight sensors were installed in each plot at different soil depths. In the sedimentary plain and high terrace, probes were installed at 0.1, 0.15, 0.2, 0.3, 0.5, 0.8, 1.2 and 1.6 m depth. In the low terrace and floodplain, probes were installed at slightly different depths - 0.1, 0.15, 0.2, 0.3, 0.4, 0.6, 0.8 and 1 m. After installing the TDR probes, the pits were covered by the removed soil, keeping the same order of soil horizons, as they were. Next to the TDR pits, eight tensiometers were vertically installed at the same depths as the TDR probes. Plots were randomly chosen and after one year of continuous measurements, the TDR sensors and tensiometers were relocated. TDR travel time measurements were translated into volumetric water content applying a calibrated regression equation deduced by Tobón (1999).

The water balance of the forest ecosystems is approached through the equation:

$$\frac{\Delta S}{\Delta t} = P_g - (E + T + \Delta \mathrm{Sff} + \Delta \mathrm{Ss} + \mathrm{Dr}\,(\mathrm{Q}) \; +)+\mathrm{L}) \qquad (1)$$

Where $P_g$ is the gross rainfall above the forest, E is the amount of water intercepted and evaporated from the forest canopy, T is the forest transpiration, ΔSff is the forest floor moisture changes, ΔSs is the soil moisture change, Dr is the drainage from the catchment and L represent the amount of water as deep percolation. Basic methods for the calculation of a water balance on a catchment basis require accurate measurements of rainfall and runoff. For the compartment approach of the water balance see Tobón (1999). The evapotranspiration (ET), namely evaporation of water intercepted by the forest canopy and transpiration, were evaluated separately through the Penman Monteith equation (Monteith, 1965).

## RESULTS AND DISCUSSION

Rainfall distribution differs between plots, within a landscape unit, when examining separate storms, although annual totals are rather similar. Mean annual precipitation was 3400 mm $yr^{-1}$ with storms ranging from 0.4 to 161.6 mm and durations between 20 minutes to 13 hours. Most showers (63%) fell during the afternoon and at night. Comparing our data on five years rainfall with data from earlier years in Colombian Amazonia (Duivenvoorden and Lips, 1995) rainfall characteristics are similar to the long-term average in the studied area.

The variability of throughfall within a subplot was large and between forest type. Larger variability was observed in small rainfall events than with major events. Throughfall percentage ranges from zero, with events below 2 mm, to 95% in storms larger than 100 mm, but mean throughfall varies from 50% to 93% depending on gross rainfall amounts and the type of forest ecosystem, being larger in the SP. Average percentage of throughfall ranges from 82 to 87% in the four ecosystems. Throughfall was calculated as the percentage of gross rainfall for five different rainfall sizes and from the totals of measured daily gross rainfall and throughfall during the study (Table 1). Fractions of throughfall depend, among others, on the rainfall size. Moreover, it is clear from the analysis of throughfall and storm size, that the high standard deviation (SD) of throughfall is the result of the large variability in rainfall classes.

**Table 1. Throughfall percentage relative to gross rainfall for 5 storm classes in four forest ecosystems and their standard deviations (SD): SP, HT, LT, FP, ColombianAmazonia**

| Rainfall ranges (mm) | Throughfall % | | | | | | | | | | | |
|---|---|---|---|---|---|---|---|---|---|---|---|---|
| | SP | SD | n | HT | SD | n | LT | SD | n | FP | SD | n |
| | | | | | | | | | | | | |
| < 5 | 58.7 | 11.4 | 41 | 56.2 | 12.1 | 34 | 52.3 | 9.6 | 32 | 47.4 | 13.2 | 27 |
| 5 – 20 | 81.4 | 6.3 | 78 | 80.5 | 5.6 | 68 | 79.8 | 6.8 | 71 | 74.5 | 6.9 | 57 |
| 20 – 40 | 88.9 | 2.8 | 39 | 87.9 | 2.6 | 36 | 87.7 | 3.4 | 41 | 83.0 | 2.5 | 32 |
| 40 – 80 | 90.6 | 2.1 | 19 | 90.0 | 1.9 | 19 | 88.8 | 2.5 | 17 | 84.6 | 3.1 | 19 |
| > 80 | 92.8 | | 1 | 92.2 | 0.8 | 3 | 92.4 | 1.0 | 2 | 88.5 | 1.2 | 5 |
| Total | 87.2 | 2.4 | 178 | 86.7 | 2.4 | 160 | 85.8 | 1.2 | 163 | 81.9 | 1.0 | 140 |

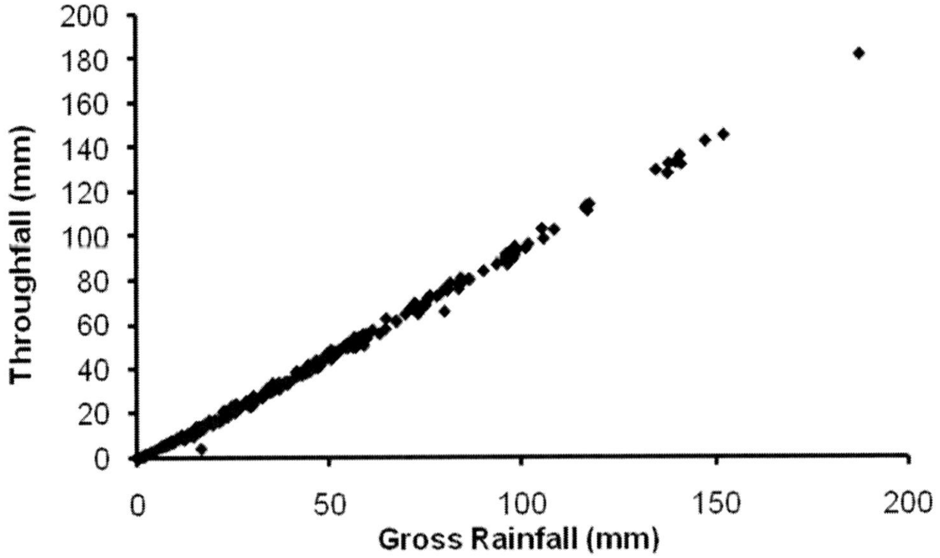


Figure 1. Trends of average throughfall amounts against gross rainfall in a forest ecosystem (SP, as an example) in Colombian Amazonia.

Regressions of throughfall versus gross rainfall were computed separately for the four forest ecosystems from the single storms. Average throughfall per plot was highly correlated with gross rainfall in all ecosystems (Figure 1). ANOVA analysis shows that the ratio of mean throughfall and gross rainfall in the FP is significantly different from the other ecosystems (significant at 0.05 level). Contribution of stemflow to the net rainfall was very low, with large differences in the amount of stemflow between forest types. Although stemflow amounts seem to be related to rainfall (Figure 2), the percentage of stemflow in all plots varied from 0.2 to 3.2% of gross rainfall. Total average percentage of stemflow relative to gross rainfall was 0.85% (± 0.46) in the SP, 0.94 (± 0.51) in the HT, 1.45 (± 0.88) in the LT and 1.12 (± 0.56) in the FP. Differences are primarily due to the higher contribution of tree palms to the total stemflow per plot.

Variability of throughfall and stemflow in undisturbed Amazonia rainforest ecosystems has been attributed to the high species diversity (Hutjes et al., 1990; Hertwitz, 1985) and forest structure, which in the present study renders the estimation of stemflow difficult to assess on an areal basis. Although the percentage of contribution of stemflow to the net rainfall is very low, it probably represents an important input of solutes, to the forest floor concentrated around the base of the tree. Moreover, results suggest that little water is stored in excess on the stems, as indicated by the very small stemflow quantities collected once rainfall has ceased.

Net precipitation at the forest floor at each forest type, is the result of throughfall plus stemflow. The net precipitation in the studied forest ranged from 6 to 47% depending on rainfall. However, average net precipitation for the studied forests ranged from 83% to 89%, indicating that the Amazonia forest interception was about 11 to 17%.

Average water storage capacity of FF's in the ecosystems studied was 16.3 mm at the SP, 7.6 mm in the HT, 8.1 mm in the LT and 4.6mm in the FP.

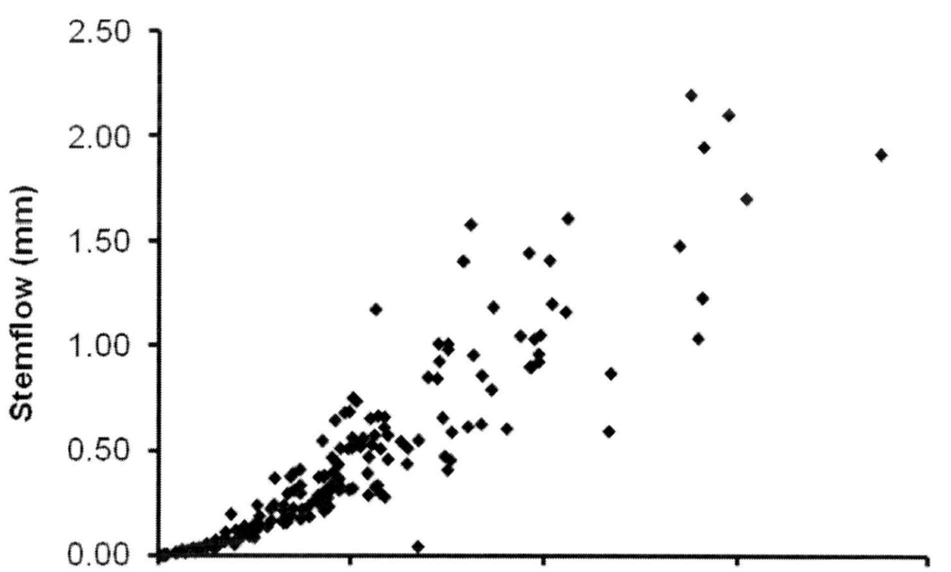


Figure 2. Average values of stemflow against gross rainfall in a forest ecosystem (SP) in Colombian Amazonia.

This gives an average of 1.51 (± 0.30) mm $kg^{-1}$, expressed as the weighted mean storage capacity per unit dry mass. The differences between ecosystems in FF water storage capacity can be accounted for by the differences in amounts of FF mass: the FF in the SP had the lowest storage per unit FF thickness (0.99 mm $cm^{-1}$), which may be explained by the loose structure of this FF and its composition, i.e. only a thin layer of fresh litter and many very fine and fine roots. The FF in the other ecosystems is rather compact and has far less fine roots, as in most part of the Amazonia (Wittmann and Parolin, 2005).

When comparing the storage capacity of the FF of Amazonia ecosystems with that of FF's in Coniferous, Douglas Fir, and Bracken forests, it appears that the value found is lower than 4.83 mm $kg^{-1}$, as reported by Pitman (1989) for Bracken forest, and slightly higher than the 1.30 mm (±0.32) found by Pradham (1973) and 0.97 mm found by Putuhena (1996) for a Eucalyptus plantation in the U.K. Noting these differences, the authors clarify that the amount of FF dry mass per unit area in the Colombian Amazonia is higher than most values reported by Pitman (1989) and Perkins *et al.* (1978) for temperate forests.

Water draining from the FF varies considerably between sites, which appeared to be related to the thickness of the litter layer and amount of litter. Although litterflow was not observed for most gross rainfall events lower than 5mm, some litterflow was collected upon small throughfall values (about 2mm). The daily average percentage of litterflow ranged from 25% to 93% of net precipitation, depending on rainfall characteristics and antecedent litter wetness (Figure 3). The trend points to a logarithmic relation between net precipitation and its interception by the FF's for small rainfall events and a linear relation for larger events are shown in Figure 3. The tendency of litter interception is related to the asymptotic nature of the wetness curve: FF's retain higher percentages of water during the earliest stage of the rainfall event.However, if the event lasts long enough, the FF storage capacity reaches its maximum value, and total drainage and uptake rates become equal to net precipitation.

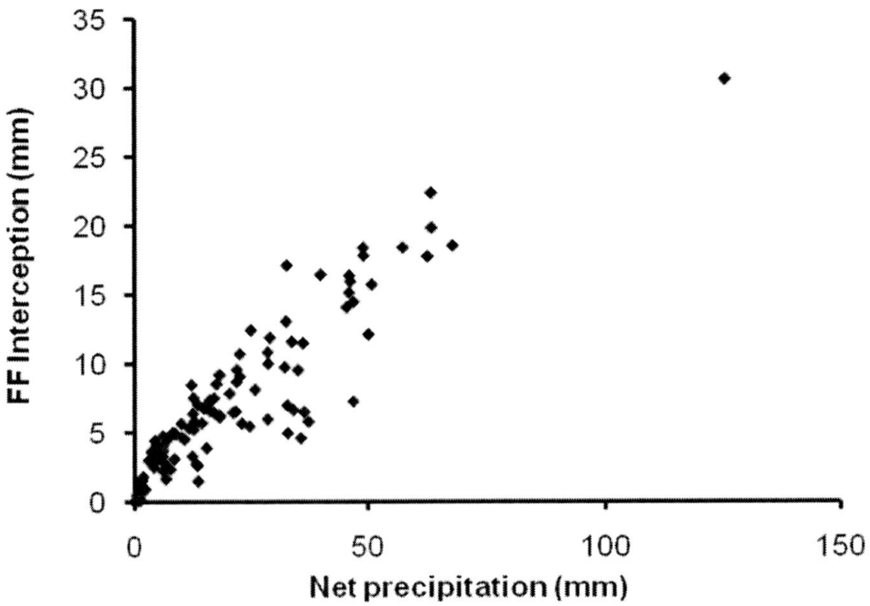


Figure 3. FF water interception, as calculated from the difference between daily measurements of net rainfall and litterflow, versus net precipitation in a forest ecosystem (SP) in Colombian Amazonia.

Figure 4 shows the measured FF water content in the four forest ecosystems. The FF in the SP exhibited the lowest water content over the whole period, while in the LT and FP water content were highest. Furthermore, the FF in the SP and HT dried out to lower values than in the FP and LT. As a general tendency, the dynamics of the FF water content in the different forest ecosystems showed a clear response to rainfall events: increases in water content in the FF were only observed after rainfall events. During the period of measurements, FF water depth ranged from 7.0 mm to 23.1 mm in the SP, 3.9 to 12.4 mm in the HT, 3.8 to 11.4 mm in the LT, and 2.7 to 8.1 mm in the FP.

Assuming that there is no evaporation from the FF, water in this compartment can either be taken up by roots or drain to the mineral soil. Water depletion from the FF's was analyzed during the two dry periods (from day 384 to 403 and from day 445 until 461), when drainage was zero since water storage in the FF's remained below the measured storage capacity, where water uptake was the only hydrological process active during these periods. In both periods and all ecosystems the water content decreased, implying that the forests took up a certain amount of water from the FF. This uptake differs between forest ecosystems and between periods. Depletion rates during the dry periods were 0.51 and 0.61 mm $d^{-1}$ in the SP, 0.20 and 0.34 mm $d^{-1}$ in the HT, 0.21 and 0.40 mm $d^{-1}$ in the LT, and 0.19 and 0.24 mm $d^{-1}$ in the FP. The relatively high uptake rate in the SP may be related to the relative high proportion of fine roots in the FF.

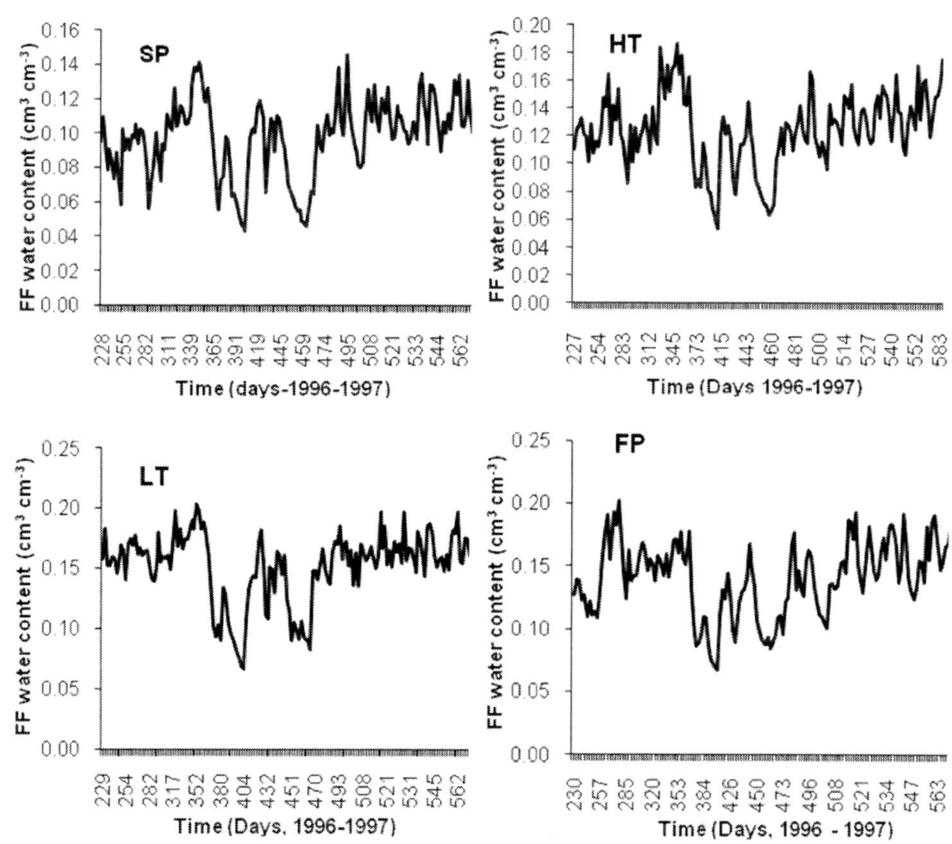


Figure 4. Temporal dynamics of measured forest floor (FF) water content (TDR) in four forest ecosystems in the Colombian Amazonia (SP, HT, LT, and FP).

Soils from northwest Amazonia show very dynamic soil moistures through space and time. Water content dynamics in the SP clearly differ from those in the other ecosystems withsmaller changes in water content and smaller differences with soil depth (Figure 5). Field observations and laboratory data indicate that these differences can be ascribed to the textural change (increasing clay content) in the soil profiles and to differences in water availability and root distribution in each ecosystem. Overall results clearly indicate that alluvial soils of an Andean river (Caquetá River) have a higher available water capacity than the soils of the Tertiary Sedimentary plain.

Soil water content dynamics in the soils are shown in Figure 6. The mineral soil had a high water content during most of the studied period in all ecosystems, with water availability relatively low, primarily in the SP. Soil moisture slightly decreased during droughts (1997 and 2005), with the largest decrease occurring in the SP ecosystem. Measurements of soil water pointed to the existence of high macroporosity, mainly in the upper part of the soil profiles in all ecosystems (Tobón, 1999). Soils developed from the SP present lower water availability and higher water content throughout the study period than soils from the alluvial system of the River Caquetá, except for those in the FP which have the highest water content. There are no considerable differences in water storage between identical depths among sites within the same ecosystem, with exception of the SP, which exhibits large differences between both sites and depths. Differences can be explained by the variations in soil texture and position of plots on the slope. The eight depths at which water content was measured allow the calculation of soil moisture depth.

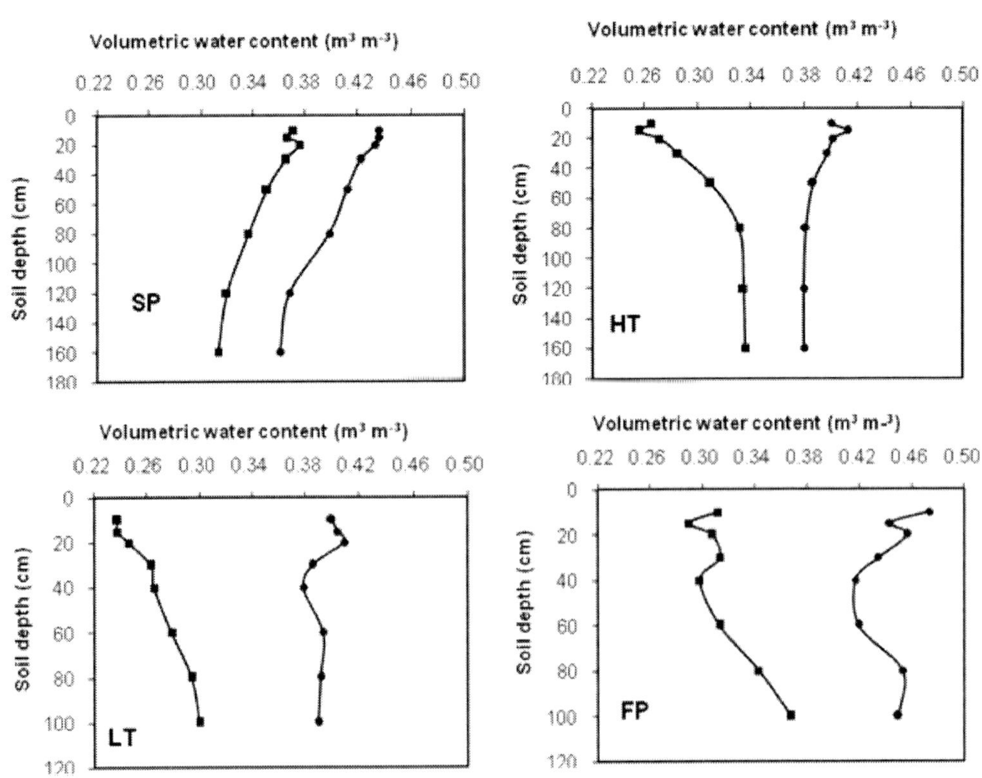


Figure 5. Profiles of maximum (■) and minimum (●) measured soil water content during the monitoring period in four forest ecosystems in Colombian Amazonia.

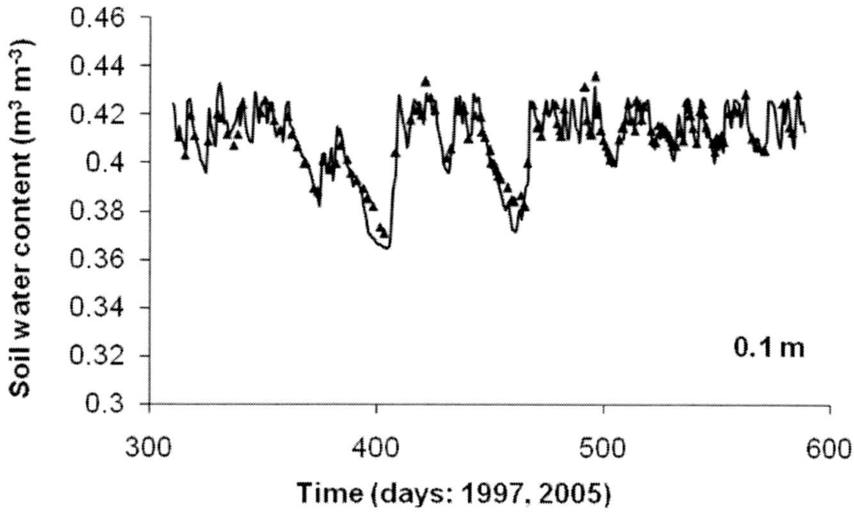


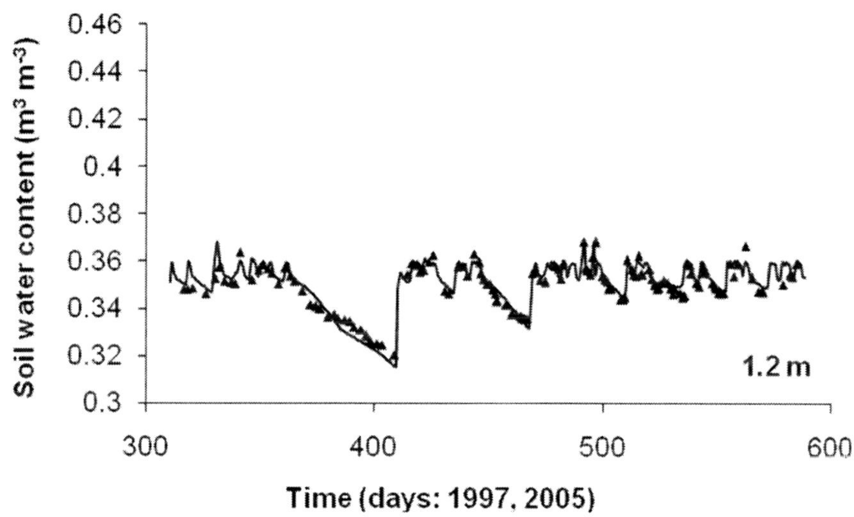


Figure 6. Soil water content dynamics at two soil depths (0.1 and 1.2 m) in a forestecosystem (SP), in Colombian Amazonia.

As an example, Figure 7 presents the temporal dynamics in soil moisture for three soil layers in the SP and HT. Water depth during the wet periods were almost similar for all ecosystems with a slightly higher values in the FP. However, during dry periods, water depth in the SP was higher than in the other ecosystems, which is consistent with the property of these soils retaining high amounts of water at low matric potential.

Changes in water uptake from the soil among forest ecosystems were found to differ: During the two dry periods (days 384 to 403 and days 445 until 461), the highest water depletion was observed in the FP with 54.1 and 65.4 mm in each period.Differences in water storage between soil layers within the same plot were observed during the dry periods when the upper part of the soil profiles dried out differently in each ecosystem. In the SP ecosystem, the upper 0.5 m of the crest profile showed the highest storage, and that in the valley bottom the lowest.

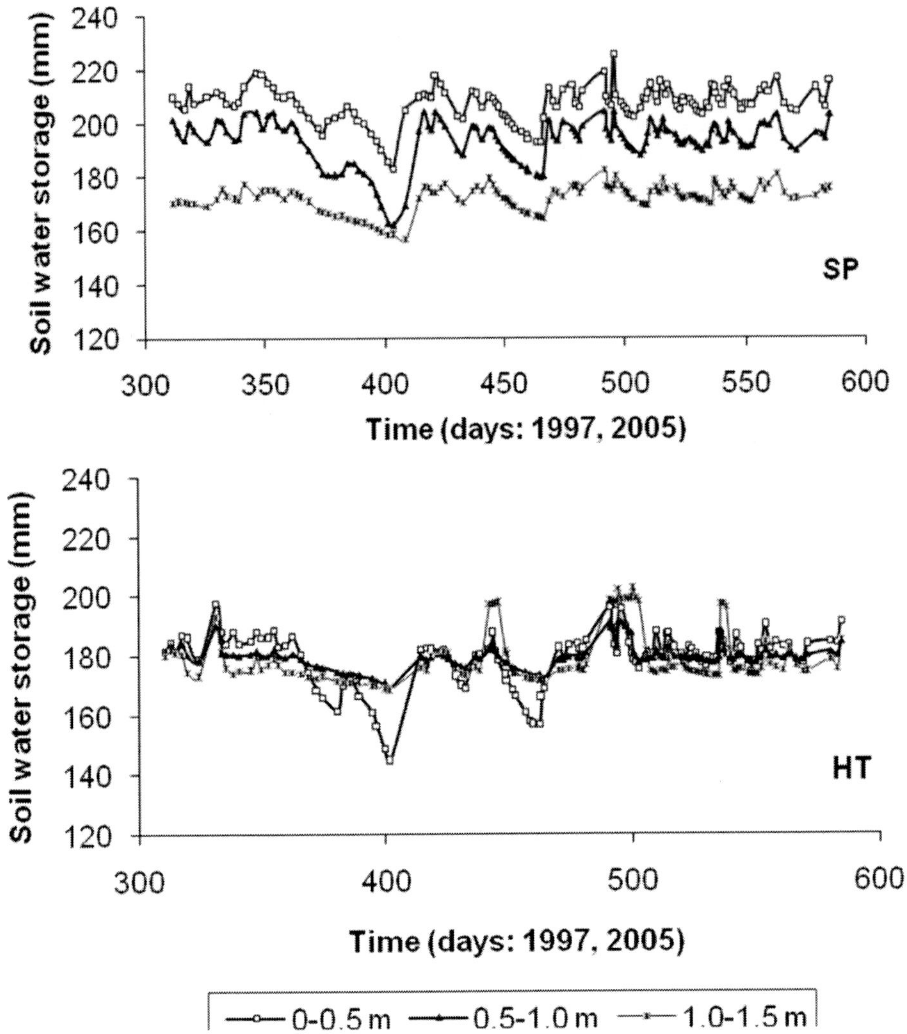


Figure 7. Profiles of soil water depth in soil layers of 0.5 m at three different depths intwo forest ecosystems (SP and HT) in Colombian Amazonia.

In contrast, storage at 1.0 to 1.5 m was higher on the slope and bottom profile than on the crest. From the soil survey and texture analysis, it is likely that differences are due to a combination of the different topography, soil texture, and water uptake by trees (Bruno et al., 2006).

Soil water fluxes were examined depending on the dynamics of soil moisture through the soil profiles. In rainfall events smaller than 10 mm, soil water fluxes were noticeable up to 0.15 m deep, while only large rainfall events (larger than 25 mm) produced fluxes throughout the soil profile. Nevertheless, small storms during wet seasons induced vertical water fluxes through the profile. In general, high fluxes were observed from the upper soil layers in all ecosystems, with the highest value in the SP, which is consistent with the observed high macroporosity in these soils (Tobón, 1999). Decreasing water fluxes were similarly observed for all ecosystems in soils layers deeper than 0.5 m, decreasing to a very low values at 3.0 m. Low upward fluxes (up to 0.002 m $d^{-1}$) were found for the soil layers between 0.1 m and 0.5

m in all ecosystems, except for the dry periods when considerable amounts of upward fluxes were observed up to 0.8 m. This shows the role of deep soil layers supplying water as the topsoil becomes dry. Upward fluxes were found to be negligible at lower depths than 0.9 m.

Calculated actual forest transpiration was of similar magnitude as reference transpiration during most of the studied period. This implies that there is no significant reduction in transpiration in the studied forest ecosystems over the period. However, a considerable reduction of the actual transpiration occurred in all ecosystems during the short dry periods, up to 60% (Figure 8). Although the soils showed to have low water availability, results indicate that the forest is supplied with sufficient water during most of the year.

During most of the studied period, the soil layers in all ecosystems presented a relative constant contribution of water uptake, which was higher from the first 0.5 m of the soil than from the deep soil layers, in agreement with the root distribution (Figure 9).

This indicates that water uptake during the wet periods strongly depended on root distribution through the soil profile. Changes were observed during the dry periods when the fraction of water uptake from deep soil layers increased, but immediately after the first rainfall following a dry period, the uptake fraction from upper layers peaks considerably, while it decreases at lower layers.

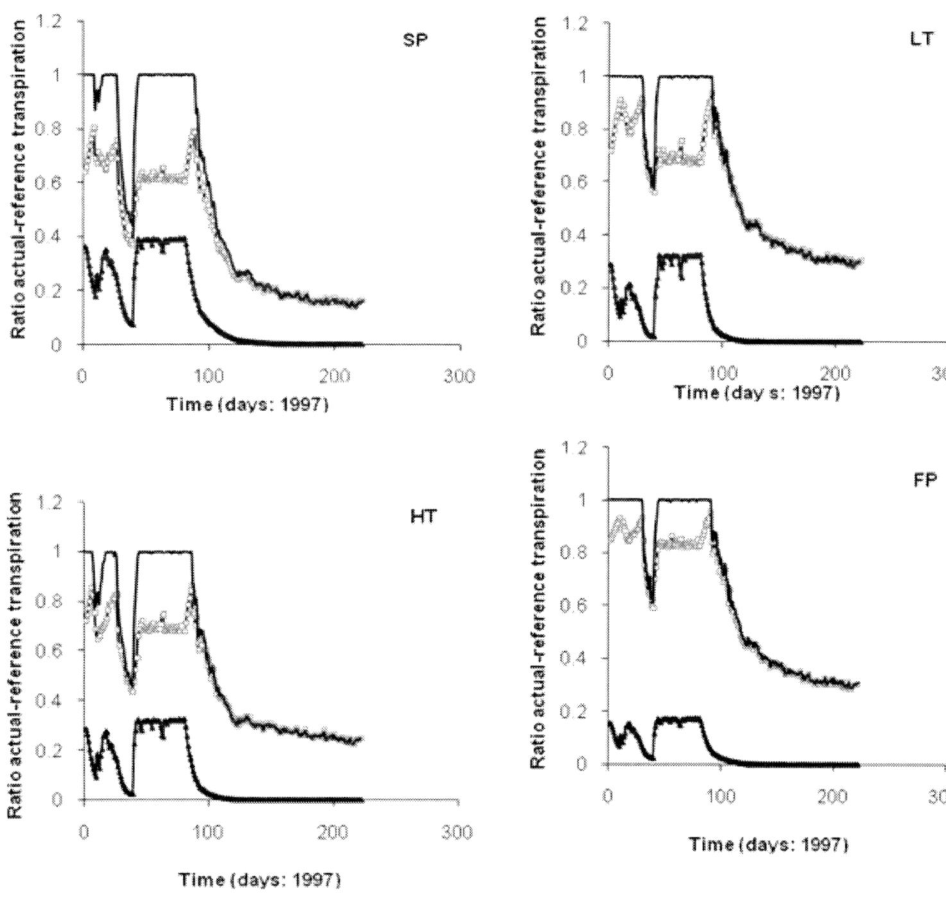


Figure 8. Temporal dynamics of the ratio between soil water uptake or actual plant transpiration and the reference transpiration (Monteith, 1965) in four forest ecosystems in Colombian Amazonia.

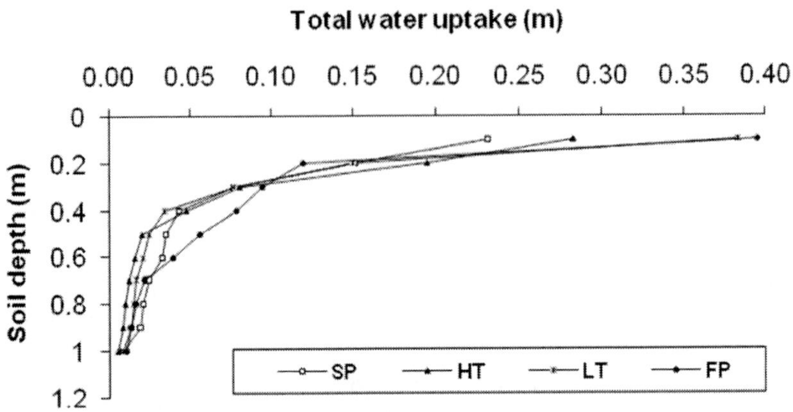


Figure 9. Vertical variation of total water uptake by the forest ecosystems (SP, HT, LT, and FP) as calculated from soil moisture depletion.

This suggests that at the start of the dry periods, the forest uptakes available water from the FF and from the surface soil horizons. Subsequently, when the water content in the upper part of the soil profile decreases beyond a certain limit, increases of water uptake from the sub-surface soil horizons, as the preferential water uptake.

During the studied period, the relative contribution of water uptake from the mineral soils to forest transpiration was 64.8% of the reference transpiration in the SP, 70.6% in the HT, 74.2% in the LT and 82.7% in the FP.

In all ecosystems, these percentages increased during the droughts and decreased immediately after the first rain storm following a dry period. These results agree with those from the analysis of water storage dynamics, which showed a higher storage in the SP during the droughts and the strongest depletion in the FP. This explains why the uptake from the mineral soil is highest in the FP: in this ecosystem the FF is very thin and contributeslittle to forest transpiration.

Mean annual transpiration values were 1193 mm in the SP, 1198 mm in the HT, 1214 mm in the LT and 1217 mm in the FP. The largest differences between actual and reference transpiration was observed in the SP, given that transpiration by the SP forest largely depends on the FF water availability (up to almost 30% of reference transpiration) when water storage capacity is low.

During the short dry seasons in 1997 and 2005, transpiration was somewhat suppressed, i.e. by a factor of 0.43 and 0.53. In other periods the depletion rate of stored soil water was significantly lower than total rainfall surplus and enabled the actual transpiration to be at the potential rate.

ET values from Colombian Amazonia are higher than those reported for some areas in Central Amazonia (Fisher et al., 2008; Lesack, 1993; Shuttleworth, 1988), including those reported by Bruijnzeel (1990) in one of the most comprehensive studies on the available information for ET in tropical lowland forests.

Moreover, results from this study concerning trends of ET contradict the conclusion by Lesack (1993) who concluded that actual ET may decline significantly during wetter than normal years. Our results show that transpiration can indeed be lower, but this is fully compensated by evaporation of intercepted water. Our results also confirm the conclusion of Bruijnzeel (1990), in the sense that values of ET can be substantially higher during wettest years or in areas with high rainfall.

Our results point to a reduction in forest transpiration during the dry periods, while available water capacity seems to be low, mainly in the SP. However, there are no clear indications for any significant physiological effects caused by water deficits such as leaf shedding or wilting. This is probably due to the short duration of the dry periods. Several authors found that Amazonia forest withered during the drought in 2005 (Tomasella, et al., 2011; Phillips, et al., 2009; Fisher, et al., 2008; Marengo et al., 2008a; Fisher, et al., 2007). However, the research results did not yield any differences in leaf shedding from studied forests, implying that in the systems studied, large soil water deficits are unlikely to occur.

Results of the quantification of the water balance for each compartment are presented in Table 2. Comparing the water balances shows differences both between compartments and among ecosystems in their water evaporation, uptake, and drainage. When considering results of each individual ecosystem, a primary observation is that the percentage of evaporation relative to gross rainfall was relatively constant over the years, although net values of evaporation differ.

An exception is the years 1997 and 2005where this percentage increased in all ecosystems. As to the forest canopy compartment, Table 2 shows that although the storage depth of this canopy is low, in terms of intercepting incoming rainfall, the net annual interception is considerably high and differs between ecosystems. Annual average evaporation of intercepted water was 366 mm in the SP, 361 mm in the HT, 405 mm in the LT, and 508 mm in the FP.

The water balance for the four forest ecosystems indicated that transpiration was relatively constant through the studied period in all ecosystems. This is not surprising since rainfall is well distributed, and FF and soil water storage was showed to be a sufficient reservoir to meet forest transpiration demands during most of the studied period. This actual transpiration appeared to be larger in the FP than in the other ecosystems, while in the SP it was lowest.

Comparison of the temporal dynamics of discharge with incoming net rainfall indicates that the fraction of discharge differs between years, with the highest value during 1996 in all ecosystems, and between ecosystems, ranging from 52% to 58% with the lowest value for the FP, in agreement with the highest interception and transpiration in this ecosystem. These percentages are higher than most values reported for the Amazonia forests (Elsenbeer and Cassel, 1991; Poels, 1987), but similar to that by Lesack (1993). However, we have to consider that gross rainfall is considerably higher in the Colombian Amazonia.

It is clear from Table 2 that the water balance surplus is almost zero, which is explained by the nature of the calculations applied, which balance inputs and outputs, and by the changes of FF and soil water storage.

Additionally, soil water storage did not change considerably over the years, while in FF relatively large changes occurred. Larger changes could be expected to occur in the forest floor and the mineral soil at the end of the dry period.

**Table 2. Summary of the water balance data for four forest ecosystems in the Middle Caquetá, Colombian Amazonia. All units are expressed as a depth of water in mm $yr^{-1}$. ET is calculated as the sum of evaporation and actual transpiration**

| Ecosystem | Year | Gross rainfall | Throughfall | Evaporation | Reference transpiration | $\Delta s_{ff}$ | Ff water uptake | Ff drainage | $\Delta s$ | Soil water uptake | Transpiration | Total drainage at 1 m | Evapo-Transpiration |
|---|---|---|---|---|---|---|---|---|---|---|---|---|---|
| **SP** | 1996 | 3308 | 3008 | 300 | 1321 | -10 | 392 | 2626 | -12 | 785 | 1177 | 1854 | 1477 |
| | 1997 | 2300 | 1981 | 319 | 1276 | -2 | 236 | 1746 | -5 | 496 | 732 | 1255 | 1052 |
| | 2004 | 3112 | 2831 | 281 | 1188 | -4 | 342 | 2478 | -6 | 743 | 1085 | 1741 | 1366 |
| | 2005 | 2480 | 1880 | 392 | 1292 | -3 | 214 | 1657 | -6 | 532 | 756 | 1131 | 1098 |
| **HT** | 1996 | 3466 | 3081 | 385 | 1321 | -4 | 307 | 2778 | -15 | 897 | 1204 | 1896 | 1589 |
| | 1997 | 2341 | 2029 | 312 | 1276 | -4 | 175 | 1858 | -4 | 541 | 716 | 1319 | 1029 |
| | 2004 | 3356 | 3023 | 333 | 1188 | -3 | 289 | 2737 | -13 | 845 | 1134 | 1995 | 1467 |
| | 2005 | 2390 | 1991 | 399 | 1292 | -1 | 192 | 1800 | -5 | 550 | 742 | 1255 | 1141 |
| **LT** | 1996 | 3404 | 3022 | 382 | 1321 | -3 | 282 | 2742 | -13 | 903 | 1185 | 1850 | 1568 |
| | 1997 | 2393 | 2046 | 347 | 1276 | -1 | 168 | 1879 | -5 | 569 | 737 | 1314 | 1084 |
| | 2004 | 3350 | 3011 | 339 | 1188 | -4 | 257 | 2758 | -11 | 922 | 1179 | 1847 | 1518 |
| | 2005 | 2355 | 1929 | 426 | 1292 | -2 | 167 | 1764 | -3 | 617 | 784 | 1150 | 1210 |
| **FP** | 1996 | 3418 | 2965 | 453 | 1321 | -2 | 162 | 2804 | -14 | 1030 | 1192 | 1789 | 1645 |
| | 1997 | 2289 | 1892 | 397 | 1276 | 0 | 95 | 1798 | -3 | 650 | 745 | 1151 | 1141 |
| | 2004 | 3220 | 2843 | 377 | 1188 | -3 | 254 | 2592 | -9 | 869 | 1123 | 1735 | 1500 |
| | 2005 | 2540 | 2042 | 498 | 1292 | -1 | 102 | 1941 | -2 | 719 | 821 | 1224 | 1319 |

# CONCLUSION

Average annual rainfall in Northwest Amazonia was found to be around 3,400 mm $yr^{-1}$, mostly falling in small showers during the afternoon and at night, with events of short duration. The hydrological functioning of forest ecosystems showed that the SP ecosystem has the highest percentage of net precipitation relative to gross rainfall, and the FP the lowest of the studied forests. The observed differences in throughfall, stemflow, and evaporation between ecosystems can partly be attributed to differences in forest structure: In the canopy this study clearly demonstrated that rainfall interception is a function of structural forest characteristics e.g. canopy cover and LAI. Ecosystem forest floors results showed similar drainage dynamics, but amounts of water retained were found to be different for each ecosystem. The FF in the SP consistently had the lowest water content and dried out to a greater extent than other ecosystems. In terms of water storage, the SP showed the highest values during the studied period, while the FP showed the lowest. Differences between ecosystems in FF water storage amounts areprimarily the result of differences in the FF thickness and to a lesser extent their moisture conditions.

Measured forest floor and soil water content dynamics indicate that there are high vertical water fluxes from the FF to the mineral soil, with the highest in the SP. On average, 34% of the SP forest transpiration was supplied by the FF, whereas in the FP, the FF contributed toonly 14% ofthe forest transpiration. The contribution of the mineral soil to the total forest transpiration during the study period differed between ecosystems, ranging from 63% to 79% of the reference transpiration. Differences are explained by the differences in the root distribution between ecosystems and the respective contribution of the forest floor to transpiration. Though rainfall reduces considerably in 2005, there was no long shortage of water for the vegetation in this part of the Amazonia. Thus, rainfall distribution is the key factor to maintain forest transpiration at the potential rate, and the FF and soil water storage capacity of undisturbed forests is high enough to meet forest ET demands during most dry periods.

Recent research demonstrated the important role played by deep roots to supply water to the forest in central Amazonia (Hodnett et al., 1996; Nepstad et al., 1994). This raises the question to whether such a phenomenon is more widespread. This research suggests that this is not the case in the study area. A truly dry period was found to not exist, and the short dry period (January to February) did not last long enough to significantly affect forest transpiration. Moreover, fine roots appeared to be concentrated in the FF and on top of the mineral soil in ecosystems studied. As to the substrate, water supply to the forests is largely determined by the moisture conditions in the FF and mineral topsoil. Therefore, neither lack of moisture stress nor lack of oxygen can explain the concentration of fine roots in the FF and top of the mineral soils. It is therefore concluded that root distribution most probably is determined by soil chemistry, notably gradients in nutrient availability rather than by moisture.

This research indicates that the impact of deforestation onchanges the hydrology of a system at the local scale remains largely unclear, and will be very site dependent. For example, trends in changes of transpiration due to land-use changes are diverse and difficult to predict without comparable data, or without making broad assumptions. Results from this research indicate that at least in the short term, removal of forest will cause increases in streamflow. For the research sites, although some bare soil evaporation can be expected, streamflow is expected to be about 1,600 mm $yr^{-1}$, larger than at the present, simplydue to the reduction in interception and transpiration. The situation, however, can be complex, as for example with the effect of large scale land-use changes on the eastern slopes of the Andes. The seasonal distribution of rainfall in the east is reflected in the fluctuations of the water level of the River Caquetá (changes of about 10m). Any change in the river discharge in the upper part of the basin of the River Caquetá, whether induced by changes in climate resulting from changes in land use in the Amazon basin, or by changes in hydrology of the Andean slopes, will directly affect the hydrological functioning of large areas in Andean rivers.

## ACKNOWLEDGMENT

Thanks to the Universidad Nacional de Colombia for the time to pursue de research and the time to write this paper. I like to thank the Dutch Tropenbos Foundation for the financial support to the project. Thanks to the *Nonuya* indigenous community in the Middle Caquetá – Colombia, for their support to the field work and hospitality. To the memory of Sebastian "tocayo" (RIP), my unforgettable field assistance during the entire field work.

## REFERENCES

Álvarez, D. E. (1993). Composición florística, diversidad, estructura y biomasa de un bosque inundable, en la Amazonia Colombiana. Ir. Dissertation, *Universidad de Antioquia, Colombia*. 550 pp.

Botero P. (1984). Relación fisiografía-suelo-aptitud de la tierra en la Amazonia colombiana. *Revista CIAF* 9 (1): 3-25. IGAC, Bogotá.

Bruijnzeel, L.A. (1990). Hydrology of moist tropical forests and effects of conversion: a state of knowledge review. UNESCO, Paris. 36 pp.

Bruno, R.; da Rocha, H.; de Freitas, H.; Goulden, M.; Miller, S. (2006). Soil moisture dynamics in an eastern Amazonian tropical forest. *Hydrological Processes,* 20:12 15. p. 2477 – 2489.

Chauvel, A., Grimaldi, M. and Tessier, D. (1991). Changes in soil pore space distribution following deforestation and revegetation: an example from the central Amazonia Basin, Brazil. *For. Ecol. and Manage*., 38: 259-271.

Cuevas, E. and Medina, E. (1988).Nutrient dynamics within Amazonian forest ecosystems. II. Fine root growth, nutrient availability and leaf litter decomposition. *Oecologia*, 76: 222-235.

Duivenvoorden, J. F. (1995). Tree species composition and rainforest-environment relationships in the Middle Caquetá area, Colombia, NW Amazonia. *Vegetatio*, 120: 1-113.

Duivenvoorden, J. M. and Lips, J. M. (1995).A land-ecological study of soils, vegetation and plant diversity in Colombian Amazonia.Ph.D. Dissertation, Landscape and Environmental Research Group, Faculty of Environmental sciences, University of Amsterdam.Tropenbos Series 12, Wageningen.438 pp.

Duivenvoorden J. F. and J. Lips.(1993). Ecología del Paisaje del Medio Caquetá. Memoria explicativa de los mapas.Tropenbos Colombia, Santafé de Bogotá. 301 pp.

Elsenbeer, H. and Cassel, D. K. (1991).The mechanisms of overland flow generation in a small catchment in western Amazonia. In: water management of the Amazon basin, edited by B.P.F. Braga and C.A. Fernandez-Jauregui,.213-220 pp, UNESCO, Montevideo, Uruguay.

Fisher, R., Williams, M., Ruivo, M., de Costa, A., Meir, P. (2008). Evaluating climatic and soil water controls on evapotranspiration at two Amazonian rainforest sites. *Agricultural and forest meteorology*: 148: 850-861.

Fisher, R. A.; Williams, M.; da Costa, L; Malhi, Y.; da Costa F.; Almeida, S.; Meir, P. (2007). The response of an Eastern Amazonian rainforest to drought stress: results and modelling analyses from a throughfall exclusion experiment. Global Change Biology vol. 13 issue 11 p. 2361-2378.

Gash, J. H. C., Nobre, C. A., Roberts, J. M. and Victoria, R. L. (1996).An overview of ABRACOS. In: Amazonian Deforestation and Climate. John Wiley and Sons, Chichester, United Kingdom.611 pp.

Golley, F. B. (1983). The abundance of energy and chemical elements. In Golley, F. B. (ed) 1983a Tropical rainforest ecosystems. Elsevier, Amsterdam, pp. 101-115.

Hertwitz, S. R. (1985). Interception storage and capacities of tropical rainforest canopy trees. *J. of Hydrol.* 77: 237-252.

Hodnett, M. G.,Oyama, M. D. and Tomasella, J. (1996). Comparisons of long-term soil water storage under pasture and forest in three areas of Amazonia. In: J. H. C. Gash, C. A. Nobre, J. M. Roberts and R. L. Victoria (Ed.). Amazonia deforestation and Climate. Pp. 57-77. John Wiley and Sons Ltd.

Hölscher, D., Sá, T. D. de A., Bastos, T. X., Denich, M. and Fölster, H. (1997).Evaporation from young secondary vegetation in eastern Amazonia. *J. of Hydrol.* 193: 293-305.

Hutjes, R. W. A., Wierda, A. and Veen, A. W. L. (1990). Rainfall interception in the Tai forest, Ivory Coast: Application of two simulation models to a humid tropical system. *J. of Hydrol.* 114: 259-275.

Jetten, V. G. (1996). Interception of tropical rainforest: Performance of a canopy water balance model. *Hydrol. Process*, Vol. 10: 671-685.

Jordan, C. F. (1989). An Amazon rainforest: the structure and function of a nutrient stressed ecosystem and the impact of slash-and-burn agriculture. Man in the Biosphere series Vol. 2. UNESCO, Paris, and the Parthenon Publ. Group.Ltd, UK.

Köppen, W. (1936).Das geographische System der Klimate. In: Handbuch der Klimatologie. Verlag Gebrüder Borntraeger, Berlin, 44 pp.

Laurance, W. F. (1998). A crisis in the making: responses of Amazonian forests to land use and climate change. Trends in Ecology and Evolution vol. 13 issue 10. p. 411-415.

Leopoldo, P. R., Franken, W., Salati, E. and Ribeiro, M. N. (1987). Towards a water balance in central Amazonia region. *Experentia* 43 (1987), BirkhäuserVerlag, Basel/Switzerland.

Leopoldo, P. R., Franken, W. and Villa Nova, N. (1995). Real evaporation and transpiration through a tropical rainforest in central Amazonia as estimated by the water balance method. *For. Ecol. Manag.* 73: 185-195.

Lesak, L. F. W. (1993). Water balance and hydrologic characteristics of a rainforest catchment in the Central Amazon Basin. *Water Resour. Res.* 29 (3): 759-773.

Lesack, L. F. W. and Melack, J. M. (1991).The deposition, composition and reference sources of major ionic solutes in rain of the central Amazon Basin. *Water Resour. Res*. 27: 2953-2978.

Lips J. M., Duivenvoorden J. F. (1990). Estudios de suelos en la Amazonia colombiana. *TROPENBOS. Colombia*.

Londoño, V. C. (1993). Análisis estructural de dos bosques asociados a unidades fisiográficas contrastantes, en la region de Araracuara (Amazonia Colombiana). Ir. Dissertation, Universidad Nacional de Colombia. 478 pp.

Longman, K. A. and Jeník, J. (1990).Tropical forest and its environment.Longman Scientific and technical Ed. 1990.347 pp.

Malhado, A; Whittaker, R.; Malhi, Y; Ladle, R.; terSteege, H; Phillips, O; Aragão, L.; Baker, T.; et al. (2010). Are compound leaves an adaptation to seasonal drought or to rapid growth? Evidence from the Amazon rainforest.Global Ecology and Biogeography vol. 19 issue 6. p. 852-862.

Marengo J. A, Nobre C. A, Tomasella J, Oyama M. D, Oliveira G. S, de Oliveira R, Camargo H, Alves L. M, Brown I. F. (2008a). The drought of Amazonia in 2005. *Journal of Climate* 21: 495–516.

Mejía, L., Botero, P., y Montoya, A. (1980). Levantamiento detallado y semidetallado de los suelos de Araracuara. Centro Interamericano de Foto interpretación CIAF, Bogotá. 213 pp.

Milliken, W; Zappi, D; Sasaki, D; Hopkins, M; P, Toby.(2010). Amazon vegetation: how much don't we know and how much does it matter?.Kew Bulletin vol. 65 issue 4 p. 691 – 709.

Nepstad, D. C., de Carvalho, C. R., Davidson, E. A., Jipp, P. H., Lefebvre, P. A., Negreiros, G. H., da Silva, E. D., Trumbore, S. E. and Vieira, S. (1994). The role of deep roots in the hydrological and carbon cycles of Amazonian forests and pastures. *Nature*, 372: 666-669.

Nobre, C. A., Sellers, P. and Shukla, J. (1991). Amazonia deforestation and regional climate change. *J. Climate*, 4: 957-988.

Pitman, J. I. (1989). Rainfall interception by bracken litter-relationship between biomass, storage and drainage rate. *J. of Hydrol*. 111: 281-291.

Phillips O. L, et al. (2009).Drought sensitivity of the Amazon rainforest. *Science* 323 (5919): 1344–1347.

Poels, R. (1987). Soils, water and nutrients in a forest ecosystem in Suriname.Ph.D. thesis, Agricultural University of Wageningen, the Netherlands.

Pradham, I. P.(1973). Preliminary study of rainfall interception through leaflitter. *Indian For*., 99: 440-445.

Putuhena, W. and Cordery, I. (1996).Estimation of interception capacity of forest floor. *J. of Hydrol*. 180: 283-299.

Sala, S; Duque, S; Núñez, M; Lamaro, A. (2002).Diatoms from the Colombian Amazonia.CryptogamieAlgologie vol. 23 issue 1. p. 75-99.

Salati, E., Dall'olio, A., Matsui, E. and Gat, J. R. (1979).Recycling of water in the Amazon Basin: an isotopic study. *Water Resour. Res*. 15: 67-71.

Salati, E. andVose, P.B. (1984). Amazon Basin: a system in equilibrium. *Science*, 225: 129-138.

Shukla, J., Nobre, C. and Sellers, P. J. 1990. Amazon deforestation and climate change. *Science*, 247: 1322-1325.

Shuttleworth, W. J. (1988). Evaporation from Amazonian rainforest. *Proc. R. Soc.* London, B233, 321-346.

Soil Survey Staff. 2010. Keys to Soil Taxonomy, 11th ed. USDA-Natural Resources Conservation Service, Washington, DC.

Tiessen, H., Cuevas, E. and Chacón, P. (1994).The role of soil organic matter in sustaining soil fertility. *Nature*, 371: 783-785.

Tobón Marin, C., Bouten, W., Sevink, J. (2000a). Gross rainfall and its partitioning into throughfall, stemflow and evaporative loss in four forest ecosystems in western Amazonia. *J. of Hydrol.* Vol. 237(1-2): 40-57.

Tobón Marin, C., Bouten and S. Dekker.(2000b). Forest floor water dynamics and root water uptake in four forest ecosystems in northwest Amazonia. *J. of Hydrol.* Vol. 237 (1-2): 169-183.

Tobón, C. 1999. Monitoring and modelling hydrological fluxes in support of nutrient cycling studies in Amazonian rainforest ecosystems. Tropenbos series17, Wageningen, the Netherlands.169 pp.

Tomasella, J., L. S. Borma, J. A. Marengo, D. A. Rodríguez, L. A. Cuartas, C. A. Nobre, and M. C. R. Prado.(2011).The droughts of 1996 1997 and 2004 2005 in Amazonia: Hydrological response in the river main-stem. *Hydrol. Processes*, 25, 1228–1242.

Ubarana, V. N. (1996). Observations and modelling of rainfall interception at two experimental sites in Amazonia.*In:* Amazonian deforestation and climate, edited by J. H. C. Gash, C. A. Nobre, J. M. Roberts and R. L. Victoria. Ed. John Wiley and Sons.611 pp.

Vasconcelos, H.; Vilhena, J.; Facure, K.; Albernaz, A. (2000).Patterns of ant species diversity and turnover across 2000 km of Amazonian floodplain. *Forest Journal of Biogeography* vol. 37 issue 3. p. 432-440.

Wittmann, F; Parolin, P. (2005).Aboveground Roots in Amazonian Floodplain Trees. *Biotropica* vol. 37 issue 4 December 2005. p. 609-619.

Wright, I. R., Gash, J. H., da Rocha, H. R., Shuttleworth, W. J., Nobre, C. A., Maitelli, G. T., Zamparoni, C. A. and Carvalho, P. (1992). Dry season micrometeorology of central Amazonian ranchland. *Q.J.R. Meteorol. Soc.* 118: 1083-1099.

# INDEX

## A

## D



## E

## F

## I



## J



## K



## L

## M

## N



## O



## P

## Q

## R



## S

## T



## U

## V



## W



## Y



## Z